VII Workshop on Electron-Volt Neutron Spectroscopy 2017

Journal of Physics: Conference Series
Volume 1055

Rome, Italy
7 - 8 November 2017

Editors:

G. Romanelli
G. Festa
M. Krzystyniak
C. Andreani
F. Fernandez-Alonso
R. Senesi

ISBN: 978-1-5108-6792-5
ISSN: 1742-6588

Printed from e-media with permission by:

Curran Associates, Inc.
57 Morehouse Lane
Red Hook, NY 12571

proceedings.com

Some format issues inherent in the e-media version may also appear in this print version.

Printed by Curran Associates, Inc. (2018)

Institute of Physics
Dirac House, Temple Back
Bristol BS1 6BE UK

Phone: 44 1 17 929 7481
Fax: 44 1 17 920 0979

techtracking@iop.org

Additional copies of this publication are available from:

Curran Associates, Inc.
57 Morehouse Lane
Red Hook, NY 12571 USA
Phone: 845-758-0400
Fax: 845-758-2633
Email: curran@proceedings.com
Web: www.proceedings.com

Table of contents

Volume 1055

VII International Workshop on Electron-Volt Neutron Spectroscopy

7–8 November 2017, Rome, Italy

Accepted papers received: 3 July 2018
Published online: 17 July 2018

Preface

Papers

The science

Conclusions

IOP Conf. Series: Journal of Physics: Conf. Series **1055** (2018) 011001 doi:10.1088/1742-6596/1055/1/011001

Neutrons matter: VII international workshop on electron-Volt neutron spectroscopy – A preface to the workshop proceedings

G Romanelli[a], G Festa[b], M Krzystyniak[a,c], C Andreani[b,d], F Fernandez-Alonso[a,e], and R Senesi[d,f]

[a] ISIS Facility, Rutherford Appleton Laboratory, Chilton, Didcot, Oxfordshire, OX11 0QX, UK

[b] Centro Fermi - Museo Storico della Fisica e Centro Studi e Ricerche "Enrico Fermi" Piazza del Viminale 1, Rome 00184 Italy

[c] School of Science and Technology, Nottingham Trent University, Clifton Campus, Nottingham, NG11 8NS, UK

[d] Università degli Studi di Roma "Tor Vergata", Dipartimento di Fisica and Centro NAST, Via della Ricerca Scientifica 1, Rome , 00133 Italy

[e] Department of Physics and Astronomy, University College London, Gower Street, London, WC1E 6BT, UK

[f] CNR-IPCF Sezione di Messina, Viale Ferdinando Stagno d'Alcontres 37, Messina, 98158 Italy

E-mail: Giovanni.romanelli@stfc.ac.uk

Abstract. We present here a collection of works reporting on the recent experimental and theoretical activities taking advantage of epithermal neutron spectroscopy, and in particular focusing on recent results presented during the *VII International Workshop on Electron-Volt Neutron Spectroscopy* held in Rome on 7-8 November 2018.

1. An introduction to epithermal neutrons

Neutrons are amongst the most exquisite probes to investigate and characterise materials [1]. These particles are referred to as cold and thermal when their energies are lower then ca. 500 meV, therefore matching the typical energy scales in the dynamics of condensed-matter systems. Cold and thermal neutrons are routinely employed for scientific investigations based on Quasi Elastic (QENS) and Inelastic Neutron Scattering (INS) [2], to probe diffusion and tunnelling, or lattice and internal molecular excitations, respectively. Similarly, as the wavelengths of such probes are commensurate with the interatomic distances in solid and liquids, their elastic scattering on a sample is employed for diffraction experiments characterising the structure of materials.

Traditionally, neutron sources have focused on the application of cold and thermal neutrons, by slowing down neutrons with energies in the MeV region via the interaction with moderating materials. As an example, Figure 1 shows the flux of moderated neutrons on the VESUVIO spectrometer [3] at the ISIS pulsed neutron and muon source [4], where neutrons with initial energies up to 800 MeV are slowed down in a room-temperature water moderator, so as to

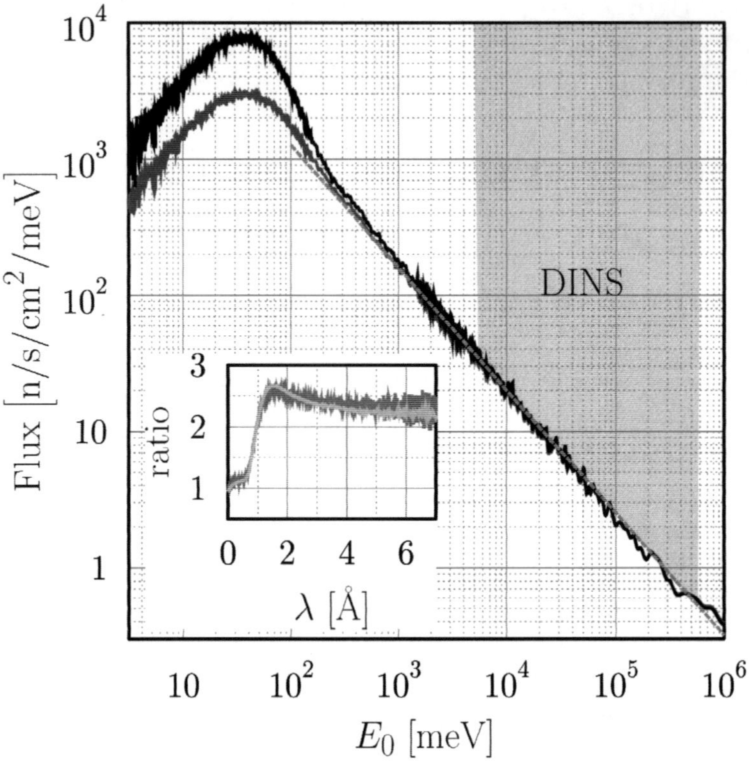

Figure 1. The incident-neutron flux on the VESUVIO spectrometer at ISIS. The red dashed line shows the epithermal tail of partially moderated neutrons, as discussed in the text. The figure is reproduced from Ref. [5].

have maximum flux in the thermal region, at about $3/2k_BT \sim 38$ meV, with k_B denoting the Boltzmann's constant, and T the temperature of the water moderator. In addition to thermal neutrons, one can notice a substantial contribution, in the shape of a long tail, of neutrons only partially moderated in the region of 1 eV – 1 keV. These neutrons are referred to as epithermal, or eV neutrons, and constitute the main subject of the articles in this collection. In particular, the flux distribution in the epithermal region is inversely proportional to the neutron energy (red dashed line in Figure 1), and is almost independent on the moderator material and geometry, as exemplified by the two overlapping spectra (blue and black lines) in the figure (see Ref. [5] for details).

Since the advent of proton-driven spallation sources in the 1970s, eV neutron spectroscopy has witnessed an ongoing development of the technique, as well as of its scientific and technological applications. The main technique based on epithermal neutrons is Deep Inelastic Neutron Scattering (DINS), also referred to as Neutron Compton Scattering (NCS) [6, 7]. The flagship instrument is the VESUVIO spectrometer [8, 9]. DINS allows for a direct measurement of nuclear kinetic energies and momentum distributions in condensed-matter systems, observables readily connected to nuclear quantum effects (NQEs), or anharmonicity in the local potential affecting a given atom. DINS is an example of MAss-selective Neutron SpEctroscopy (MANSE), where the signals from different elements are spectroscopically separated depending on their mass. Moreover, epithermal neutrons can be absorbed by some elements, usually isotopes of heavy elements such as Au, Ag, In, I, Ta, and others, in a nuclear resonant process. The study of these

Figure 2. Group picture at Centro Fermi.

resonances, namely Neutron Resonance Capture Analysis (NRCA) [10], can provide valuable information about the elemental composition of a sample, such as in archaeological or cultural-heritage investigations [11], as well as it serves as a quantum thermometer for local effective temperatures affecting atoms [12]. The cross section for such resonances can be several orders of magnitude higher than the neutron scattering cross section, and the value of the resonant energy is isotope-specific.

2. Broadening horizons

Over the past decades, a number of thematic workshops on the use of epithermal neutrons have helped defining the way forward. After a first scientific meeting in Los Alamos (USA) in 1984 [13], workshops on eV spectroscopy were held in Abingdon (UK) in 1995 and 1998, in Santa Fe (USA) in 2005, Oak Ridge National Laboratory (USA) in 2006, Rome (Italy) in 2010, and Abingdon in 2014 [14]. In the last episode of this story, an always larger and more diverse community met in Rome on 7 8 November 2017 for the VII International Workshop on electron-Volt Neutron Spectroscopy [15].

In recent years, DINS has provided, *inter alia*, a unique benchmark for the development of state-of-the-art computer simulations to describe NQEs in the hydrogen dynamics [16], such as path-integral molecular dynamics calculations [17, 18]. Yet, the 2014 edition of the workshop underlined how the realms of applicability of DINS go well beyond hydrogen [19], with an increasing number of studies focusing on moderate-weight masses such as oxygen [20], or lithium and fluorine [21]. This change in strategy of data analysis promoted MANSE as a unique technique where the momentum distribution of several atoms in a system can be analysed concurrently and independently.

In 2017, the horizon was moved well beyond MANSE. The new trend appears to be the

IOP Conf. Series: Journal of Physics: Conf. Series **1055** (2018) 011001 doi:10.1088/1742-6596/1055/1/011001

Table 1: Programme of the first day of the workshop.

	Tuesday 7 November 2017 – Villa Wolkonsky
11:30	Registration & lunch
	Opening
13:00	Welcome address: Ken O'Flaherty (British Embassy) and Andrew D. Taylor (STFC)
13:10	Carla Andreani – Occhialini prize lecture ("Tor Vergata")
	Hydrogen dynamics in stable and metastable water by deep inelastic neutron scattering
	Session – Fundamental systems; chair: Felix Fernandez-Alonso (STFC)
13:30	Alexander I. Kolesnikov (Oak Ridge National Laboratory)
	Inelastic and deep inelastic neutron scattering study of water under ultra-confinement
14:00	Francesco Mallamace (Università di Messina)
	Nuclear Compton scattering and NMR, a common way to understand the local order
14:30	Daniele Colognesi (CNR)
	Self dynamics in simple liquids: connections between the Gaussian approximation and the asymptotic impulsive regime
15:00	Coffee break
	Session – Simulation and modelling; Chair: Roberto Senesi ("Tor Vergata")
15:30	Roberto Car (Princeton University)
	The momentum distribution of the nuclei as an environmental probe
16:00	Angelos Michaelides (University College London)
	Quantum effects in adsorption and diffusion at solid surfaces
16:30	Michele Ceriotti (École Polytechnique Fédérale de Lausanne)
	Theoretical benchmarks of deep inelastic neutron scattering. From the kinetic energy to the full particle momentum distribution
17:00	Poster session
20:00	Conference dinner

combination of several concurrent measurements available, for example, on the VESUVIO spectrometer using NRCA, neutron diffraction, and neutron transmission. Moreover, a more systematic comparison with parent techniques, such as INS, or first-principles simulations, such as DFT-based phonon calculations, are now the new standard of eVS scientific output, as for example in Ref. [22].

To demonstrate the broad range of topics discussed, as well as the increasing interest from an always larger community in the use of eV spectroscopy, we report below the list of the registered participants in the workshop, and the list of presented posters, while Tables 1 and 2 show the programme of the two days of the workshop.

2.1. List of registered attendees

- Prof Carla Andreani – Università degli Studi di Roma "Tor Vergata", Italy
- Ms Laura Arcidiacono – University College London, United Kingdom
- Dr Jeff Armstrong – Science & Technology Facilities Council, United Kingdom
- Prof Fabio Bruni – Università di Roma 3, Italy
- Ms Elena Budennaia – Nottingham Trent University, United Kingdom
- Prof Roberto Caciuffo – European Commission, Joint Research Centre, Germany
- Prof Roberto Car – Princeton University, United States
- Prof Michele Ceriotti – École Polytechnique Fédérale de Lausanne, Switzerland

IOP Conf. Series: Journal of Physics: Conf. Series **1055** (2018) 011001 doi:10.1088/1742-6596/1055/1/011001

Table 2: Programme of the second day of the workshop.

	Wednesday 8 November 2017 – Villa Wolkonsky
	Session – Beyond fundamental systems; Chair: Roberto Car (Princeton University)
8:00	Gregory Chasse (Queen Mary University of London)
	Atomistic and vibrational source of mechanical toughness in glass bio-cements and portland cement during setting
8:30	Fengge Gao (Nottingham Trent University)
	Better understanding of self-cross linking behaviour of smectite nanoclays through VESUVIO
9:00	Christoph Salzmann (University College London)
	Ice - a miraculous material
09:30	Coffee break & group photo
	Session – Methods and instrumentation; Chair: Giuseppe Gorini (University Milano Bicocca)
10:00	Anton Tremsin (University of California at Berkeley)
	High resolution neutron transmission imaging with single pixel spectroscopic resolution
10:30	Erik Schooneveld (STFC)
	Detection of epithermal neutrons on Vesuvio
11:00	Rolando Granada (Comisión Nacional de Energía Atómica)
	Neutron moderation: Some recent problems and results
11:30	Transfer to Centro Fermi
12:00	Welcome address: Luisa Cifarelli (President of Centro Fermi)
	The lost notebook of Enrico Fermi and the discovery of neutron induced radioactivity
12:30	Rosario Nania (Centro Fermi)
	The Centro Fermi in Rome: the projects, the future
12:45	Lunch & group picture
14:00	Ceremony – “Neutrons Matter” prize
16:00	Closing remarks

- Dr Gregory Chass – Queen Mary University of London, United Kingdom
- Prof Luisa Cifarelli – Museo Storico della Fisica e Centro Studi e Ricerche “Enrico Fermi”, Italy
- Dr Daniele Colognesi – Consiglio Nazionale delle Ricerche, Italy
- Dr Carmelo Corssaro – Consiglio Nazionale delle Ricerche, Italy
- Mr Alex Di Giulio – Università degli Studi di Roma “Tor Vergata”, Italy
- Dr Kacper Druzbiki – University Adam Mickewicz, Poland
- Prof Felix Fernandez-Alonso – Science & Technology Facilities Council, United Kingdom
- Dr Giulia Festa – Museo Storico della Fisica e Centro Studi e Ricerche “Enrico Fermi”, Italy
- Prof Fengge Gao – Nottingham Trent University, United Kingdom
- Prof Giuseppe Gorini – Università di Milano Bicocca, Italy
- Ms Ekaterina Gousseva – Queen Mary University of London, United Kingdom
- Dr Rolando Granada – Comisión Nacional de Energía Atómica CNEA, Argentina
- Mr Brandon Hewer – Science & Technology Facilities Council, United Kingdom
- Dr Maria Grazia Izzo – Università di Roma, La Sapienza, Italy
- Mr Alexander Jurgens – Università degli Studi di Roma “Tor Vergata”, Italy
- Mr Venkat Kapil – École Polytechnique Fédérale de Lausanne, Switzerland
- Prof Maths Karlsson – Chalmers University of Technology, Sweden
- Dr Somayeh Khazaei – Martin-Luther University, Germany
- Dr Alexander Kolesnikov – Oak Ridge National Laboratory, United States

- Dr Maciej Krzystyniak – Science & Technology Facilities Council, United Kingdom
- Ms Sveva Lungo – University of Messina, Italy
- Mr Mustafa Mahmoud – Queen Mary University of London, United Kingdom
- Dr Domenico Mallamace – CSGI, Center for Colloid and Surface Science, Italy
- Prof Francesco Mallamace – University of Messina, Italy
- Dr Adriana Mamede – University of Coimbra, Portugal
- Prof Rossana Marra – Università degli Studi di Roma "Tor Vergata", Italy
- Prof Maria Paula Marques – University of Coimbra, Portugal
- Ms Louise McCann – Science & Technology Facilities Council, United Kingdom
- Ms Sara Miceli – Università degli Studi di Roma "Tor Vergata", Italy
- Prof Angelos Michaelides – University College London UCL, United Kingdom
- Dr Rosario Nania – Museo Storico della Fisica e Centro Studi e Ricerche "Enrico Fermi", Italy
- Mr Matteo Nardini – Università degli Studi di Roma "Tor Vergata", Italy
- Ms Dalila Onorati – Università degli Studi di Roma "Tor Vergata", Italy
- Dr Stewart Parker – Science & Technology Facilities Council, United Kingdom
- Dr Roberto Pinna – Università di Milano Bicocca, Italy
- Dr Rosina Ponterio – Consiglio Nazionale delle Ricerche, Italy
- Ms Sara Postorino – Università degli Studi di Roma "Tor Vergata", Italy
- Prof Paolo Radaelli – University of Oxford, United Kingdom
- Mr James Rawlings – Nottingham Trent University, United Kingdom
- Dr Giovanni Romanelli – Science & Technology Facilities Council, United Kingdom
- Dr Alexander Rosu-Finsen – University College London, United Kingdom
- Mr Mohammed Salha – Queen Mary University of London, United Kingdom
- Dr Christoph Salzmann – University College London, United Kingdom
- Mr Valerio Scacco – Università degli Studi di Roma "Tor Vergata", Italy
- Dr Claudia Scatigno – Università degli Studi di Roma "Tor Vergata", Italy
- Mr Giuseppe Scionti – Università della Calabria, Italy
- Dr Erik Schooneveld – Science & Technology Facilities Council, United Kingdom
- Prof Roberto Senesi – Università degli Studi di Roma "Tor Vergata", Italy
- Ms Zainab Sharif – University College London, United Kingdom
- Dr Goran Škoro – Science & Technology Facilities Council, United Kingdom
- Mr Fu Song – Queen Mary University of London, United Kingdom
- Dr Corrado Spinella – Consiglio Nazionale delle Ricerche, Italy
- Dr Andrew Taylor – Science & Technology Facilities Council, United Kingdom
- Dr Kun Tian – Università degli Studi di Roma "Tor Vergata", Italy
- Dr Anton Tremsin – University of California, Berkeley, United States
- Mr Pierfrancesco Ulpiani – Università degli Studi di Roma "Tor Vergata", Italy
- Mr Muhammed Umarji – Queen Mary University of London, United Kingdom
- Dr Cirino Vasi – Consiglio Nazionale delle Ricerche, Italy
- Ms Aurora Zaganelli – Università degli Studi di Roma "Tor Vergata", Italy

2.2. List of presented posters

- L. Arcidiacono *et al.* – University College London; *Byzantine gold coins analysed using time resolved prompt gamma activation analysis*
- L. Arcidiacono *et al.* – University College London; *Time-resolved prompt gamma activation analysis at ISIS Spallation Neutron Source*
- J. Armstrong *et al.* – Science & Technology Facilities Council; *How do rotations alter the local potential of nanoscopic molecules? A* C_{60} *case study*
- E. Budennaia *et al.* – Nottingham Trent University; *Using Vesuvio to gain better understanding of self cross- linking behaviour of smectite clays*
- C. Corsaro *et al.* – Consiglio Nazionale delle Ricerche; *The similarities between Neutron Compton Scattering and NMR spectroscopy applied to investigate the degradation of cellulosic materials*
- A. Di Giulio *et al.* – Università degli Studi di Roma "Tor Vergata"; *A McStas simulation of the incident beam in the VESUVIO spectrometer*

IOP Conf. Series: Journal of Physics: Conf. Series **1055** (2018) 011001 doi:10.1088/1742-6596/1055/1/011001

- K. Drubicki *et al.* – University Adam Mickewicz; *Hydrogen dynamics in molecular solids: Insight from simulations with quantum colored-noise thermostat*
- E. Gousseva *et al.* – Queen Mary University of London; *Resolving interfacial structures and dynamics in hybrid organic - inorganic cements*
- B. Hewer *et al.* – Science & Technology Facilities Council; *Advances in the reduction and analysis of data from epithermal neutron experiments using MANTID*
- V. Kapil *et al.* – É'cole Polytechnique Fé'dérale de Lausanne; *Probing the momentum distribution of protons in ab inito water*
- S. Khazaei *et al.* – Martin-Luther university; *Rotation tunnelling of coupled methyl quantum rotors in 4methylpyridine: single rotor potential versus coupling interaction*
- M. Mahmoud *et al.* – Queen Mary University of London; *Graphene nanoflake reinforced cements*
- D. Mallamace *et al.* – Center for Colloids and Surface Science; *The local structure of hydrogen bond in water studied in a wide temperature range by Neutron Compton scattering and NMR spectroscopy*
- A. Mamede *et al.* – University of Coimbra; *Shining the beam on bones Human Burned Skeletal Remains Probed by Neutron Spectroscopy*
- M. P. Marques *et al.* – University of Coimbra; *International water – a secondary target in chemotherapy? Impact of Pt/Pd anticancer drugs in breast cancer cell probed by neutron techniques*
- M. Nardini *et al.* – Università degli Studi di Roma "Tor Vergata"; *Neutron diffraction and resonance capture analysis from Sumerian pottery from the citystate of Abu Tbeirah*
- D. Onorati *et al.* – Università degli Studi di Roma "Tor Vergata"; *Optimization of gamma detection methods for neutron energy analysis on eV neutron spectrometers*
- S. Parker *et al.* – Science & Technology Facilities Council; *Observation of the stretch mode in H_2 and D_2 by INS spectroscopy*
- A. Parmentier *et al.* – Università degli Studi di Roma "Tor Vergata"; *Hydrogen mean force and anharmonicity in polycrystalline and amorphous ice*
- J. Rawlings *et al.* – Nottingham Trent University; *Towards a better understanding of the self-cross-linking behaviour of smectite clays*
- G. Romanelli and M. Krzystyniak – Science & Technology Facilities Council; *Recent upgrades on the VESUVIO spectrometer*
- A. Rosu-Finsen *et al.* – University College London; *Benchmarking acidic and basic dopants with respect to facilitating the ice V to XIII and ice VI to XV hydrogen-ordering phase transitions*
- M. Salha *et al.* – Queen Mary University of London; *Supreme toughness in Aluminium enriched cements*
- V. Scacco *et al.* – Università degli Studi di Roma "Tor Vergata"; *Analysis of neutron activation decay on the VESUVIO beam line at ISIS*
- Z. Sharif *et al.* – University College London; *Exploring new areas of the ice phase diagram for metastable ice V with varying degrees of hydrogen-order*
- F. Song *et al.* – Queen Mary University of London; *Neutron scattering on atomic structure, stabilities and dynamics of cement*
- M.B. Umarji *et al.* – Queen Mary University of London; *Water gelation as a structural scaffold in cementitious materials*
- P. Ulpiani *et al.* – Università degli Studi di Roma "Tor Vergata"; *Enhancing the YAP detectors discrimination threshold installed in the VESUVIO spectrometer at the ISIS spallation neutron source*

3. An auspice

A detailed programme and outcome of the "Neutrons Matter: VII International Workshop on electron-Volt Neutron Spectroscopy" was recently presented in Ref. [15]. Also, a copy of the flyer is shown in Figure 3 for future reference. The development of the eV spectroscopy with neutrons on the VESUVIO spectrometer has been a clear example of a successful international collaboration in science. Over the past decades, a solid agreement between the British Science and Technology Facilities Council and the Italian Consiglio Nazionale delle Ricerche has provided a fertile ground for the development of the technique, that has facilitated the growth of a world-wide scientific community. This workshop, as well as other past editions, was partially co-funded by the two councils, and for the second time it took place in Rome, the eternal city.

So as to underline this long-standing agreement, the scientific sessions of the workshop were hosted in the British Embassy in Rome, a very suggestive venue. Moreover, the time of the workshop coincided with the anniversaries of the discovery of the neutron by the British scientist

J. Chadwick, and the patent for neutron moderation by the Italian physicist E. Fermi. The group picture in Figure 2 was taken during a visit to the Museo Storico della Fisica e Centro Studi e Ricerche "Enrico Fermi" in the city centre, on the very stairs where a number of Nobel laureates were photographed in 1931 during the first international conference in nuclear physics. May this precedent be an auspice for the future of spectroscopy with epithermal neutrons, and for the underpinning international communities.

Acknowledgements

We would like to thank the British Embassy in Rome for their hospitality in hosting a substantial part of the workshop, as well as for supporting travel expenses for early-career researchers. A special thanks to Prof Cifarelli and Dr Nania for the visit to the Museo Storico della Fisica e Centro Studi e Ricerche "Enrico Fermi". We acknowledge the financial support of the Science & Technology Facilities Council, of the School of Neutron Scattering "F. P. Ricci", of the Museo Storico della Fisica e Centro Studi e Ricerche "Enrico Fermi", and of the Consiglio Nazionale delle Ricerche.

References

[1] Fernandez-Alonso F and Price D L 2013 *Neutron Scattering – Fundamentals* (New York: Academic Press)
[2] Mitchell P C H 2005 *Vibrational spectroscopy with neutrons: with applications in chemistry, biology, materials science and catalysis* vol 3 (World Scientific)
[3] Last accessed on June 2018 URL `http://www.isis.stfc.ac.uk/Pages/VESUVIO.aspx`
[4] Last accessed on June 2018 URL `http://www.isis.stfc.ac.uk`
[5] Romanelli G, Krzystyniak M, Senesi R, Raspino D, Boxall J, Pooley D, Moorby S, Schooneveld E, Rhodes N J, Andreani C and Fernandez-Alonso F 2017 *Measurement Science and Technology* **28** 095501
[6] Andreani C, Krzystyniak M, Romanelli G, Senesi R and Fernandez-Alonso F 2017 *Advances in Physics* **66** 1–73
[7] Andreani C, Senesi R, Krzystyniak M, Romanelli G and Fernandez-Alonso F 2017 *Neutron Scattering - Applications in Biology, Chemistry, and Materials Science* (*Experimental Methods in the Physical Sciences* vol 49) ed Fernandez-Alonso F and Price D L (Academic Press) pp 403 – 457
[8] Mayers J and Reiter G 2012 *Measurement Science and Technology* **23** 045902
[9] Seel A G, Krzystyniak M and Fernandez-Alonso F 2014 *Journal of Physics: Conference Series* **571** 012006
[10] Postma H and Schillebeeckx P 2006 *Neutron Resonance Capture and Transmission Analysis* (John Wiley & Sons, Ltd) ISBN 9780470027318
[11] Festa G, Arcidiacono L, Pappalardo A, Minniti T, Cazzaniga C, Scherillo A, Andreani C and Senesi R 2016 *Journal of Instrumentation* **11** C03060
[12] Mayers J, Baciocco G and Hannon A 1989 *Nuclear Instruments and Methods in Physics Research Section A: Accelerators, Spectrometers, Detectors and Associated Equipment* **275** 453 – 459
[13] Last accessed on June 2018 URL `https://www.osti.gov/biblio/5858602`
[14] Seel A G, Senesi R and Fernandez-Alonso F 2014 *Journal of Physics: Conference Series* **571** 011001
[15] Romanelli G, Krzystyniak M, Festa G, Andreani C, Senesi R and Fernandez-Alonso F 2018 *Neutron News* **29** 4–6
[16] Andreani C, Senesi R, Krzystyniak M, Romanelli G and Fernandez-Alonso F 2018 *La Rivista del Nuovo Cimento* **41** 291–340
[17] Flammini D, Pietropaolo A, Senesi R, Andreani C, McBride F, Hodgson A, Adams M A, Lin L and Car R 2012 *The Journal of Chemical Physics* **136** 024504
[18] Ceriotti M 2014 *Journal of Physics: Conference Series* **571** 012011
[19] Krzystyniak M, Seel A G, Richards S E, Gutmann M J and Fernandez-Alonso F 2014 *Journal of Physics: Conference Series* **571** 012002
[20] Romanelli G, Fernandez-Alonso F and Andreani C 2014 *Journal of Physics: Conference Series* vol 571 p 012003
[21] Seel A G, Ceriotti M, Edwards P P and Mayers J 2012 *Journal of Physics Condensed Matter* **24** J5401
[22] Krzystyniak M, Druzbicki K, Romanelli G, Gutmann M J, Rudic S, Imberti S and Fernandez-Alonso F 2017 *Physical Chemistry Chemical Physics* **19**(13) 9064–9074

NEUTRONS MATTER

VII International Workshop on Electron-volt Neutron Spectroscopy

85th anniversary - J. Chadwick (1932)
80th anniversary - Enrico Fermi (1937)

7th – 8th November 2017
Villa Wolkonsky and Centro Studi e Ricerche Enrico Fermi
Rome, Italy

Neutrons Matter, the seventh edition of the *International Workshop on Electron-volt Neutron Spectroscopy*, celebrates the 85th anniversary *of the discovery of the neutron* by Chadwick, and the 80th anniversary *of the Patent issued by the Ministero delle Corporazioni* on 27th October 1937 to Enrico Fermi for the use of neutrons to enhance artificial radioactivity. The latest discovery took place at the Istituto di Fisica di Via Panisperna at the very heart of Rome. Following a successful sixth edition of this workshop series in the United Kingdom in 2014, the worshop aim to bring to the fore and highlight the most recent (and quite exciting) advances in the field, with an emphasis on the use of the technique across condensed matter research, as well as the concurrent use of contemporary computational materials modelling. The workshop will be held in Rome, in the beautiful setting of Villa Wolkonsky and at the Museo Storico della Fisica e Centro Studi e Ricerche Enrico Fermi.

PROGRAMME

Programme includes invited oral presentations and posters

Dates & Venue

7th Nov 2017 All day: Villa Wolkonsky.
Evening: conference dinner
8th Nov 2017 Morning: Villa Wolkonsky
Lunch and Afternoon: Museo Storico della Fisica e Centro Studi e Ricerche Enrico Fermi

International Advisory Committee

Prof Carla Andreani, Università di Roma Tor Vergata, Italy - Chair
Prof Felix Fernandez Alonso, STFC, UK - Secretary
Prof Roberto Car, Princeton University, US
Prof Michele Ceriotti, University of Lausanne, Switzerland
Dr Javier Dawidowski, Centro Atómico Bariloche, Argentina
Prof Giuseppe Gorini, Università degli Studi di Milano–Bicocca, Italy
Prof Francesco Mallamace, Università di Messina, Italy
Prof George Reiter, University of Houston, USA
Prof Roberto Senesi, Università di Roma Tor Vergata, Italy
Prof Corrado Spinella, CNR, Italy

Supporting Organizations

- British Embassy, Rome, Italy
- Consiglio Nazionale delle Ricerche (CNR), Italy
- Centro Studi e Ricerche Enrico Fermi, Italy
- School on Neutron Scattering "Francesco Paolo Ricci" (SoNS), Italy
- Science & Technology Facilities Council (STFC)
- UK Government Science and Innovation Network, UK
- Università degli Studi di Roma Tor Vergata, Italy

Organising Committee

Dr Giulia Festa, Centro Fermi, Italy - Chair
Dr Alessandra Ferraris, British Embassy, Italy
Dr Matthew Krzystyniak, STFC, UK
Dr Laura Nuccilli, British Embassy, Italy
Dr Dalila Onorati, Università di Roma Tor Vergata, Italy
Dr Giovanni Romanelli, STFC, UK

APPLICATION ABSTRACT HOTEL REGISTRATION

Attendance 60-70 participants, early-career researchers welcome

Topics

- New Materials for Energy and Functional Materials
- Fundamental Systems
- Information & Communication Technologies
- Simulation and Modelling
- Applications to cultural heritage
- Recent instrumentation and methodological developments resulting from joint collaborative efforts across the Consiglio Nazionale delle Ricerche and the UK Science & Technology Facilities Council.

"NEUTRONS MATTER" Prize

A prize will be awarded to the best poster presented by an early-career researcher. Poster session is scheduled on the 7th November at Villa Wolkonsky. The Prize ceremony will take place on the 8th November at Museo Storico della Fisica e Centro Studi e Ricerche Enrico Fermi. Poster Abstract.

Proceedings

Workshop Proceedings will be published in the *Journal of Physics*, UK Institute of Physics. Contributions will be reviewed by the International Advisory Committee. Proceedings Abstract.

Scientific Secretariat

Sara Miceli, Università di Roma Tor Vergata, Italy

Important dates

Request of hotel reservation extended deadline	**25th September 2017**
Proceedings abstract extended deadline:	**27th September 2017**
Notification of abstract acceptance	**15th October 2017**
Deadline for Poster Abstract:	**25th September 2017**

For questions related to the science programme:
Carla Andreani and Felix Fernandez Alonso
For questions related to the conference proceedings:
Giulia Festa and Giovanni Romanelli
For any other question:
Sara Miceli

British Embassy Rome · Centro Fermi · Consiglio Nazionale delle Ricerche · Science & Technology Facilities Council ISIS · UK Science & Innovation Network · SoNS

Figure 3. The flyer of the workshop.

IOP Conf. Series: Journal of Physics: Conf. Series **1055** (2018) 011002 doi:10.1088/1742-6596/1055/1/011002

Peer review statement

All papers published in this volume of *Journal of Physics: Conference Series* have been peer reviewed through processes administered by the proceedings Editors. Reviews were conducted by expert referees to the professional and scientific standards expected of a proceedings journal published by IOP Publishing.

Observation of the stretch mode in H_2 and D_2 by inelastic neutron scattering spectroscopy

S F Parker,[1] R A Ewings,[1] F Fernandez-Alonso[1,2] and D Colognesi[3]

[1]ISIS Facility, STFC Rutherford Appleton Laboratory, Chilton, Didcot, Oxon OX11 0QX, UK
[2]Department of Physics and Astronomy, University College London, Gower Street, London WC1E 6BT, UK
[3]Consiglio Nazionale delle Ricerche, Istituto dei Sistemi Complessi, via Madonna del Piano 10, I-50019 Sesto Fiorentino, (FI), Italy
E-mail: stewart.parker@stfc.ac.uk

Abstract. The interaction of H_2 with materials is relevant to areas such as catalysis, hydrogen storage and fuel cells. Inelastic neutron scattering has been extensively used for such studies, which have generally utilised the J 0→1 rotational line at 14.5 meV of parahydrogen. Unfortunately, this is very difficult to model. The problem would be mitigated by studying the fundamental H–H stretch vibration at 516 meV, because the calculation becomes a conventional lattice dynamics problem, which is tractable with density functional theory. Here, we have chosen to investigate solid and liquid H_2 and D_2 as a test of whether it is possible to observe the H–H and D–D stretch. This is only possible with a direct geometry spectrometer at a spallation neutron source, since only this combination of facilities gives access to high energy neutrons with a wide range of momentum transfer. The measurements were successful and, in addition to several rotational states of the ground vibrational state, a feature recoiling from an origin value of 516 meV for H_2 and from 370 meV for D_2 is observed in both the solid and liquid states. These are assigned to the H–H and D–D stretch respectively.

1. Introduction

The study of molecular hydrogen (H_2, dihydrogen) in, or on, materials is relevant to areas such as catalysis [1], hydrogen storage [2] and fuel cells [3]. Inelastic neutron scattering (INS) is well-suited to studies of such systems because the large ^{1}H cross section and the transparency of most materials to neutrons means that most of the information obtained directly relates to the adsorbed dihydrogen. These studies are almost invariably carried out with the J 0→1 rotational line at 14.5 meV of parahydrogen as the probe of choice [4]. The reasons for this are that not only its transition energy [4] but also its underlying line shape [5,6] are sensitive probes of the local environment and the instruments generally have their best resolution in this energy region.

The disadvantage of studying the rotational transition is that it is very difficult to model. This has been done with either a phenomenological model of how the rotational energy levels behave in an applied potential [4] or have required very sophisticated multi-dimensional *ab initio* calculations [7]. These difficulties could be reduced by studying the fundamental H–H stretch vibration at 516 meV, because the calculation is now a conventional lattice dynamics problem, which is tractable with density functional theory [8].

As a first step towards studying adsorbed hydrogen in materials, we have chosen to investigate solid and liquid H_2 and D_2 as a test of whether it is possible to observe the H–H/D–D stretch directly. Such a measurement is only possible with a direct geometry spectrometer at a spallation neutron source, since only this combination of facilities gives access to high energy neutrons with a wide range of momentum transfer (Q, Å^{-1}) [9]. The only previous attempt [10] at this measurement, that we are aware of, was inconclusive.

2. Experimental

The INS measurements were carried out with the direct geometry spectrometer MAPS [9, 11] at the ISIS Facility [12]. The high resolution "A" chopper package was used with incident energies, E_i, in the range 30 – 2000 meV. The sample, H_2 or D_2, was condensed into a thin-walled aluminium can with a path length of 1 mm. Parahydrogen was prepared in an apparatus that has been previously described [13].

3. Results and discussion

3.1. Solid and liquid H_2

The left side of figure 1 shows the $S(Q,\omega)$ map of solid H_2 at 7 K recorded with E_i = 2000 meV. The H_2 was nominally ~100% parahydrogen (p-H_2). The data consists of a series of apparent steps originating at zero energy transfer, in addition there is a continuous sweep of intensity originating at ~500 meV.

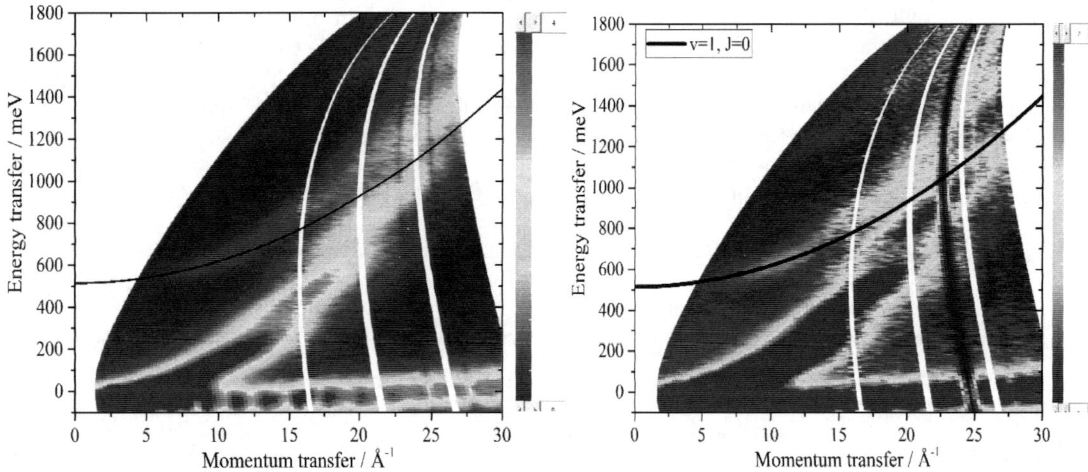

Figure 1. $S(Q,\omega)$ map of H_2 at 7 K recorded on MAPS with E_i = 2000 meV. Left: solid state at 7 K, right: liquid state at 18 K. The black curve shows the recoil trajectory of an H_2 molecule with one quantum of vibrational excitation.

When a light particle is struck by a neutron it recoils with an energy, E_{total} (meV), given by [4,8]:

$$E_{\text{total}} = \hbar\omega_{v,J'} + E_r = \hbar\omega_{v,J'} + \frac{\hbar^2 Q^2}{2M} = \hbar\omega_{v,J'} + \frac{2.09\,Q^2}{M} \quad (1)$$

where $\hbar\omega_{v,J'}$ is the energy (meV) of the molecule in vibrational level v (= 0, 1…) and rotational level J', E_r is the recoil energy (meV) and M is the molecular mass (amu). The black curve in figure 1 is the trajectory of an H_2 molecule with $\omega_{v,J'} = \omega_{1,0}$ *i.e.* in the first vibrational level. It can be seen that this accounts for the broad swath of intensity, particularly at the lower momentum transfers. At higher energy and momentum transfers, deviations from this prediction indicate that the ultimate scattering response tends to approach the atomic-recoil limit, readily accessible via the use of so-called deep inelastic neutron scattering techniques [14]. Figure 2 shows a series of Q-cuts in the region of the ~500 meV excitation. At the lowest Q available, 4.4 Å^{-1}, the peak is observed at an energy of 523 meV, in reasonable agreement with that predicted by equation (1) of 536 meV $(= \hbar\omega_{1,0} (= 516) + E_r(= 20))$, where 516 meV is the gas phase value of the H–H stretch [15]. We conclude that the upper band of intensity originating at ~500 meV is indeed the H–H stretch of solid H_2. We also note that, from the outset, this spectral feature exhibits prominent recoil associated with the entire H_2 molecule. This behaviour seems to violate the well-known Sachs-Teller tensor mass theory [16], which would classically predict an m = 4/3 amu recoil, instead of the observed m = 2 amu. This can be explained on the basis of the characteristic and quite disparate energy scales of centre-of-mass H_2 motions in the solid and liquid phases (about 5 meV) relative to the quantum H_2 rotations and the intramolecular H–H motions (516 meV, the highest-known vibrational energy of a fundamental in condensed matter). The excitation of the latter vibrational mode requires incident neutron energies such that the overall scattering response is essentially that of a free H_2 molecule. Compared to other hydrogen-containing systems studied to date via INS spectroscopy, H_2 may therefore be considered an interesting and special case characterised by two extreme limits in terms of the energy scales associated with intramolecular and intermolecular modes.

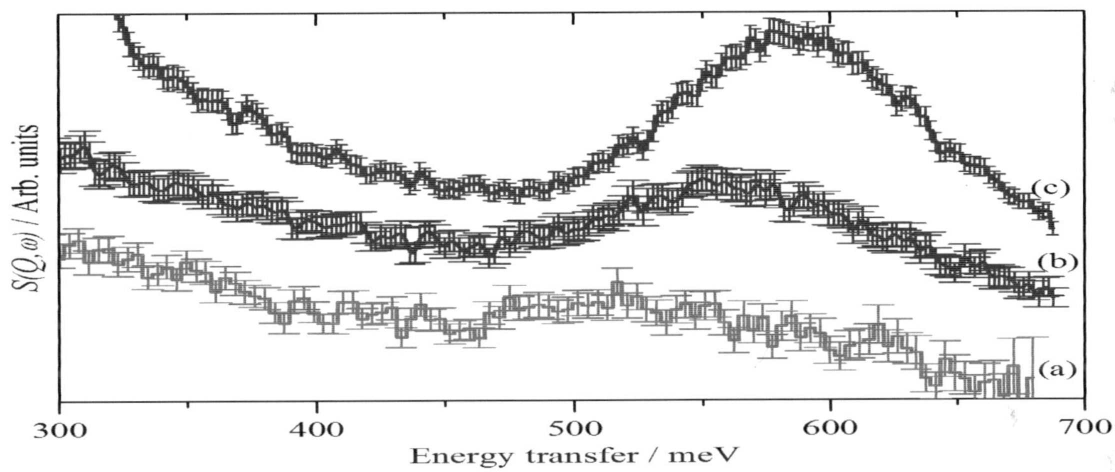

Figure 2. INS spectra of solid H_2 at 7 K recorded on MAPS with E_i = 2000 meV in the region of the H–H stretch. The spectra shown are generated from figure 1 by summing over the range: (a) $2 \leq Q \leq 6$, (b) $4 \leq Q \leq 8$ and (c) $6 \leq Q \leq 10$ Å^{-1}.

The series of steps in figure 1 are the recoil from rotational states of p-H_2 in the vibrational ground state: $\omega_{v,J'} = \omega_{0,1}, \omega_{0,3}, \omega_{0,5}\ldots$ The rotational states have energies, E_{rot}, given by:

$$E_{rot} = J(J+1)B \tag{2}$$

where B is the rotational constant (= 7.35 meV). For p-H_2, $J = 0$ is the ground state and only transitions to J = odd are intense – *i.e.* proportional to the incoherent-scattering cross section of hydrogen. These were previously observed by Langel *et al* [10] and, based on the Young and Koppel model [17], they showed that the rotational states exhibit a strong Q-dependence, such that they are only observable over particular Q-ranges, as seen here. There is also intensity at $Q > 15$ Å^{-1} that lies above the trajectory of the recoiling H–H stretch, it is possible that this is from rotationally excited states of the stretch mode: $\omega_{v,J'} = \omega_{1,3}, \omega_{1,5}\ldots$ (the $\omega_{1,1}$ state is indistinguishable from the stretch mode at this resolution). As noted [10], the rotational states are still observable in the liquid phase, right side of figure 1. Somewhat surprisingly, the H–H stretch mode is also still observable.

Figure 3 shows the region of the J 0→1 rotational line for the solid, 3a, and liquid states, 3b. The translational modes of the H_2 molecule (peak at 5 meV and the combination with the rotational mode at

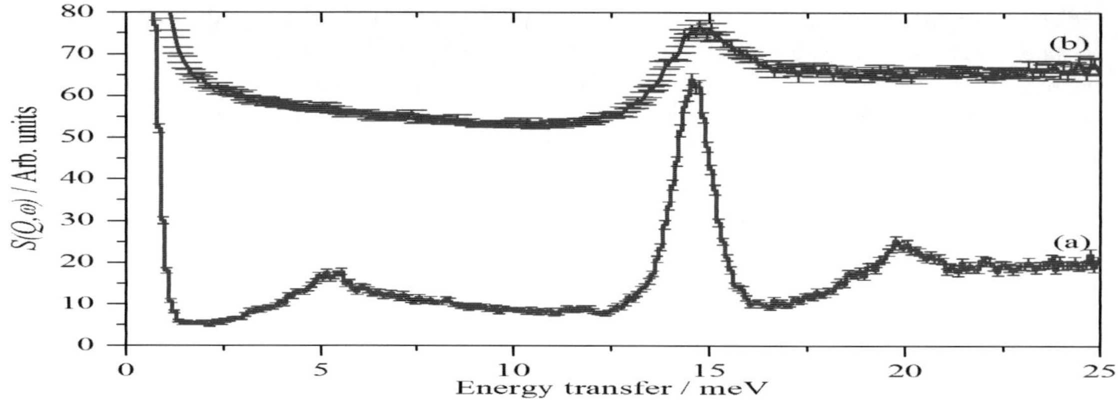

Figure 3. INS spectra of (a) solid H_2 at 7 K and (b) liquid H_2 at 18 K recorded on MAPS with E_i = 50 meV. The spectra shown are summed over the available Q range: 0.2 – 3.8 Å^{-1}.

20 meV) seen for the solid state, figure 3a, as expected, do not lead to well-defined spectral signatures in the liquid state.

3.2. Solid and liquid D_2

Figure 4 shows the *S(Q,ω)* map of solid D_2 at 7 K recorded with E_i = 1400 meV. In the left side figure, intensity recoiling from ~370 meV (the gas phase value of the D–D stretch [15]), this is more clearly seen when the same data is plotted on a logarithmic intensity scale. The solid curve shows the expected recoil line for D_2 with $\omega_{v,J'} = \omega_{1,0}$ and the feature is so assigned.

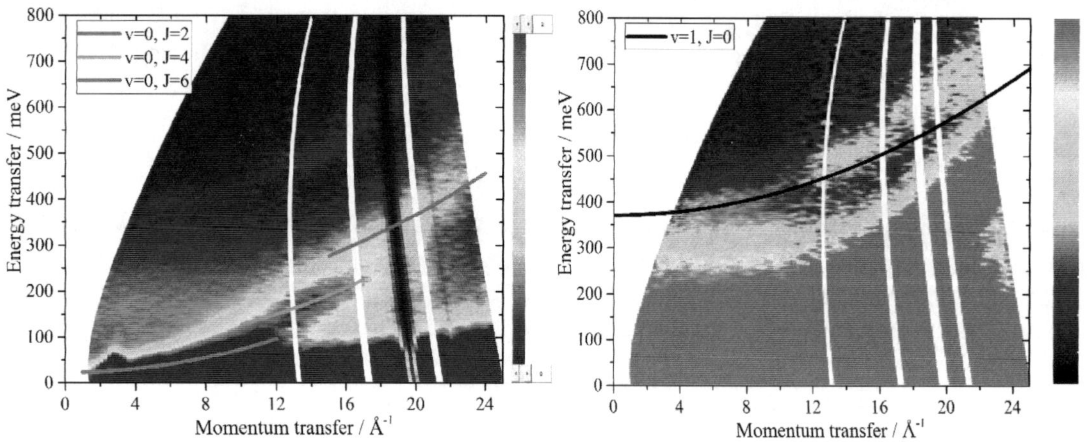

Figure 4. *S(Q,ω)* map of solid D_2 at 7 K recorded on MAPS with E_i = 1400 meV. Left: intensity plotted on a linear scale and right: intensity plotted on a logarithmic scale. The superimposed curves on the left side figure correspond to: $\omega_{v,J'} = \omega_{0,2}, \omega_{0,4}, \omega_{0,6} \ldots$, the black curve on the right side figure corresponds to $\omega_{v,J'} = \omega_{1,0}$.

As with H_2, the D–D stretch mode is also visible in the liquid state. The left side of figure 5 shows the *S(Q,ω)* map of the liquid at 18 K and the right side shows a series of *Q*-cuts in the region of the stretch mode.

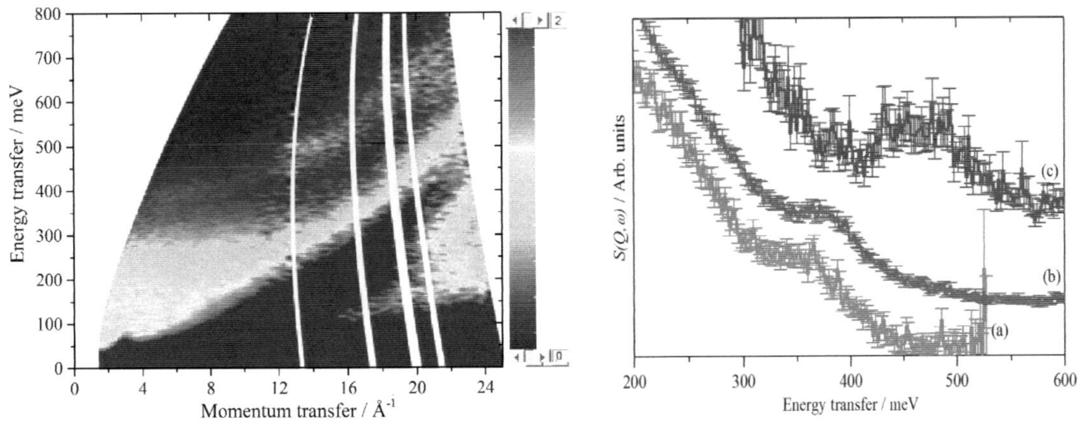

Figure 5. Left: $S(Q,\omega)$ map of liquid D_2 at 18 K recorded on MAPS with E_i = 1400 meV. Right: The region of the D–D stretch. The spectra shown are generated from the left part of this figure by summing over the range: (a) $2 \leq Q \leq 6$, (b) $6 \leq Q \leq 10$ and (c) $10 \leq Q \leq 16$ Å^{-1}.

D_2 is also a mix of ortho, o-D_2, and para, p-D_2, spin states, although because of the different nuclear spin, o-D_2 is the lowest energy state. For deuterium, the coherent, b_{coh}, and incoherent, b_{inc}, scattering lengths are similar: 6.671 and 4.04 fm respectively (1 fm = 1×10^{-15} m), thus incoherent transitions (*i.e.* involving a change in the total nuclear spin of the molecule) in both the o-D_2 and p-D_2 manifolds are possible [18]. For o-D_2 transitions of the type $\omega_{v,J'} = \omega_{0,2}, \omega_{0,4}, \omega_{0,6}...$ the cross-section is: $\{4(b_{coh})^2 + 2.5(b_{inc})^2\}$ = 2.19 barn and transitions of the type $\omega_{v,J'} = \omega_{0,1}, \omega_{0,3}, \omega_{0,5}...$ have a cross-section of $\{1.5(b_{inc})^2\}$ = 0.24 barn. Thus for o-D_2, only $J = 0 \rightarrow J'$ = even transitions would be expected to be visible in our spectra.

For p-D_2, $J = 1 \rightarrow J'$ = odd transitions have a cross section of: $\{4(b_{coh})^2 + (b_{inc})^2\}$ =1.94 barn, while transitions to an even state are given by: $\{3(b_{inc})^2\}$ = 0.49 barn. So one would expect to see the $J = 0 \rightarrow J' = 0, 2, 4, 6...$ as strong lines originating from o-D_2 and the $J = 1 \rightarrow J' = 1, 3, 5, 7...$ transitions originating from p-D_2 as weaker. The former will be located at ω =0, 6B, 20B, 42B ... (plus the recoil), while the latter are at ω = 0, 10B, 28B, 54B... (plus the recoil), with the rotational constant B = 3.774 meV.

For these measurements D_2 was condensed from the gas phase. At room temperature, D_2 has an ortho content of 66.7% [18], thus this would be the minimum o-D_2 content of our sample. However, the (at least) twofold larger quantity of o-D_2, means that these are the more readily observed transitions and, as may be seen from the left side of figure 4, the first three transitions are apparent and have the expected Q-dependence. Unfortunately, we are unable to resolve the p-D_2 transitions from the stronger o-D_2 transitions because the separation between the features is less than the instrument resolution, ~30 meV, with these operating conditions.

The left part of figure 6 shows the $S(Q,\omega)$ map in the low energy region. In contrast to H_2, there is a strongly dispersive mode. This goes to zero at $Q = 2.1 \approx 4\pi/(a\sqrt{3})$ Å^{-1}, where a is the basal lattice parameter, 3.596 Å, it has a maximum energy of ~9 meV, indicating that it corresponds to the longitudinal acoustic phonon [19]. The right side of figure 6 shows a series of Q-cuts: 6a is at the peak of the dispersion curve, Q ~1 Å^{-1}, 6b is where the mode goes to zero, Q ~2 Å^{-1} and 6c is summed across the entire Q-range, this shows reasonable agreement with the density of states calculated from single crystal measurements [20]. From 6b, modes are evident at 4.8 and 7.3 meV, in agreement with [18], these are assigned as a transverse longitudinal acoustic phonon [19] and the J 0→1 transition respectively. The latter is more clearly seen in figure 6c, which also shows weak additional transitions at 14 and 23 meV. These are assigned as the J 1→2 transition of p-D_2 and the J 0→2 transition of o-D_2 respectively.

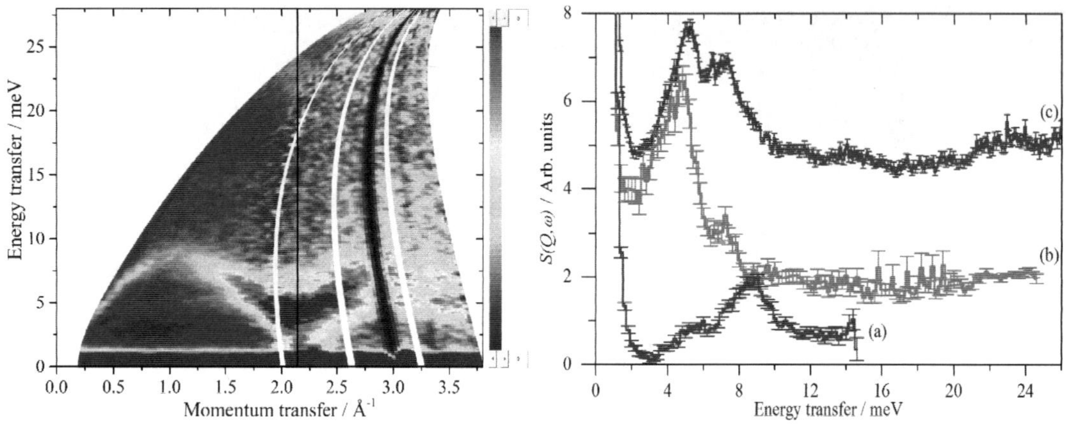

Figure 6. Left: $S(Q,\omega)$ map of solid D_2 at 7 K recorded on MAPS with E_i = 30 meV. Right: spectra shown generated from the left part by summing over the range: (a) $0.9 \leq Q \leq 1.1$, (b) $2 \leq Q \leq 2.2$ and (c) $0.2 \leq Q \leq 3.8$ Å^{-1}.

4. Conclusions

The motivation for this work was to test if it was possible to observe the H–H stretch mode of solid H_2. As figures 1 and 4 show, this is not only possible for H_2 but also for D_2. The cell that was used contained ~0.1 mole of H_2 or D_2. Given that the D–D stretch was still observable in the latter, together with the difference in the cross section of H and D, this suggests that the stretch mode in as little as 0.01 mole of H_2 should be detectable. As described in the Introduction, this will open new ways to investigate adsorbed H_2 in technologically interesting materials.

Acknowledgement
The STFC Rutherford Appleton Laboratory is thanked for access to neutron beam facilities.

References

[1] Christmann K 1988 *Surf. Sci. Rep.* **9** 1-163.
[2] Ramirez-Cuesta A J, Jones M O and David W I F 2009 *Materials Today* **12** 54-61.
[3] Mitchell P C H, Parker S F, Tomkinson J and Thompsett D 1998 *J. Chem. Soc. Faraday Trans.* **94** 1489-1493.
[4] Mitchell P C H, Parker S F, Ramirez-Cuesta A J and Tomkinson J 2005 *Vibrational Spectroscopy With Neutrons* (Singapore, World Scientific) chapter 6 pp 219-283.
[5] Fernandez-Alosno F, Bermejo F J, Cabrillo C, Loutfy R O, Leon V and Saboungi ML 2007 *Phys. Rev. Lett.* **98** 215503.
[6] Fernandez-Alonso F, Cabrillo C, Fernández Perea R, Bermejo F J, González M A, Mondelli C and Farhi E 2012 *Phys. Rev.* B **98** 215503.
[7] Celli M, Powers A, Colognesi D, Xu M, Bačić Z and Ulivi L 2013 *J. Chem. Phys.* **139** 164507.
[8] Clark S J, Segall M D, Pickard C J, Hasnip P J, Probert M J, Refson K and Payne M C 2005 *Z. Krist* **220** 567-570.
[9] Parker S F, Lennon D and Albers P W 2011 *Applied Spectroscopy* **65** 1325-1341.
[10] Langel W, Price D L, Simmons R O and Sokol P E 1988 *Phys. Rev.* B **38** 11275-11283.
[11] https://www.isis.stfc.ac.uk/Pages/maps.aspx
[12] https://www.isis.stfc.ac.uk/Pages/home.aspx
[13] Kibble M G, Ramirez-Cuesta A J, Goodway C M, Evans B E and Kirichek O 2014 *J. Phys.: Conf. Series* **554** 012006.
[14] Andreani C, Krzystyniak M, Romanelli G, Senesi R and Fernandez-Alonso F 2017 *Adv. Phys.* **66** 1-73.

IOP Conf. Series: Journal of Physics: Conf. Series **1055** (2018) 012001 doi:10.1088/1742-6596/1055/1/012001

[15] Herzberg G 1939 *Spectra of Diatomic Molecules* (New York, Van Nostrand Reinhold Company) p533.
[16] Mitchell P C H, Parker S F, Ramirez-Cuesta A J and Tomkinson J 2005 *Vibrational Spectroscopy With Neutrons* (Singapore, World Scientific) chapter 2 pp 61-64.
[17] Young J A and Koppel J U 1964 *Phys. Rev.* **135** A603-A611.
[18] Frei A, Gutsmiedl E, Morkel C, Müller A R, Paul S, Urban M, Schober H, Rols S and Unruh T 2009 *Phys. Rev.* B **80** 064301.
[19] Nielsen M 1973 *Phys. Rev.* B **7** 1626-1635.
[20] Schmidt J W, Nielsen M and Daniels W B 1984 *Phys. Rev.* B **30** 6308-6319.

IOP Conf. Series: Journal of Physics: Conf. Series **1055** (2018) 012002 doi:10.1088/1742-6596/1055/1/012002

Inelastic and deep inelastic neutron spectroscopy of water molecules under ultra-confinement

A I Kolesnikov[1,*], G F Reiter[2], T R Prisk[3], M Krzystyniak[4], G Romanelli[4], D J Wesolowski[5,#], and L M Anovitz[5]

[1]Neutron Scattering Division, Oak Ridge National Laboratory, Oak Ridge, Tennessee 37831, USA

[2]Physics Department, University of Houston, Houston, Texas 77204, USA

[3]Center for Neutron Research, National Institute of Standards and Technology, Gaithersburg, Maryland 20899-6100, USA

[4]ISIS Facility, Rutherford Appleton Laboratory, Chilton, Didcot, Oxfordshire OX11 0QX, UK

[5]Chemical Sciences Division, Oak Ridge National Laboratory, Oak Ridge, Tennessee 37831, USA

[*]Corresponding author: kolesnikovai@ornl.gov
[#]Retired from Chemical Sciences Division ORNL

Abstract. Water confined within sub-nanometer channels of silicate minerals presents an extreme case of confinement, where the restricted molecules are situated in channels whose diameter is not much larger than the water molecule itself. Recently, we discovered a new quantum tunneling state of the water molecule confined in 5 Å channels in the mineral beryl, characterized by extended proton and electron delocalization. Several peaks were observed in the inelastic neutron scattering (INS) spectra which were uniquely assigned to water quantum tunnelling. In addition, the water proton momentum distribution measured with deep inelastic neutron scattering (DINS) at 4.3 K directly showed coherent delocalization of the water protons in the ground state. The obtained average kinetic energy (E_K) of the water protons was found to be ~30% less than it is in bulk liquid water and ice phases. In the current work we present INS and DINS study of water in single crystal beryl in wider temperature range, T=5−260 K, where we observed significant increase of E_K of the confined water protons with temperature increase. The obtained INS data also indicate that with increasing temperature water molecules are progressively involved in hydrogen bonding (HB) with the beryl cage, while HB is almost absent at low temperatures.

IOP Conf. Series: Journal of Physics: Conf. Series **1055** (2018) 012002 doi:10.1088/1742-6596/1055/1/012002

1. Introduction

Water has many unusual properties in bulk states and under confinement [1-8]. The behavior of water confined within the nanopores with very weak water to pore surface interaction presents special interest because the reduced dimensionality could result in new phenomena, e.g. anomalously soft dynamics of water in quasi one-dimensional single-wall carbon nanotubes (SWNT) [9-12]. There are extreme cases when water is confined within the pores of a few Angstrom size, which can accommodate only one water molecule, such as water incapsulated in C_{60}-fullerenes [13], or water in mineral beryl [14]. Due to weak interaction with the walls, the confined water displays significant quantum behavior. In case of fullerenes water is almost free and quantum transitions between the para and ortho states of water in C_{60}-fullerenes were observed in inelastic neutron scattering (INS) experiments [15, 16], and quantum tunneling of water was suggested to explain the fine structure of the optical spectra of beryl [17-19]. Recently, using INS, deep inelastic neutron scattering (DINS) and *ab initio* calculations we discovered new quantum tunneling state of water in beryl, characterized by extended proton and electron delocalization [20]. The observed number of peaks in the INS spectra were uniquely assigned to water quantum tunneling, and the DINS study directly revealed that in the ground state water protons are coherently delocalized and have anomalously small average kinetic energy, E_K.

In the current work we present results of INS and DINS study of water in single crystal beryl in wide temperature range, T=5–260 K, to understand the vibrational dynamics of the water at higher temperatures, when all split ground states of water protons are well populated. We show that at high temperatures E_K of the water protons significantly increases (due to observed broadening of the proton momentum distribution).

2. Experimental

In this study we used low-alkali beryl single crystals (~0.1 wt.% of Na, see ref. 21 for details). The beryl structure ($Be_3Al_2Si_6O_{18}$, space group *P6/mcc*) consists of six-membered rings SiO_4 tetrahedra with open channels along the *c*-axis [14]. The channels form cavities 5.1 Å in diameter (which may contain alkali cations and water) separated by narrow (~2.8 Å) "bottlenecks". In alkali-poor beryl the water dipole moment is in the *ab*-plane and water protons are distributed over 6-fold equivalent positions across the channel, above and below the oxygen site. The used samples contain 0.6 water molecules per beryl formula unit, or about 2 wt.% water. Neutron spectroscopy is a preferred technique to study dynamics of the hydrogen containing materials, because neutron scattering cross-section for hydrogen (~82 barn for thermal neutrons, see [22]) is much larger than for other atoms, particularly this is right for beryl (7.63, 1.503, 2.167 and 4.232 barn for Be, Al, Si and O atoms, respectively), and INS can measure all vibrational modes.

2.1. Inelastic neutron scattering

The INS measurements were performed using the fine-resolution Fermi chopper spectrometer SEQUOIA at the Spallation Neutron Source at the Oak Ridge National Laboratory [23, 24]. To observe all possible vibrations with a good energy resolution ($\Delta E/E \approx$ 1-2% in the energy range of interest) we used five incident neutron energies, E_i=25, 50, 160, 250 and 800 meV. The measurements were conducted for two sample orientations, with neutron momentum transfer **Q** approximately parallel and perpendicular to the beryl crystal *c*-axis (see details in refs. [21, 25]). Most of the data were collected at seven temperatures, T=5, 25, 45, 80, 120, 200 and 260 K, which were controlled by closed-cycle helium refrigerator. The INS spectra from the empty container at the similar conditions were also measured and subtracted from the sample data. The collected INS data were corrected for the detectors efficiency and transformed from the time-of-flight and instrument coordinates to the dynamic structure factor $S(Q,E)$ by using the MANTID [26] and DAVE [27] software packages.

2.2. Deep inelastic neutron scattering

The DINS (also known as Neutron Compton scattering) measurements [28, 29] were done using the VESUVIO spectrometer [30] at the ISIS pulsed neutron source. VESUVIO is an inverse-geometry

IOP Conf. Series: Journal of Physics: Conf. Series **1055** (2018) 012002 doi:10.1088/1742-6596/1055/1/012002

machine based on neutron capture resonances and its technical details can be found in [30, 31]. In the current experiment, we use 64 detectors, arranged with scattering angles from 37 to 67 on both sides of the beam. The data are corrected for multiple scattering [32] and a γ-ray background [33].

3. Results and discussion

3.1. Vibrational spectra and tunnelling modes

Figures 1 to 4 show the INS spectra for water in beryl at different temperatures. We used in this study different single crystals of beryl than in our previous work [20], but the results are well reproduced. The spectrum obtained at 5 K with E_i=25 meV (which provides good energy resolution data for energy transfer in the range 2 to 22 meV) for orientation **Q**⊥*c*-axis shows strong peaks at 8.4, 12.7 and 14.7 meV, which significantly decrease their intensities on heating to 25 K, and become invisible at higher temperatures. These peaks in the INS spectrum were assigned [20] to be due to the transitions between the lower and upper split ground state of water due to water tunneling between the six symmetrically equivalent positions around the *c*-axis in the channels of the beryl. This is understandable, because with temperature increase the lower ground state is depopulated due to transition to the higher split ground state according to Boltzmann law. The spectra for the orientation **Q**||*c*-axis show only one strong peak at around ~11 meV, which can be assigned to translational vibrations of water molecules along the *c*-axis. This peak is well observed up to the highest temperature 260 K. What is interesting, the position of the peak shows almost linear blue shift with temperature, from 10.9 meV at 5 K to 11.7 meV at 260 K. The peak width (full width at half maximum) also increases linearly from 0.71 meV at 5 K to 1.53 meV at 260 K. The large shift (7.3%) of the peak position to higher energy can be explained by negative thermal expansion of the beryl along the *c*-axis at temperatures below 600 K observed in previous x-ray diffraction results [34]. At room temperature the thermal expansion of beryl is $+2.6\cdot10^{-6}$ K^{-1} and $-2.9\cdot10^{-6}$ K^{-1} along the *a*- and *c*-axis, respectively, which is consistent with the values obtained in other studies [35, 36]. With temperature increase the intensity of this peak first increases (due to increase of Bose population factor) and then decreases (at T>150 K) due to suppression by the Debye-Waller factor, which we discussed in [20].

The spectra measured with E_i=50 meV (Figure 2, left panel) show a prominent band between 18 and 28 meV with two maxima at 21 and 24.5 meV for both orientations, with intensity of the band for **Q**⊥*c*-axis being about 1.6 times larger. We attribute this band to librational motions of the ultra-confined water molecules. This is consistent with the pioneering INS investigations of beryl [37], Raman spectroscopy studies [38], and first-principles calculations [39]. The band observed between 36 and 55 meV (Figure 2) is probably due to two-phonon neutron scattering involving the librational band at 18-28 meV. Other peaks are observed at 85, 96 and 151 meV for **Q**⊥*c*-axis orientation, and at 86, 114 and 128 meV for **Q**||*c*-axis (see Figure 2, right panel and Figure 3). The intensities of these phonon modes change a little with temperature increase from 5 to 260 K, and can be assigned to beryl cage vibrations [40], Al-O (85-86 meV), Be-O (96 meV) and Si-O (114, 127, 151 meV).

The peak at 197.5 meV is strong for **Q**||*c*-axis and very weak for **Q**⊥*c*-axis orientation (see Figure 3), and this peak can be assigned to water intramolecular bending H-O-H mode [20, 21], as we described in our previous study, in agreement with numerous literature data [17-19, 38-40]. The peaks at about 208.5, 219.3 and 230.1 meV can be well described as combination of the bending mode and the translational vibrations of water molecule along the *c*-axis (at 10.9 meV at 5 K).

The water intramolecular stretching O-H modes are observed at ~465 meV as strong and weak peaks for **Q**||*c*-axis and **Q**⊥*c*-axis orientations, respectively (see Figure 4). The obtained very large energy for the intramolecular stretching modes, and small value for the intramolecular bending mode (which are almost the same as for free water molecule) are in agreement with the assumption of absence of hydrogen bonds acting on water molecules in beryl, in accord with previous studies [17-19, 38-40]. The intensity of the peak at 465 meV for **Q**⊥*c*-axis orientation is much smaller than it should be based only on the

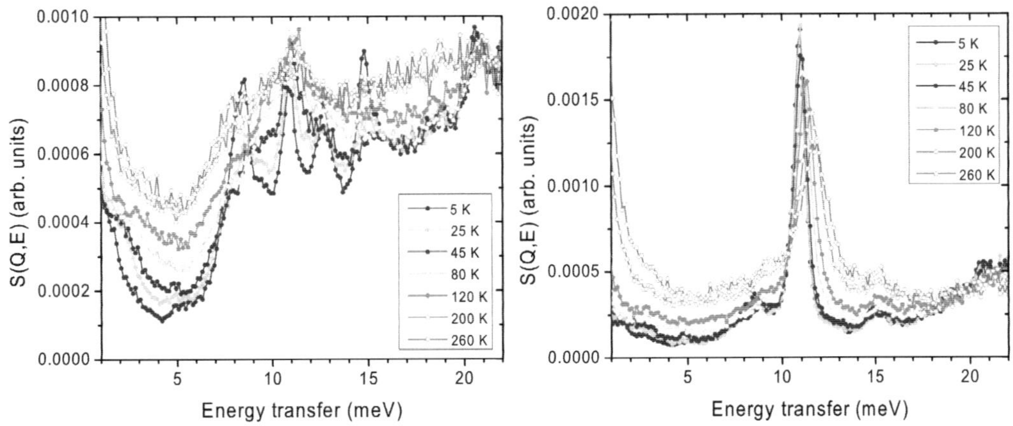

Figure 1. INS spectra of water in single crystal beryl at different temperatures, 5 to 260 K, measured with E_i=25 meV for neutron momentum transfer being approximately perpendicular and parallel to the crystal *c*-axis (left and right panels, respectively).

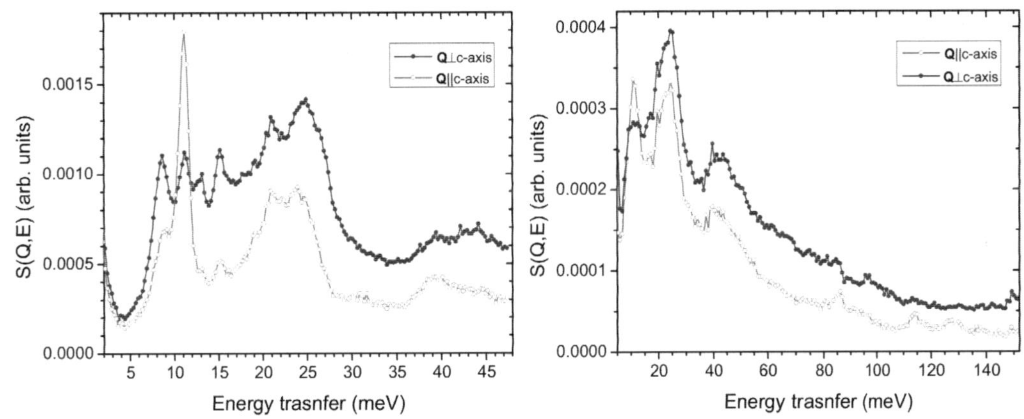

Figue 2. INS spectra of water in single crystal beryl at 5 K, measured with E_i=50 meV and 160 meV (left and right panels, respectively)

projections of the hydrogen displacement due to the stretching modes on the *ab*-plane and along *c*-axis, therefore an additional suppression of the peak should be involved due to the Debye-Waller factor, $DW=exp[-<(\mathbf{u}_H\mathbf{Q})^2>]$, where $\mathbf{u}_H$ is vibrational displacement of water protons. Hence, the data suggest larger mean squared displacement (MSD) of water protons in *ab*-plane. With temperature increase, the maximum value of the peak decreases (due to increase of the MSD of water protons), and a prominent excess of intensity is observed at the left side of the peak, between 405-440 meV. One might be tempted to attribute the increase in intensity on the left-hand side of the peak to multiphonon neutron scattering. Here the multiphonon process would consist of the simultaneous creation an O-H stretching mode quantum and the annihilation of low energy (translational and/or librational) water quanta. However, in that scenario, the effect would be proportional to the Bose population factor governing the low energy modes, $n(E_{LE},T)$. This implies that there should be an accompanying even larger increase in the scattering on the high-energy side of the peak, proportional to $n(E_{LE},T)+1$, which is inconsistent with the

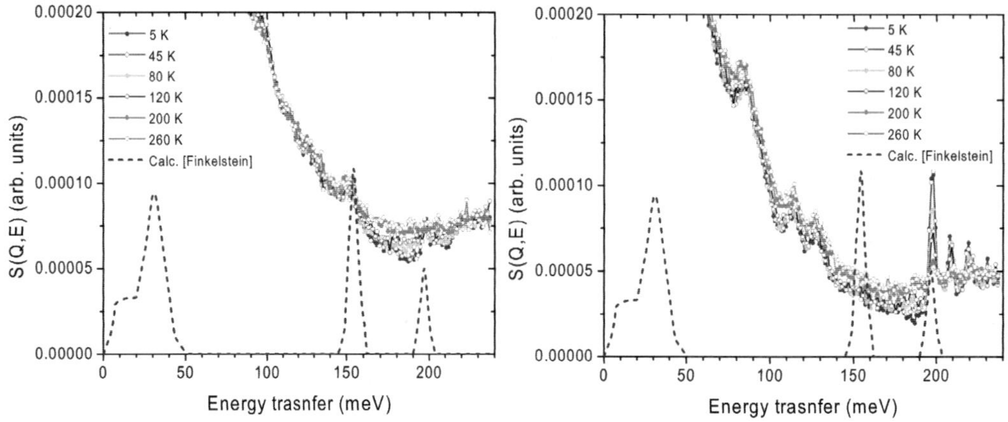

Figure 3. INS spectra of water in single crystal beryl at different temperatures, 5 to 260 K, measured with E_i=250 meV for neutron momentum transfer being approximately perpendicular and parallel to the crystal *c*-axis (left and right panels, respectively). Dashed lines show the calculated spectrum from [45].

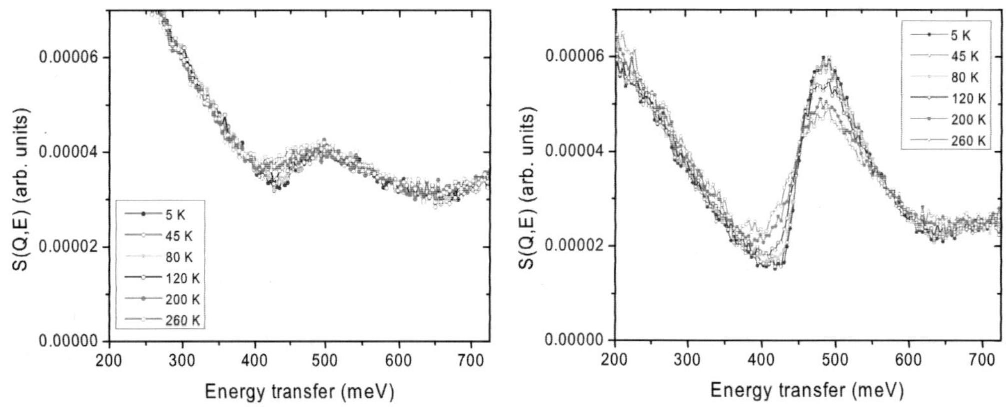

Figure 4. INS spectra of water in single crystal beryl at different temperatures, 5 to 260 K, measured with E_i=600 meV for neutron momentum transfer being approximately perpendicular and parallel to the crystal *c*-axis (left and right panels, respectively).

observed behavior. Rather, the additional intensity on the low energy side of the O-H stretching peak indicates significant softening of the stretching modes for at least some of the water molecules at higher temperatures (already at 120 K and above). Such softening is likely due to hydrogen bonding between the water molecules and the beryl cage.

This result sheds light on a discrepancy in our previous study [21] between the low temperature INS data, which suggested that no hydrogen bonding is present, and the high temperature (T=233-363 K) dielectric spectroscopic (DS) results, which proposed the presence of hydrogen bonds for water in beryl. Therefore, based on the INS and DS [21] results we propose that water in beryl does not form hydrogen bonds at low temperatures, and the hydrogen bonds progressively appear for some water molecules at higher temperatures.

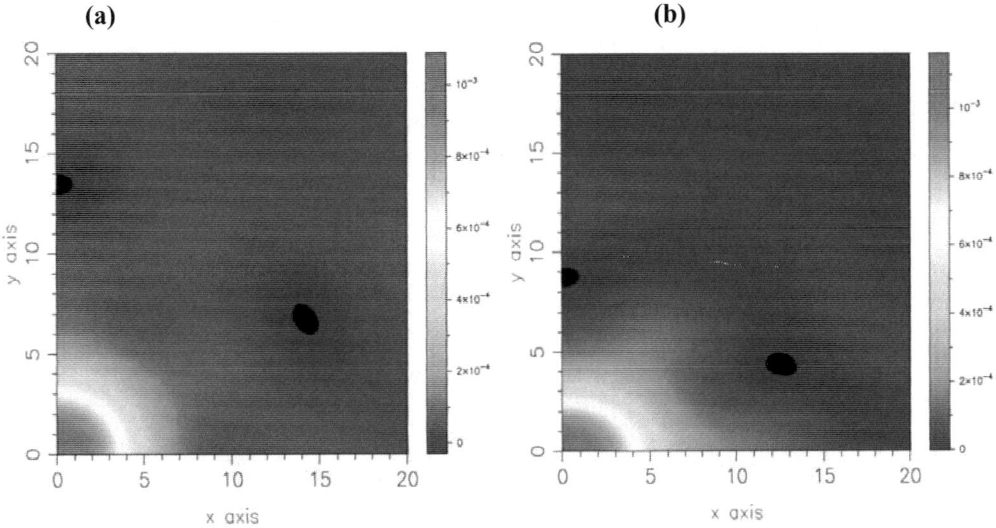

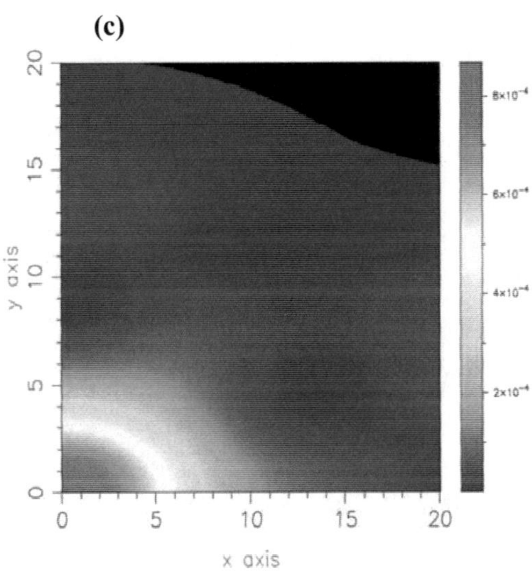

Figure 5. Contour plot of projection of water proton momentum distribution in beryl onto the *xy*-plane, obtained from the DINS data collected at **(a)** T=5 K, **(b)** T=60 K, and **(c)** T=260 K. Black spots in the (a) and (b) plots are due to slightly negative values of the fitting results of the momentum distribution.

3.2. Water protons momentum distribution and kinetic energy

The DINS spectra for beryl were collected for two orientations of beryl singles crystals, with their *c*-axis being vertical (perpendicular to the incoming neutron beam), and with *c*-axis being aligned along the incoming neutron beam. The data were treated with the use of MANTID software package [41]. The data for both configurations were fitted simultaneously to a 3D momentum distribution that is the average distribution of the two protons of the water molecule. We used a basis of Hermite polynomials of tenths order (with the terms consistent with the sixfold symmetry of the beryl channel) and spherical harmonics [12]. The *z*-axis of the fitting was chosen along the crystal *c*-axis, and the *xy*-plane was perpendicular to *z*-axis (parallel to the *ab*-plane). Figure 5 shows the contour plots of projection of water proton momentum distribution in beryl onto the *xy*-plane at different temperatures, and Figure 6 shows the curves fitted to the measured Compton profiles along the *x*-, *y*- and *z*-axis, and the corresponding parameters σ_x, σ_y and σ_z, that determine the width of the momentum distribution along the corresponding

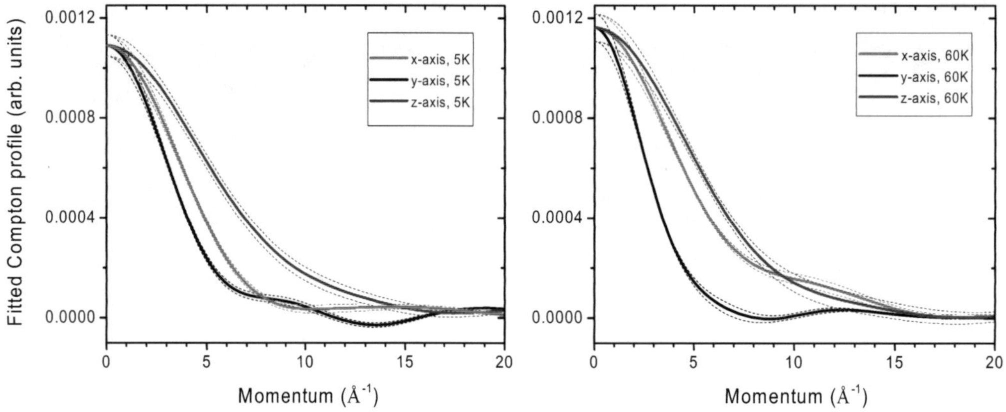

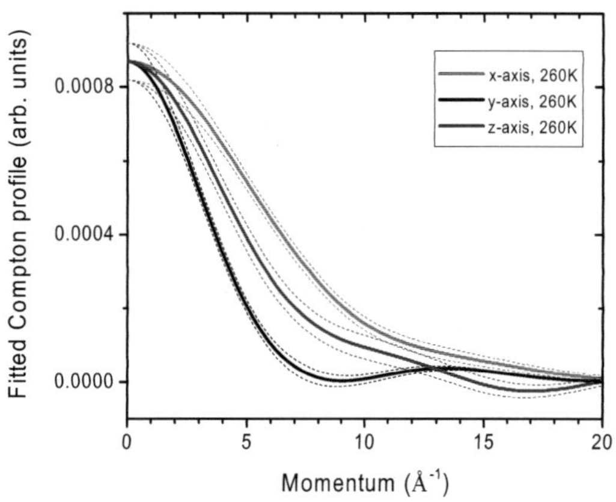

Figure 6. The fitted curves to the measured Compton profiles for water in beryl along *x*-, *y*- and *z*-axis at 5 K, 60 K and 260 K.

axis, are presented in Table 1. We notice that the fitting results in Figures 5 and 6 show slightly negative values of the momentum distribution. It can be extremely difficult to impose non-negativity constraints on the intensity of the momentum distribution when several Hermite polynomials are considered. As non-constrained coefficients can affect the value of the mean kinetic energy, one should pay attention to such aspect. In this light, the article by Krzystyniak in these proceedings [42] discusses the analytic constraints amongst the fitting parameters in the case of polynomials up to the fourth order. In our results, the slight negativity of the momentum distribution could be justified by including the fitting error bars, therefore making the result overall compatible with the non-negativity constraints.

The average kinetic energy of proton E_K is proportional to the second moment of momentum distribution and can be expressed as:

$$E_K = \frac{\hbar^2}{2m}(\sigma_x^2 + \sigma_y^2 + \sigma_z^2), \quad (1)$$

where m is mass of proton. The obtained E_K are 104.0 meV, 207.7 meV and 210.3 meV for 5, 60 and 260 K, respectively. The widths of momentum distribution E_K at 5 K is in very good agreement with our old data at 4.3 K measured on different beryl sample. We anticipated increase of average kinetic energy of water proton at higher temperatures due to occupation not only the lower split ground state (as at 5K) but also other split ground states, which should result in broadening of the momentum distribution. The

Table 1. The widths (Å^{-1}) of momentum distribution for water proton in beryl.

Temperature, K	σ_x	σ_y	σ_z
4.3 (ref. [20])	3.66	3.61	4.98
5	4.13	3.40	4.65
60	5.83	3.51	7.35
260	6.77	3.90	6.37

observed two times increase of E_K already at 60 K is very unusual, but so large values for proton kinetic energy were documented before for water confined in double wall carbon nanotubes of 16 Å inner diameter (198 meV at 120 K, and 221 meV at 170 K), in Nafion-1120* (245 meV at 300 K), and Dow-858* (262 meV at 300 K) [43]. So, at low temperature (5 K) water in beryl shows coherent quantum delocalization (tunneling) over six equivalent positions across the channel, and water protons are localized in inverse (momentum) space, having narrow momentum distribution and low average kinetic energy. With temperature increase, water still tunnel between these positions, and it progressively occupies other split ground states, which results in increase of E_K.

In the semi-empirical (SE) harmonic approximation (see e.g. [44]) due to very large values of the O-H stretching modes of water in beryl (~465 meV) the estimated E_K, with the use of only water intramolecular modes, should be 132.5 meV [20, 44]. The measured energy of the stretching modes is the energy difference between the first excited and ground states, and the energy of the ground state is assumed to be half of this difference. Probably, the energy of the ground state of the tunneling particle is much lower than that for particle in single harmonic potential.

Recently, DFT based first-principles calculations were carried out for beryl [45], where the obtained generalized vibrational density of states (GVDOS) for water protons (shown as dashed lines in the Figure 3) well describes our INS spectra: the intramolecular stretching and bending modes are in excellent agreement with the experiment. The authors [45] did not make assignments for the low-energy modes, but the part of the spectrum below 20 meV could be probably related to the water translational modes (there is a peak in this range of GVDOS for oxygen atoms, which are mostly involved in the translational vibrations of water), therefore the peak around 30 meV should be due to water librational vibrations, which is in accord with INS data. The calculated (high intensity) peak at 154 meV is the only one, which does not correspond to the peak from water protons in the measured spectrum. There is actually a peak in the INS data (obtained for **Q**⊥*c*-axis orientation) but the temperature dependence of the intensity of the peak and other literature data indicate that this peak is due to vibrations of beryl cage, as discussed above. Using the DFT derived GVDOS for water proton the authors [45] calculated the mean kinetic energy (E_K=104.4 meV) which is in remarkable agreement with the 5 K DINS data. The big discrepancy between the new calculated value and the old one, estimate in SE approach, was explained by different distribution of GVDOS of protons, most probably due to coupling with vibrational modes of the beryl cage. We think that, unfortunately, there is a mistake in the calculated GVDOS for water in beryl. From the values of energy fractions shared by the H and O atoms in beryl presented in Table VI of the ref. [45], it follows violation of one from two known sum rules, which are known as "the orthonormality and closure conditions" (see e.g. [46]):

$$\sum_{i=1}^{3} |\mathbf{e}(i|\mathbf{k}j)|^2 = 1 \tag{2}$$

and

$$\sum_{j=1}^{9} |\mathbf{e}(i|\mathbf{k}j)|^2 = 1, \tag{3}$$

where *i* is the atomic number in the primitive cell (*i*=1 to 3, for 2 hydrogen atoms and 1 oxygen), ***k*** is a wave vector, and *j* stands for phonon branches (*j*=1 to 9, for 3 translational, 3 librational and 3 intramolecular modes for H_2O). The squared phonon eigenvectors in the above equations directly relate to energy fractions used in [45]. The Eq. (3) is well satisfied, while the Eq. (2) is not: the sum over all atoms is about twice larger for the lowest energy mode, and the sum is twice smaller for intramolecular modes, than they should be. The proposed explanation [45] that this (very large) violation is due to coupling of the water intramolecular vibrations with the hosting beryl nano-cage does not look acceptable because in this case the energy of the intramolecular modes should be strongly changed, while they almost coincide to those for the free water molecule (with no HB). Therefore, the very low average kinetic energy of water protons in beryl at 5 K obtained from the DINS data is still awaiting a quantitative explanation.

4. Conclusion

We studied dynamics of single crystal beryl in wide temperature range, from 5 to 260 K, by using INS and DINS technique. The obtained data confirm our previous study [20] that at 5 K water molecule exhibits quantum tunneling behavior, and protons have very small average kinetic energy (104 meV). At 60 K, and also at 260 K, the E_K increases about two times. In the INS spectra we observed also appearing of the low energy feature of the water O-H stretching peak with temperature increase (at 120 K and above) which indicates progressive appearance of the hydrogen bonding of some water molecules with the beryl cage.

Acknowledgements

The neutron scattering experiment at Oak Ridge National Laboratory's Spallation Neutron Source was sponsored by the Scientific User Facilities Division, Office of Basic Energy Sciences, U.S. Department of Energy. This research was sponsored by the Division of Chemical Sciences, Geosciences, and Biosciences, Office of Basic Energy Sciences, U.S. Department of Energy. The STFC Rutherford Appleton Laboratory is thanked for access to neutron beam facilities. Some of the beryl crystals used in our experiments were cut by Bradley S. Wilson of Coast-to-Coast Rarestones, International.

The identification of any commercial product or trade name does not imply endorsement or recommendation by the National Institute of Standards and Technology.

References

[1] Ball P 2008 Water–An enduring mystery *Nature* **452** 291
[2] Finney J L 2004 Water? What's so special about it? *Phil. Trans. R. Soc. B* **359** 1145
[3] Angell C A 2008 Insights into phases of liquid water from study of its unusual glass-forming properties *Science* **319** 582
[4] Debenedetti P G and Stanley H E 2003 Supercooled and glassy water *Phys. Today* **56** 40
[5] Wang H-W, DelloStritto M J, Kumar N, Kolesnikov A I, Kent P R C, Kubicki J D, Wesolowski D J, and Sofo J O 2014 Vibrational density of states of strongly H-bonded interfacial water: Insights from inelastic neutron scattering and theory *J. Phys. Chem. C* **118** 10805
[6] Gainaru C, Agapov A L, Fuentes-Landete V, Amann-Winkel K, Nelson H, Köster K, Kolesnikov A I, Novikov V N, Richert R, Böhmer R, Loerting T, and Sokolov A P 2014 Anomalously large isotope effect in the glass transition of water *Proc. Natl. Acad. Sci.* **111** 17402-17407
[7] Reiter G F, Deb A, Sakurai Y, Itou M, Krishnan V G, and Paddison S J 2013 Anomalous ground state of the electrons in nanoconfined water *Phys. Rev. Lett.* **111** 036803
[8] Algara-Siller G, Lehtinen O, Wang F C, Nair R R, Kaiser U, Wu H A, Geim A K, and Grigorieva I V 2015 Square ice in graphene nanocapillaries *Nature* **519** 443
[9] Hummer G, Rasaiah J C, and Noworyta J P 2001 Water conduction through the hydrophobic channel of a carbon nanotube *Nature* **414** 188-190
[10] Koga K, Gao G T, Tanaka H, and Zeng X C 2001 Formation of ordered ice nanotubes inside

carbon nanotubes *Nature* **412** 188-190
[11] Kolesnikov A I, Zanotti J-M, Loong C-K, Thiyagarajan P, Moravsky A P, Loutfy R O, and Burnham C J 2004 Anomalously soft dynamics of water in a nanotube: a revelation of nanoscale confinement *Phys. Rev. Lett.* **93** 035503
[12] Reiter G, Burnham C, Homouz D, Platzman P M, Mayers J, Abdul-Redah T, Moravsky A P, Li J C, Loong C-K, and Kolesnikov A I 2006 Anomalous zero point motion of the protons in water in carbon nanotubes *Phys. Rev. Lett.* **97** 247801
[13] Kurotobi K and Murata Y 2011 A single molecule of water encapsulated in fullerene C_{60} *Science* **333** 613
[14] Diego Gatta G, Nestola F, Bromiley G D, and Mattauch S 2006 The real topological configuration of the extra-framework content in alkali-poor beryl: A multi-methodological study *Am. Mineral.* **91** 29
[15] Beduz C, Carravetta M, Chen J Y-C, Concistrè M, Denning M, Frunzi M, Horsewill A J, Johannessen O G, Lawler R, Lei X, Levitt M H, Li Y, Mamone S, Murata Y, Nagel U, Nishida T, Ollivier J, Rols S, Rõõm T, Sarkar R, Turro N J, and Yang Y 2012 Quantum rotation of *ortho* and *para*-water encapsulated in a fullerene cage *Proc. Natl. Acad. Sci. U.S.A.* **109** 12894-12898
[16] Goh K S K, Jiménez-Ruiz M, Johnson M R, Rols S, Ollivier J, Denning M S, Mamone S, Levitt M H, Lei X, Li Y, Turro N J, Muratae Y, and Horsewill A J 2014 Symmetry-breaking in the endofullerene $H_2O@C_{60}$ revealed in the quantum dynamics of ortho and para-water: a neutron scattering investigation *Phys. Chem. Chem. Phys.* **16** 21330-21339
[17] Gorshunov B P, Zhukova E S, Torgashev V I, Lebedev V V, Shakurov G S, Kremer R K, Pestrjakov E V, Thomas V G, Fursenko D A, and Dressel M 2013 Quantum behavior of water molecules confined to nanocavities in gemstones *J. Phys. Chem. Lett.* **4** 2015-2020
[18] Gorshunov B P, Zhukova E S, Torgashev V I, Motovilova E A, Lebedev V V, Prokhorov A S, Shakurov G S, Kremer R K, Uskov V V, Pestrjakov E V, Thomas V G, Fursenko D A, Kadlec C, Kadlec F, and Dressel M 2014 THz-IR spectroscopy of single H_2O molecules confined in nanocage of beryl crystal lattice *Phase Transitions* **87** 966-972
[19] Zhukova E S, Torgashev V I, Gorshunov B P, Lebedev V V, Shakurov G S, Kremer R K, Pestrjakov E V, Thomas V G, Fursenko D A, Prokhorov A S, and Dressel M 2014 Vibrational states of a water molecule in a nano-cavity of beryl crystal lattice *J. Chem. Phys.* **140**, 224317
[20] Kolesnikov A I, Reiter G F, Choudhury N, Prisk T R, Mamontov E, Podlesnyak A, Ehlers G, Seel A, Wesolowski D J, and Anovitz L M 2016 Quantum tunneling of water in beryl: A new state of water molecule *Phys. Rev. Lett.* **116** 167802
[21] Anovitz L M, Mamontov E, ben Ishai P, and Kolesnikov A I 2013 Anisotropic dynamics of water ultra-confined in macroscopically oriented channels of single-crystal beryl: A multi-frequency analysis *Phys. Rev. E* **88** 052306
[22] Sears V F 1992 Neutron scattering lengths and cross sections *Neutron News* **3** 26-37
[23] Granroth G E, Kolesnikov A I, Sherline T E, Clancy J P, Ross K A, Ruff J P C, Gaulin B D, and Nagler S E 2010 SEQUOIA: a newly operating chopper spectrometer at the SNS *J. Phys.: Conf. Ser.* **251** 012058
[24] Stone M B, Niedziela J L, Abernathy D L, DeBeer-Schmitt L, Ehlers G, Garlea O, Granroth G E, Graves-Brook M, Kolesnikov A I, Podlesnyak A, and Winn B 2014 A comparison of four direct geometry time-of-fight spectrometers at the Spallation Neutron Source *Rev. Sci. Instrum.* **85** 045113
[25] Kolesnikov A I, Anovitz L M, Mamontov E, Podlesnyak A, and Ehlers G 2014 Strong anisotropic dynamics of ultra-confined water *J. Phys. Chem. B* **118** 13414-13419
[26] Arnold O *et al.* 2014 Mantid–data analysis and visualization package for neutron scattering and μSR experiments *Nucl. Instrum. Methods A* **764** 156-166
[27] Azuah R T, Kneller L R, Qiu Y, Tregenna-Piggott P L W, Brown C M, Copley J R D, and Dimeo R M 2009 DAVE: A comprehensive software suite for the reduction, visualization, and

analysis of low energy neutron spectroscopic data *J. Res. Natl. Inst. Stand. Technol.* **114** 341-358

[28] Andreani C, Colognesi D, Mayers J, Reiter G F, and Senesi R 2005 Measurement of momentum distribution of light atoms and molecules in condensed matter systems using inelastic neutron scattering *Adv. Phys.* **54** 377-469

[29] Andreani C, Krzystyniak M, Romanelli G, Senesi R, and Fernandez-Alonso F 2017 Electron-volt neutron spectroscopy: beyond fundamental systems *Adv. Phys.* **66** 1-73

[30] Mayers J and Adams M A 2011 Calibration of an electron volt neutron spectrometer *Nucl. Instrum. Methods A* **625** 47-56

[31] Romanelli G, Krzystyniak M, Senesi R, Raspino D, Boxall J, Pooley D, Moorby S, Schooneveld E, Rhodes N J, Andreani C, and Fernandez-Alonso F 2017 Characterisation of the incident beam and current diffraction capabilities on the VESUVIO spectrometer *Meas. Sci. Technol.* **28** 095501

[32] Mayers J, Fielding A, and Senesi R 2002 Multiple scattering in deep inelastic neutron scattering: Monte Carlo simulations and experiments at the ISIS eVS inverse geometry spectrometer *Nucl. Instrum. Methods A* **481** 454-463.

[33] Mayers J 2011 Calculation of background effects on the VESUVIO eV neutron spectrometer *Meas. Sci. Technol.* **22** 015903

[34] Morosin B 1971 Structure and thermal expansion of beryl *Acta Cryst.* **B28** 1899-1903

[35] Schlenker L, Gibbs G V, Hill E G, Crews S S, and Myers R H 1977 Thermal expansion coefficients for indialite, emerald, and beryl *Phys. Chem. Minerals* **1** 243-255

[36] Fan D, Xu J, Kuang Y, Li X, Li Y, and Xie H 2015 Compressibility and equation of state of beryl ($Be_3Al_2Si_6O_{18}$) by using a diamond anvil cell and in situ synchrotron X-ray diffraction *Phys. Chem. Minerals* **42** 529–539

[37] Boutin H, Prask H, and Safford G J 1965 Low frequency motions of H_2O molecules in beryl from neutron inelastic scattering data *J. Chem. Phys.* **42** 1469-1470

[38] Kolesov B A and Geiger C A 2000 The orientation and vibrational states of H_2O in synthetic alkali-free beryl *Phys. Chem. Minarals* **27** 557-564

[39] Belyanchikov M A, Zhukova E S, Tretiak S, Zhugayevych A, Dressel M, Uhlig F, Smiatek J, Fyta M, Thomas V G, and Gorshunov B P 2017 Vibrational states of nano-confined water molecules in beryl investigated by first-principles calculations and optical experiments *Phys. Chem. Chem. Phys.* **19** 30740

[40] Adams D M and Gardner I R 1974 Single-crystal vibrational spectra of beryl and dioptase *J. Chem. Soc., Dalton Trans.* **0** 1502-1505

[41] Jackson S, Krzystyniak M, Seel A G, Gigg M, Richards S E, and Fernandez-Alonso F 2014 VESUVIO data analysis goes MANTID *J. Phys.: Conf. Ser.* **571** 012009

[42] Krzystyniak M *et al.* 2018 Model selection in neutron Compton scattering – A Bayesian approach with physical constraints *J. Phys.: Conf. Ser.* (in this volume)

[43] Reiter G F, Kolesnikov A I, Paddison S J, Platzman P M, Moravsky A P, Adams M A, and Mayers J 2012 Evidence for an anomalous quantum state of protons in nanoconfined water *Phys Rev B* **85** 045403

[44] Finkelstein Y and Moreh R 2016 Proton dynamics in hydrogen-bonded systems *Mol. Phys.* **114** 2108-2114

[45] Finkelstein Y, Moreh R, Shang S L, Wang Y, and Liu Z K 2017 Quantum behavior of water nano-confined in beryl *J. Chem. Phys.* **146** 124307

[46] Born M and Huang K 1954 *Dynamical theory of crystal lattices* (Oxford: Clarendon Press)

IOP Conf. Series: Journal of Physics: Conf. Series **1055** (2018) 012003 doi:10.1088/1742-6596/1055/1/012003

Hydrogen dynamics in solid formic acid: insights from simulations with quantum colored-noise thermostats

K Drużbicki[a,b], **M Krzystyniak**[c,d], **D Hollas**[e], **V Kapil**[f], **P Slavíček**[e], **G Romanelli**[c], **F Fernandez-Alonso**[c,g]

[a] Faculty of Physics, Adam Mickiewicz University, Umultowska St. 85, 61-614, Poznan, Poland
[b] Frank Laboratory of Neutron Physics, Joint Institute for Nuclear Research, Dubna, JoliotCurie 6, 141980, Russia
[c] ISIS Facility, Rutherford Appleton Laboratory, Chilton, Didcot, Oxfordshire OX11 0QX, United Kingdom
[d] School of Science and Technology, Nottingham Trent University, Clifton Campus, Nottingham, NG11 8NS, United Kingdom
[e] Department of Physical Chemistry, University of Chemistry and Technology, Prague, Technická 5, 166 28, Czech Republic
[f] Laboratory of Computational Science and Modeling, Institute of Materials, École Polytechnique Fédérale de Lausanne, 1015 Lausanne, Switzerland
[g] Department of Physics and Astronomy, University College London, Gower Street, London, WC1E 6BT, United Kingdom

E-mail: `kacper.druzbicki@amu.edu.pl`

Abstract. With an increase of computational capabilities, *ab initio* molecular dynamics becomes the natural choice for exploring the nuclear dynamics of solids. As based on classical mechanics, the validity of this approach is, in-principle, limited to the high-T regime, whilst low-temperature simulations require inclusion of quantum effects. The methods commonly used to account for nuclear quantum effects are based on the path-integral formalism, which become, however, particularly time consuming when high accuracy methods are used for calculating forces. Recently, new efficient alternative approaches to account for quantum nature of nuclei have been proposed, using so-called quantum thermostats. In this work, we examine the simulations performed with the quantum colored-noise thermostat introduced by Ceriotti [Phys. Rev. Lett., 103:030603, 2009]. We present the tests of portable implementation of the quantum thermostat in the ABIN program, which has been extended to periodic systems through the interface to CASTEP, a leading spectroscopy-oriented plane-wave density functional theory code. The range of applicability of quantum-thermostatted molecular dynamics simulations for the interpretation of neutron scattering data was examined and compared to classical molecular dynamics and lattice-dynamics simulations, using solid formic acid case as a test bed. We find that the approach is particularly useful for the modeling of low-temperature inelastic neutron scattering spectra as well as provides some theoretical estimate for the low-limit of the mean kinetic energy. While finding the quantum-thermostat to seriously affect the dynamic properties of the title system, we illustrate to which extent the unperturbed response can be successfully recovered.

1. Introduction

Molecular dynamics (MD) has evolved into a powerful numerical method to investigate structural and dynamical properties of condensed matter [1]. For an atom, these calculations are, however, only valid in the classical limit, i.e. for temperatures safely above the Debye temperature that signals the onset of quantum effects. The Debye temperature is often of the order of few hundred Kelvins and nuclear quantum effects (NQEs) are thus indeed important over a wide range of temperatures [2, 3].

Quantum fluctuations stemming from the Heisenberg uncertainty principle make the energy of a particle at 0 K higher than a the potential energy minimum, leading to a concept of zero point energy (ZPE) (see Fig. 1). While ZPE reflects the quantum nature of atomic motion, it can be approximately accounted for by lattice-dynamics calculations, which, by definition, are based on the quantum harmonic oscillator model.

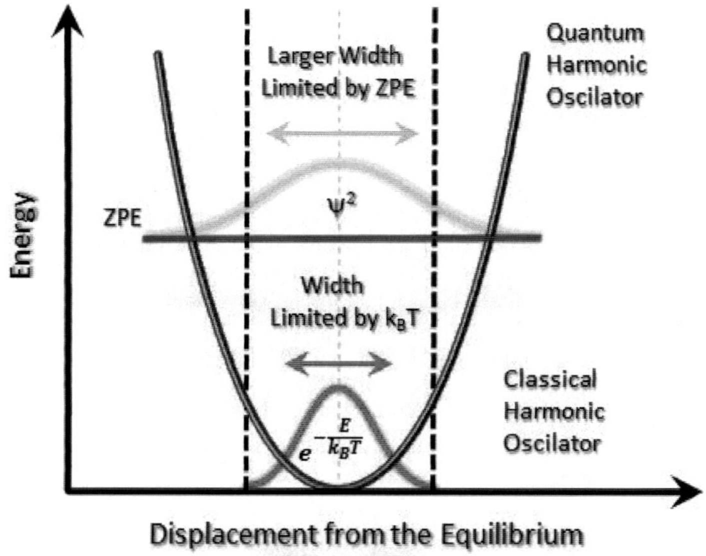

Figure 1. Schematic representation of the energy and probability of a mechanical and quantum particle in a harmonic energy well.

In order to capture anharmonic effects from first-principles, a number of static theoretical methods have recently been devised. These approaches are based on calculating vibrational excited states either from vibrational self-consistent field (VSCF) [4–6], the vibrational interaction of configurations (VCI) [7–9] or from perturbation theory (second-order vibrational perturbation theory, VMP2, for example) [10, 11]. Despite of a substantial activity in the field [12, 13], the use of static methodology in modeling of anharmonic spectra of solids is still practically unworkable for polyatomic molecular systems because of its complexity.

While classical MD simulations in principle do not account for ZPE, they do naturally account for mechanical anharmonicity effects, since no assumption about the shape of the underlying potential energy surface (PES) is imposed. However, if the ZPE is very large (e.g. hydrogenous materials) then the oscillator looks as reaching a higher temperature than the classical spring and so it enters the anharmonic part of the PES. However, in such a case, the low-temperature classical MD simulations do not properly account for anharmonic effects. That is because the particle distribution widths are narrow (see Fig. 1) and the simulations stay near the potential energy minima all the time. In such a way, only the harmonic part of the PES is probed.

Indeed, it has been illustrated that low-temperature *ab initio* Born-Oppenheimer Molecular Dynamics (BOMD) simulations provide vibrational spectra identical to the harmonic ones calculated through the Hessian matrix diagonalization (see references [14] and [15] for an illustration). Above the low-temperature limit, it is necessary to perform the simulations at higher, effective temperature conditions, possibly adjusted to mean square atomic displacements (MSD) found experimentally. For hydrogen atoms this may lead to very high effective temperatures. The problems associated with such a crude approach have been nicely illustrated by Williams et al. [16]. It has been shown that selection of classical effective temperature is a serious problem, stemming from the fact that the measured MSDs only represent a probable upper limit that may be allowed in the BOMD simulation for each atom in the same molecule, whereas the mapping of the quantum system onto the classical one could only be done provided that the temperature is set differently for each mode (or mass). As shown by Van-Oanh et al. [17], ZPE contributions can be somehow alleviated by preparing the initial conditions for MD simulations in the microcanonical ensemble (NVE) according to semi-classical quantization based on a normal-mode representation. Such an approximation, however, was found to be inefficient for strongly anharmonic cases, suffering from energy relaxation effects and from so-called 'ZPE leakage', i.e. a cross-talk between the high-energy and low-energy modes.

An alternative strategy to incorporate NQEs is to place the system in contact with a quantum thermal bath (QTB), where the dynamics follows a generalized Langevin equation (GLE) and the vibrational delocalization is mimicked by a quantum thermostat (QT). In such an approach, the otherwise white noise spectrum of the Fourier Transform of the stochastic force correlation function is filtered (colored) in such a way that it ends up resembling the spectral density of the correlation function for a quantum harmonic oscillator with ZPE contribution included [2,18–20].

More strictly, in GLE, the particle of unit mass, which position is described by the generalized coordinate q, with a conjugate momentum p, and being subject to a potential $V^{'}(q)$, yields the following equations of motion:

$$\dot{q(t)} = p(t)$$

$$q(t) = -V^{'}(q) - \int_{-\infty}^{t} K(t-s)p(s)ds + \zeta(t) \tag{1}$$

where $\zeta(t)$ is a time-correlation function of the random force and $K(t)$, is the memory kernel, introducing the dependence of the system evolution on its history in terms of a non-Markovian stochastic differential equation [21]. Such definition, however, requires the use of integro-differential equations, leading to a serious numerical challenge. Alternatively, one may map the non-Markovian dynamics onto an equivalent, Markovian dynamics in an extended space n, by introducing auxiliary fictitious momenta s, linearly coupled to the physical momentum p:

$$\dot{q(t)} = p(t) \tag{2}$$

$$\begin{pmatrix} \dot{p} \\ \dot{\mathrm{s}} \end{pmatrix} = \begin{pmatrix} -V^{'}(q) \\ \mathbf{0} \end{pmatrix} - \begin{pmatrix} a_{pp}\mathbf{a}_p^T \\ \bar{\mathrm{a}}_p\mathbf{A} \end{pmatrix} \begin{pmatrix} p \\ \mathrm{s} \end{pmatrix} + \begin{pmatrix} b_{pp}\mathbf{b}_p^T \\ \mathbf{b}_p\mathbf{B} \end{pmatrix} \begin{pmatrix} \xi \end{pmatrix}$$

where s is the vector of n additional degrees of freedom; ξ denotes a n+1 dimensional vector of uncorrelated Gaussian numbers [21]; a_{pp} stands for a friction coefficient and b_{pp} denotes the intensity of the random force, both linked by the so-called fluctuation-dissipation theorem ($b_{pp}^2 = 2a_{pp}mk_{\mathrm{B}}T$).

The matrices in Eq. (2) are related by the so-called drift matrix $\mathbf{C}$:

$$\mathbf{A}_p\mathbf{C}_p + \mathbf{C}_p\mathbf{A}_p^T = \mathbf{B}_p\mathbf{B}_p^T \tag{3}$$

where $\mathbf{A}_p = \begin{pmatrix} a_{pp} & \mathbf{a}_p^T \\ \bar{\mathbf{a}}_p & \mathbf{a} \end{pmatrix}$ and $\mathbf{B}_p = \begin{pmatrix} b_{pp} & \mathbf{b}_p^T \\ \bar{\mathbf{b}}_p & \mathbf{b} \end{pmatrix}$. This allows one to predict the response properties analytically.

For a quantum thermostat, it is, however, required to relax the fluctuation-dissipation theorem, which leads to non-equilibrium GLE dynamics. This may be realized by noting that the kinetic energy distributions of both classical and quantum harmonic oscillator take the form of a Gaussian, with the widths differing only in the standard deviation. While the effective temperature can be defined as:

$$T^*(\omega) = \frac{\hbar\omega}{2k_B}\coth\frac{\hbar\omega}{2k_BT} \tag{4}$$

one can take an advantage of GLE dynamics by thermostatting the different vibrational modes present in the system at different effective temperatures T^*. In such a way, the kinetic energy distribution with deviation $\sigma_q^2 = \frac{\hbar}{2m\omega} coth\frac{\hbar\omega}{2k_BT}$ can be imposed by using purely classical simulations.

Quantum thermostatting has proven to be very efficient for the control and evaluation of static properties in MD simulations. Nevertheless, its use for prediction of dynamical properties of materials, including vibrational spectra or diffusion coefficients, has been found to be limited due to a possible disturbance of these properties induced by the presence of the thermostat. The problem is of particular importance for anharmonic systems, where a strongly coupled thermostat is required to avoid the ZPE leakage.

A remedy to this deficiency has been recently provided by Rossi, Kapil and Ceriotti [21], delivering an analytical scheme to extract the microcanonical velocity-velocity autocorrelation function (VACF) from the quantum-thermostated (GLE) trajectory. It is based on the idea that the exact VACF for a harmonic oscillator of frequency ω_0, coupled to a GLE thermostat is:

$$\Theta(\omega, \omega_0) = C_{qp}(\omega_0)_{pp}^{-1} \left[\frac{A_{qp}(\omega_0)}{A_{qp}(\omega_0) + \omega^2}\right]_{pp} \tag{5}$$

where A and C are the drift and the diffusion matrices of the generalized Langevin equation. For the details of the nomenclature and the matrices elements ($[\cdot]_\cdot$), we refer the reader to section 2 of [21]. For a system with a density of states given by $g(\omega)$, the VACF can be written in the harmonic limit as:

$$C_{VV}(\omega) = \int d\omega g(\omega)\,\Theta(\omega, \omega_0) \tag{6}$$

Though the result has been derived under a harmonic approximation, the predictions happen to be in excellent agreement with the exact VACF for system as anharmonic as liquid water [21]. It was shown that the post-processing deconvolution of the perturbed spectra can properly recover the unperturbed dynamic properties even in the case of heavily thermostatted cases, i.e. for considerable anharmonic systems [21].

Modeling of anharmonic vibrational dynamics along with nuclear quantum effects is no doubt of great importance for neutron spectroscopy. On the other hand, the aforementioned state-of-the-art *ab initio* MD approaches are hardly used by the community in the modeling of neutron spectroscopy observables. To meet these needs, we examine here a possible applicability of the quantum-thermostatted molecular dynamics simulations for the prediction of inelastic-neutron-scattering (INS) spectra and neutron Compton scattering (NCS) observables in hydrogen-bonded molecular materials using solid formic acid (FA) as a testbed.

The present paper is organized as follows. First we present the test of quantum thermostat implementation in a molecular dynamics code ABIN, developed by Hollas et al. [22], which we recently interfaced to a leading solid-state plane-wave density functional theory code, CASTEP [23, 24]. We then examine the range of applicability of quantum-thermostatted *ab initio* molecular dynamics simulations for the modeling and analysis of neutron-scattering data, with solid FA acting as the reference [25].

2. Research Methodology

2.1. Experimental

The experimental data presented in this paper along with the initial structural models have been published recently in Ref. [25]. In brief, we refer to the low-temperature INS spectrum measured on the inverted-geometry TOSCA spectrometer, located at the ISIS Facility [26]. We further supplement the analysis with the variable-temperature INS spectra collected with the direct-geometry spectrometer HRMECS (Intense Pulsed Neutron Source, Argonne National Laboratory) [27]. NCS experiments were performed at 180 and 240 K on the VESUVIO spectrometer, also located at the ISIS Facility [28–31]. The experimental and data-analysis protocol for VESUVIO measurements is explained in more detail elsewhere [28–35].

2.2. Computational

The ABIN code is a Fortran program performing BOMD [22]. It was designed specifically to deal with NQEs. It can perform path integral MD (PIMD) simulations and also utilizes the recently developed quantum thermostat by Ceriotti [18–20]. ABIN itself handles the propagation of the system according to the Newton equations of motion, while the forces and energies are taken from the external electronic structure programs. The call to the chosen external program is handled via a shell script. Writing a new interface can be done without any changes introduced to ABIN or to the electronic structure code. Therefore, ABIN is very flexible and ideally suited to work with essentially any molecular modelling code. In this communication we present an extension of ABIN including periodic boundary conditions (PBC). To this end, ABIN has been interfaced to CASTEP [23, 24] and QUANTUM ESPRESSO [36, 37].

In line with our previous study, we employed periodic plane-wave density functional theory (PW-DFT) calculations, using the same numerical settings as defined in Ref. [25]. As a starting point, we used supercells with Z = 24, where the cell volumes were kept fixed to those obtained with the NPT runs at a given effective temperature, that is the temperature relative to the melting point, rather than to the absolute (classical) temperature of the material (see Ref. [25] for more details). The results of the quantum-thermostatted ABIN calculation with CASTEP, were benchmarked by comparison with the results of the BOMD simulations obtained using the CP2K code with its own quantum thermostat implementation [38]. The CP2K calculation used a mixed Gaussian and plane-wave approach (GPW) [39]. In both cases exchange and correlation were treated with the Perdew-Burke-Ernzerhof (PBE) functional [40], augmented with semi-empirical dispersion corrections (SEDC). Since both codes employ different SEDC schemes, Tkatchenko and Scheffler (TS) dispersion corrections (PBE-TS) were employed in the CASTEP simulations [41], while the Grimme-D3 scheme with Becke-Johnson damping [PBE-D3(BJ)] was used in the CP2K simulations [42–44]. For CASTEP calculations a plane-wave kinetic-energy cutoff of 900 eV was used along with nonlocal norm-conserving pseudopotentials (NCPPs) [45]. For the CP2K simulations, we employed a valence double-zeta-quality basis set optimized from molecular calculations (DZVP-MOLOPT-SR-GTH) along with the corresponding Goedecker-Teter-Hutter pseudopotentials. The plane-wave cutoff for the finest level of the multi-grid was set to 400 Ry.

The BOMD simulations were performed with an SCF convergence of 5×10^{-8} eV threshold per atom. Due to the large size of the real space supercell, the k-point grid sampling was reduced to the Γ point in both CP2K and CASTEP simulations.

A time step of 0.5 fs was used all throughout the MD runs. While the use of a smaller integration step is recommended for the simulations employing QTs, the use of the standard 0.5 fs step was dictated by a very large computational cost. Use of such an integration step may potentially result in unphysical blue-shifting of the highest predicted frequencies, estimated at the order of 50 cm^{-1}, according to molecular simulations of a FA molecule presented elsewhere [46]. Such a discrepancy might be potentially corrected a posteriori according to Tikhonov [46].

Both the ABIN (CASTEP) and CP2K simulations were run at 50, 180 and 240 K using a strongly-coupled QTs. The thermostats were defined with $N_s = 6$ (the dimension of the drift and the diffusion matrices); $\omega^{-}h/\mathrm{kT}$=100 and$N_s = 6$, $\omega^{-}h/\mathrm{kT}$=50, for the low- and higher-temperatures simulations, respectively [47]. After a successful equilibration, whose length depended on the temperature and the definition of the thermostat used, production trajectories of 12.5 ps were collected in both the quantum-thermostatted canonical (QT-NVT) and microcanonical (NVE) ensembles. The trajectories were analyzed using the TRAVIS code [48]. Furthermore, direct deconvolution of the QT-NVT trajectory into the microcanonical one (QT-NVT $\rightarrow$ NVE) was performed by the UNWIND code, written by Kapil [49].

A semi-quantitative analysis of the pairwise interactions stabilizing the crystal structure was performed with the help of the Crystal Explorer program [50], employing its internal quantum-chemistry engine, TONTO [51], and following the methodology described in details in Refs. [52] and [53].

3. Benchmarking of quantum-thermostatted molecular dynamics simulations

3.1. The Case Study of Solid Formic Acid

The structure of solid FA, has been extensively investigated both theoretically and experimentally. It has been originally shown to crystallize in an orthorhombic, $Pna2_1$ structure (see Fig. 2). The structure consists of infinite hydrogen-bonded chains, with FA molecules existing in the *trans* conformation [54]. The hydrogen-bonded chains are tightly packed in layers. In order to shed more light into the nature of the intermolecular interactions stabilizing the orthorhombic structure of FA, we have performed the analysis of the intermolecular forces within the crystallographic cell [54]. To this end we have used a recently-developed analysis of the pairwise interactions, i.e. energy frameworks scheme, which provides an insight into the anisotropy in the strengths of the crystal packing [52, 53]. The analysis of the intermolecular interactions responsible for the $Pna2_1$ structure reveals three types of attracting forces with a stabilizing energy greater than 5 kJ/mol, namely: the OH$\cdots$O hydrogen-bond between the FA molecules; the stacking forces tightly packing the chains into the layers; and the inter-chain contacts between the neighboring stacks (see Fig.2). The results of this quantitative analysis are provided in Table 1, where the interaction energies are decomposed into the electrostatic, polarization, dispersion and repulsive terms, reported for two different electronic structure models, namely Møller–Plesset perturbation theory (MP2) and DFT (B3LYP-D2) with the same, 6-31G** basis set. While the calculations refer to the crystallographic structure, and thus to the off-equilibrium geometries with respect to both MP2 and B3LYP-D2 levels of theory, the calculated energy contributions have been scaled appropriately, according to the procedure of Ref. [52]. The scaling factors have been derived from a large molecular training set explored using an CCSD(T)/CBS standard [52].

The energy decomposition derived from both approaches follows the same hierarchy of the intermolecular forces. While standard MP2 (without spin-component scaling) is generally well known for its over-binding tendency [55], one may accept the values provided by the scaled B3LYP-D2 to be more reliable [52]. The stabilization energy of more than 30 kJ/mol classifies the analyzed hydrogen-bond as rather strong, placing the chain attraction well below this energy scale.

Polymorphism in FA has been an object of some controversy, with several reports that could not be further confirmed. For instance, a first-order structural phase transition at 218 K was postulated by Zelsmann et al. [56], based on infrared spectroscopy and differential thermal

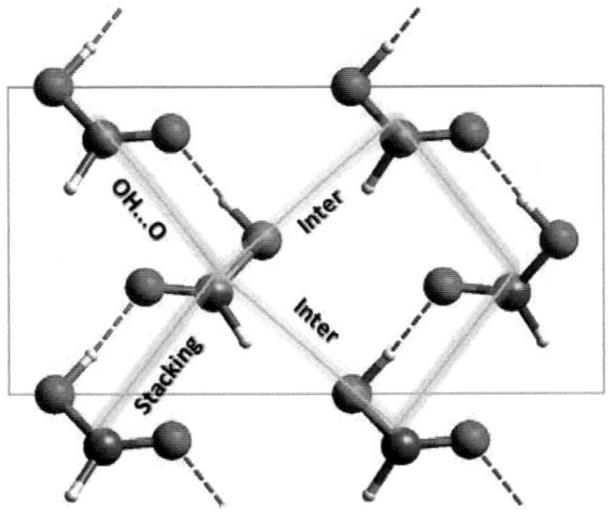

Figure 2. Crystal structure of formic acid at 4.5 K (dideutero form), as solved by Albinati [54], with an energy framework projection of the intermolecular interactions present (see Table 1 for a quantitative analysis).

analysis. However, neutron diffraction studies have established that the $Pna2_1$ phase is stable in the 98–278 K range [57]. Recently, using a combination of neutron spectroscopy and *ab initio* modelling, we have tested the hypothesis, put forward by Zelsmann, on a proton-transfer-driven phase coexistence at high temperatures [25]. Our study has revealed a temperature-induced loosening of the crystal packing, which promotes shearing of the hydrogen bonded chains. In light of the above mentioned energy decomposition, such a scenario seems to be the most plausible one [25].

Table 1. Interaction energies for dideutero-formic acid for the crystal structure solved at 4.5 K by Albinati [54]. R is the distance between the molecular centroids. Decomposition of the total energy into electrostatic, polarization, dispersion and repulsive terms is reported for two energy models (MP2 and DFT+D) [50, 51]. The energy terms are scaled appropriately according to Ref. [52]

Interaction (Symmetry)	R (Å)	Density	Energy Decomposition [kJ/mol]				
			E_{ele}	E_{pol}	E_{dis}	E_{rep}	E_{tot}
OH···O $(x+1/2, y+1/2, z+1/2)$	3.63	MP2	-68.4	-10.3	-6.7	41.9	-43.5
		B3LYP-D2	-62.8	-8.3	-7.0	46.7	-31.4
Stacking $(-x+1/2, y+1/2, z+1/2)$	3.63	MP2	-8.3	-0.9	-4.7	2.8	-11.1
		B3LYP-D2	-5.6	-0.7	-4.9	2.7	-8.5
Inter $(-x, -y, z+1/2)$	4.73	MP2	-5.4	-0.4	-3.0	1.5	-7.3
		B3LYP-D2	-4.2	-0.3	-3.1	1.5	-6.1

In our previous work [25], we explored the hydrogen dynamics by employing an artificially elevated classical temperature. The primary aim of this work was to re-examine the system with the help of quantum-thermostatted simulations, which is the subject of the next section.

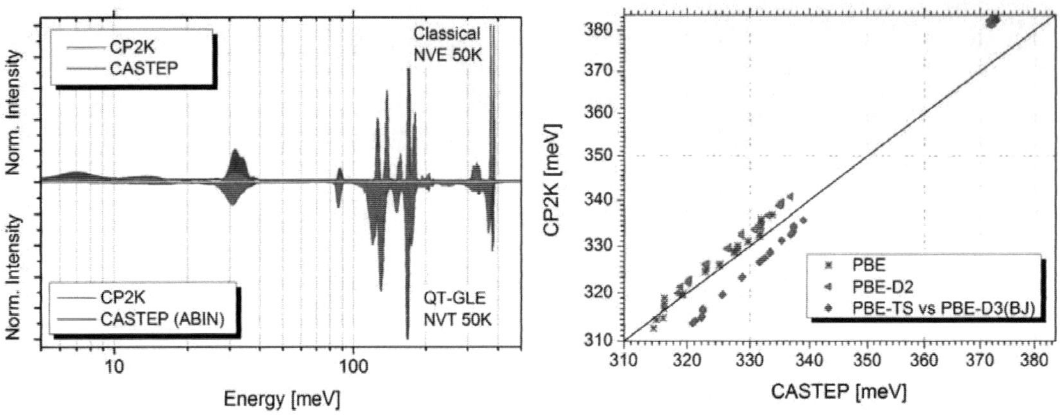

Figure 3. The left panel reports the comparison of hydrogen-projected phonon density of states derived from the classical BOMD simulations in the NVE ensemble (Upper part: NVE) and the quantum-thermostatted MD simulations in the NVT ensemble (Bottom part: QT-GLE), both performed with CASTEP (ABIN) and CP2K codes (see the text for numerical details). The panel on the right depicts a comparison of the highest harmonic frequencies calculated with both codes, with and without inclusion of the semiempirical dispersion corrections.

3.2. The Test of Quantum Thermostat Implementation for Condensed Phase

We have tested the correctness of the QT-NVT BOMD simulations performed using ABIN by comparing the production runs from CASTEP (ABIN) and CP2K simulations. To this end, we have focused on the low-temperature production runs, collected with a target temperature of 50 K. Virtually identical behavior was observed for both CP2K and CASTEP, while the CP2K simulations were characterized by a better performance, owing to the efficient implementation of the wavefunction/density propagation. The low-temperature simulations suffered from a slow-convergence, where at least 25 ps were needed to equilibrate the system with a mean temperature of ca. 450 K (with ± 50 K fluctuations). Using 324 processors (Intel Xeon 2.5 GHz) at the PL-Grid high-performance computing facility PROMETHEUS, these calculations took about 3 months per trajectory (40 ps) at each temperature.

To test the consistency of the predicted dynamic properties we have used the hydrogen-projected density of states obtained with both codes. The left panel of Fig. 3 illustrates the power spectra obtained by Fourier-transform of the velocity-velocity autocorrelation function (VACF) for both classically-initiated microcanonical (Upper - NVE) and quantum-thermostatted canonical ensembles (Bottom - QT-GLE). The results obtained from both codes are indeed consistent, with very similar intensity distributions throughout the whole frequency range, particularly in the quantum-thermostatted case.

One may note a visible difference between the CASTEP (ABIN) and CP2K data, associated with the band positions in the highest-frequency range. Such a difference can be clearly understood through the analysis of the right panel of Fig. 3, presenting harmonic phonon frequencies in the highest-energy spectral edge. This difference stems mainly from different basis-set definitions used in both codes, rather than different SEDC schemes used (see PBE-TS vs the PBE-D3). By comparing the phonon frequencies calculated with the same functional/SEDC scheme (i.e., PBE; PBE-D2, [58]) one may note that the highest-frequencies predicted by CASTEP are much softer (by ca. 10 meV). Such an effect has been extensively studied by Tosoni et al. [59], and proven to stem from the incompleteness of the basis-sets employed. Such a discrepancy might be reduced by extending the basis set size. Unfortunately, the use of any triple-zeta quality basis set in CP2K along with very high plane-wave kinetic energy cutoff for CASTEP calculations (> 1500 eV) was not possible due to a significant increase in the computational cost anticipated. Nevertheless, as shown earlier, the level of theory selected for the CASTEP simulations ensures an excellent description of the neutron scattering vibrational data [25].

4. Analysis of the Experimental Data

To examine the range of applicability of the quantum-thermostatted molecular dynamics simulations for the interpretation of neutron scattering data, we refer to INS and NCS experiments.

4.1. Inelastic Neutron Scattering

In our previous paper [25], we have extensively explored the proton vibrational dynamics in FA performing harmonic lattice dynamics (HLD) calculations, along with classical BOMD simulations, performed at elevated temperatures. As already mentioned [16], such a crude approach suffers from the fact that the effective nuclear temperature may be different for each different atom. Here, we discuss a possible advantage of employing quantum thermostatting for simulations of INS data.

Figure 4 compares the experimental and theoretical spectra (all the results from CASTEP/PBE-TS/900eV calculations) across the energy transfer range of 2 - 425 meV. One may note a limitation of an indirect-geometry instrument (TOSCA) to explore the range above 225 meV, due to a fixed $Q(E)$ dependence [60, 61]. As a consequence, only large momentum transfers are accessible in this range, which suppresses the total intensity via the enhanced Debye-Waller factor and makes the higher-order transitions dominating [60, 61]. The latter factor makes the intensities of fundamentals nearly unobservable in this range. Therefore, we supplement the analysis using HRMECS data [27] recorded in a low-momentum transfer regime.

The theoretical spectrum from HLD at 0K was derived from the second derivatives of the energy and simulated with the abINS code [62,63], accounting for the characteristics of TOSCA spectrometer. In the presented BOMD spectra, the intensity across the whole energy transfer range has been suppressed exponentially to mimic the intensity profile induced by an indirect-geometry instrument [60–63]. The HLD spectrum accounting only for the first-order transitions (fundamentals) is presented as a solid line, while the total spectrum accounting for higher-order events (overtones and combination bands) is presented in the logarithmic intensity scale (see the dotted line). The HLD spectrum was discussed in detail in Ref. [25]. A sufficient agreement with the experimental data has been achieved throughout the whole spectral range and will be, hence, treated as a reference.

The classical MD power spectrum refers to the production run (NVE), initiated from the classically-thermostatted (Nosé-Hoover) simulations at low-temperature (50K). The intensity of the resulting power spectrum has been quantum-corrected by means of standard (SA) and harmonic (HA) approximations [64]. In terms of frequency, the 50K classical *ab initio* BOMD method delivers a virtually identical spectrum to the one from HLD even for highest frequency vibrations. Furthermore, neither the SA-corrected nor HA-corrected spectra can properly reproduce the intensity relations in the lowest-energy transfer range.

On the contrary, the quantum-thermostatted NVT simulations, QT-GLE-NVT, are considerably red-shifted with respect to the classical BOMD simulations, indicating that the simulations enters the anharmonic part of the PES. Since the QT-GLE simulations account for the ZPE contributions, no further quantum corrections were applied here.

One should not forget that a better reproduction of the experimental frequencies by HLD is just a fortuitous effect, thanks to a well-known cancelation of errors due to anharmonicity by pure generalized gradient approximation (GGA) functionals, i.e. PBE-TS. Such a behavior has

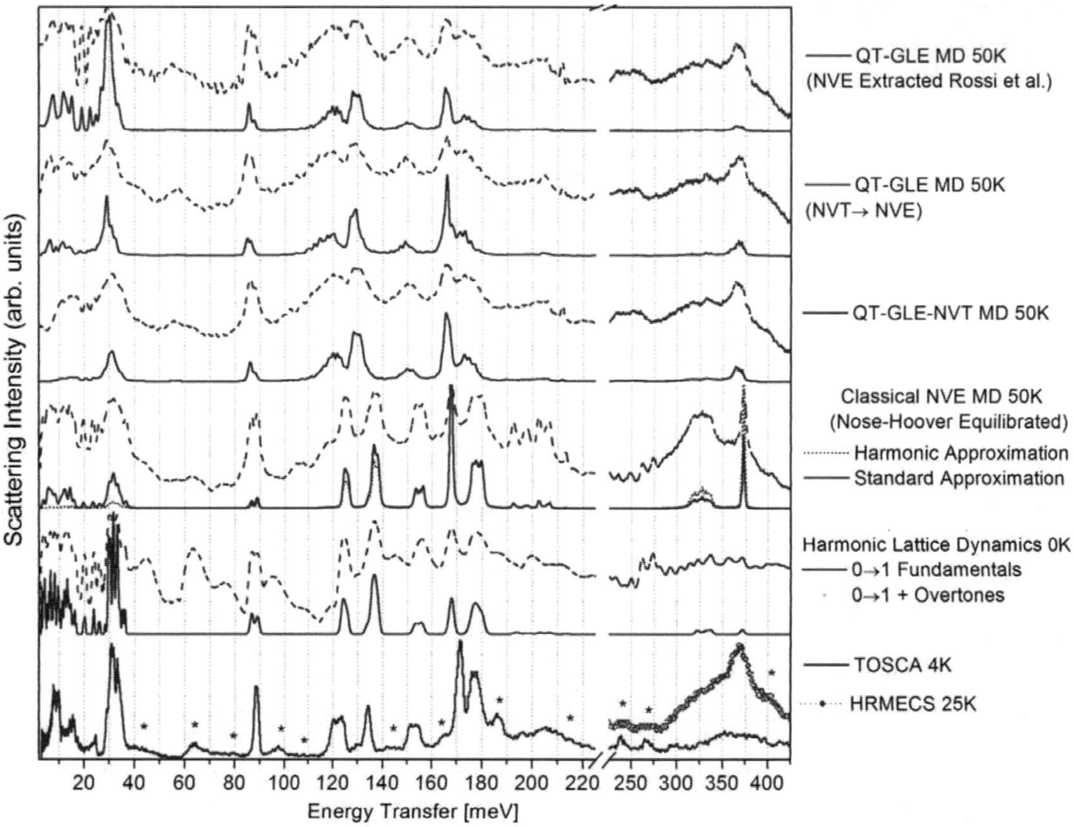

Figure 4. Low-temperature experimental spectra of formic acid recorded with the TOSCA and HRMECS ($E_0 = 600$ meV) spectrometers at 4K and 25K, respectively,[24,26] along with the set of theoretical ones (CASTEP/PBE-TS/900eV). The theoretical spectra were derived from HLD calculations and from a series of different low-temperature *ab initio* BOMD simulations (hydrogen-projected power spectra, see the text for details). The dotted lines depict the theoretical spectra presented in a logarithmic intensity scale, in order to emphasize weak overtone features and combination bands (see the asterisks in the experimental spectra).

been extensively studied, e.g. by Lin et al. [65], and has been related to improper description of the curvatures of potential energy surfaces. Recently, Pestana et al. have demonstrated similar effects in BOMD simulations of liquid water by comparing the performance of dispersion-corrected GGA with high quality meta-GGA functionals [66]. More proper modeling of the vibrational response would, hence, call for the use of more advanced approximations to the exchange-correlation functional, possibly by employing non-local hybrid functionals. However, in terms of plane-wave DFT at the present stage of development, employing non-local approximations to the exchange-correlation functional is very inefficient, making the use of codes employing atom-centered basis sets more favorable.

One may note a kind of spectral broadening with respect to the classical BOMD power-spectra, presumably due to the thermal fluctuations induced by the quantum-thermostat. This has been confirmed by performing the NVE simulations initiated from the QT-GLE-NVT trajectory (see the NVT NVE panel in Fig. 4), using a single phase-space configuration (see e.g. the bands in the range 100-200 meV). Recently, Tikhonov et al have examined the potential

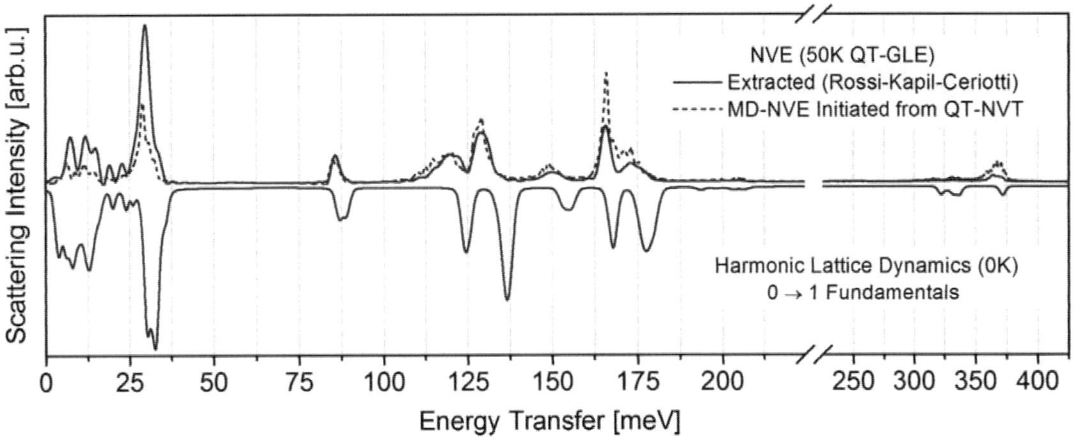

Figure 5. A mirror-like comparison of the INS spectra of solid formic acid delivered by the simulations accounting for ZPE. The top panel compares the results of simulations in the microcanonical ensemble (MD-NVE, initiated from the quantum-thermostat NVT simulations), with the NVE spectrum directly deconvoluted from the QT-GLE-NVT trajectory [21]. The bottom panel shows the 0K harmonic spectrum derived from the HLD calculations. For the sake of clarity, the band shape in the harmonic spectrum has been adjusted to the one from MD simulations.

ability of the QT-GLE-NVT BOMD simulations to predict the vibrational spectra of small gas-phase molecules at elevated temperatures, finding it inapplicable for this purpose [67]. Such a failure does not hold for the crystal-phase simulations presented here, where the thermal fluctuations do not alter the spectrum so much.

Uncoupling the system from a quantum-thermostat provides some enhancement of the spectral intensity in the lowest-part of the spectrum. This can be understood by a large-friction of the strongly coupled thermostat, which was designed to disturb all the low-frequency modes so as to prevent the ZPE leakage, the effect certainly expected for such a considerably anharmonic solids as FA. As a consequence, in the QT-GLE-NVT simulation the terahertz part of the spectrum is completely smeared out, with all the information on the external modes being lost.

Alternatively, the microcanonical VACF function has been extracted from the quantum-thermostated NVT trajectory according to the analytical scheme introduced by Ross et al. [21]. The deconvolution process corrects at the same time for dynamical disturbances and the frequency-dependent occupations of different normal modes. Such a correction has been applied to deconvolute the QT-GLE-NVT trajectory (see the top panel of Fig. 4). The deconvolution has not reduced the band broadening, but it dramatically affected the intensity relations throughout the whole energy range. Importantly, in the terahertz range all the smeared bands, found in both HLD and classical BOMD spectra, have now been clearly restored. On the whole, this study has shown that such performance opens an exciting possibility of simulating terahertz-INS spectra beyond the harmonic approach.

To summarize the performance of quantum-thermostatted BOMD simulations in the prediction of the INS spectra we refer to Fig. 5. Here, the results of the NVE simulations initiated from the quantum-thermostatted NVT (with only a single phase-space configuration taken) are compared to the ones deconvoluted directly from the QT-GLE-NVT trajectory [21]. While both hydrogen-projected VACF spectra are very similar in terms of anharmonic frequencies, the results differ considerably in terms of the INS intensities, particularly for the bands below 40 meV and around 165 meV. The most intense spectral feature, predicted at around 35 meV is due to external modes involving hydrogen-bridge stretching modes, O$\cdots$O, mixed with the out-of-plane librations of FA molecules. The bands at around 165 meV are due to the in-plane CH bending modes, CH (lower band), and due to in-plane hydrogen-bonds deformations, OHO (upper band with a lower intensity). One of the reasons of such a divergence might be the fact that only a single NVT phase-space configuration was used in the subsequent NVE run. The second reason might be a partial loss of the initial population of vibrational modes across the NVE run, an effect particularly probable in the case of an anharmonic system, with a number of coupled modes. Also, it is possible that in the QT-NVT simulations the ZPE leakage could not be completely prevented. However, by noting a reasonable overall distribution of the spectral intensities with respect to the HLD reference (see the bottom panel), the latter effect seems not to be the case here.

To further test the usefulness of the quantum-thermostatted simulations in the modeling of INS observables we continue the discussion by turning into the NCS experiments.

4.2. Neutron Compton Scattering

NCS experiments provide a direct experimental access to the mean kinetic energy $\langle E_K \rangle$ of nuclei, irrespective to the nature (harmonic or not) of their underlying dynamics. In brief, NCS operates within the limits of the so-called impulse approximation (IA) [69], with hundreds of electron-Volt energy transfer $\hbar\omega$ from incident neutrons and up to tens of Å^{-1} of momentum transfer, Q. The consequence of the IA is that a time-of-flight (TOF) spectrum measured on VESUVIO consists of separate recoil peaks, each stemming from the longitudinal nuclear momentum distribution (NMD) of each respective nucleus present in the target system, $J_{IA}\left(y, \hat{Q}\right)$ [69]. Moreover, the intensity of each NCS recoil peak is proportional to the product of the total bound scattering cross-section and the number of atoms under the beam, thus rendering NCS, unlike INS, completely free from the neutron weighting problem, which greatly facilitates the theoretical interpretation of the spectra.

Considering a momentum transfer of a nucleus in direction $\hat{Q}$, the NMD can be defined in terms of Gaussian approximation (GA) as:

$$J_{IA}\left(y, \hat{Q}\right) = \frac{1}{\sqrt{2\pi\sigma^2}} exp\left(\frac{-y^2}{2\sigma^2}\right) \tag{7}$$

where σ^2 is the second moment of the NMD and y is the so-called West-scaling variable, defined as the projection of an associated radial nuclear momentum $\vec{p}$ onto the direction of the momentum transfer $\hat{Q}$ [69].

For a given nucleus, σ^2 can be, within the harmonic approximation, further associated with the atom-projected density of vibrational states (AVDOS), $g(\omega)$, according to the following relation:

$$\sigma^2 = \frac{M}{\hbar} \int g(\omega) \frac{\omega}{2} \coth\left(\frac{\omega}{2k_B T}\right) d\omega, \tag{8}$$

$g(\omega)$, is determined as the contribution from a given nucleus to the total phonon density of states and defined by the following sum over all q points in the first Brillouin zone (BZ), N_q, and over all phonon bands N_λ:

$$g_n(\omega) = \frac{1}{3N_q} \sum_{q \in 1BZ} \sum_{\lambda=1}^{N_\lambda} e_n(\lambda, q)^2 \delta(\omega - \omega(\lambda, q)) \tag{9}$$

where $e_n(\lambda, q)$ are the polarization vectors for phonon modes λ of energy $\omega(\lambda, q)$. [69]

IOP Conf. Series: Journal of Physics: Conf. Series **1055** (2018) 012003 doi:10.1088/1742-6596/1055/1/012003

With the second moment of the NMD equal to σ^2, the mean kinetic energy, $\langle E_K \rangle$, can be defined as:

$$\langle E_K \rangle = \frac{3\hbar^2}{2M}\sigma^2 \tag{10}$$

Equation 8 has been successively used for the analysis of nuclear-dynamics in a number of materials using HLD calculations [69]. Despite the fact that Eq.(8) is derived using the quantum mechanical theory of the harmonic oscillator it can still present a good starting point for the discussion of anharmonic systems. Notably, the average kinetic energy $\langle E_K \rangle$ is equal to the half of the total energy $\langle E \rangle$ only in the harmonic approximation. To this end, two cases of an anharmonic oscillator can be considered: symmetric and asymmetric [70]. In the case of the asymmetric oscillator, described by a potential of the form $V(x) = 1/2kx^2 - 1/3ax^3$, to first order in the anharmonic parameter, $\delta = Aa/k$, the result is the same as in the case of purely harmonic oscillator [70]. In the second case, described by a potential of the form $V(x) = 1/2kx^2 - 1/4bx^4$, to second order in the anharmonicity parameter, $\delta = bA^2/k$ with A being the amplitude of the unperturbed harmonic motion, we get:

$$\frac{\langle E_K \rangle}{E} = \frac{1 - 12/16\delta}{2(1 - 9/16\delta)} \tag{11}$$

where, perhaps more intuitively, one can introduce a non-dimensional parameter ϵ, equal to first order in b with $-1/32\delta$, where ϵ is the correction to the classical turning point, A, in the spirit of the Wentzel-Kramers-Brillouin (WKB) approximation, i.e., $A \rightarrow A(1+\epsilon)$ with $A = \sqrt{\frac{\hbar\omega}{k}} = \sqrt{\frac{\hbar}{\mu\omega}}$ with ω the characteristic frequency of vibration, μ reduced mass, and k the force constant for a particular vibration under consideration. Equation (11), in the limit of large anharmonicity of, say, δ=0.1, yields the average kinetic energy being 0.49 of the total energy, thus rendering the anharmonic correction very small for this class of potentials. Thus on the whole, in case of anharmonic symmetric local effective Born-Oppenheimer potential, felt by nuclei which NMDs are under consideration, one can always re-scale the kinetic energy below the harmonic limit, in a spirit which is similar in nature to the application of scaling to $g_n(\omega)$ curves in order to mimic anharmonicity.

As a complementary approach, one may also calculate the mean-kinetic energy for an anharmonic simulation directly from the analysis of the velocity distribution function (VDF). Figure 6 provides some insight into the influence of quantum thermostats on the vibrational energy distribution. By looking at the temperature evolution of the raw QT-NVT spectra (depicted as the solid lines) one may note clearly pronounced blue-shifts throughout the whole energy transfer range, a definitely unexpected effect with increase in temperature. For comparison, the microcanonical spectra deconvoluted from the NVT trajectories are presented as dotted lines. The spectral deconvolution shifts the first moment towards the lower energy transfers leading to correct reproduction of the band positions with respect to the experimental HRMECS reference (see Fig. 7) [27]. While the presence of noise induced by larger thermostat fluctuations at elevated temperatures should not be ignored, the calculations provide an overall reasonable reproduction of the direct-geometry spectrum even close to room-temperature conditions.

Let us now comment on the performance of the QT calculations for the description of the NCS observables. The results of the mean kinetic energy calculations are depicted in Fig. 8 and collected in Table 2, respectively, as referring to classical (HLD) and semi-classical (QT-NVT; QT-NVE) simulations. The results are compared to the experimental data which, however, are only available at 180K and 240 K.

In Fig. 8, the HLD results may serve as a reference, being only slightly above the experimentally observed mean kinetic energy of 147 eV at 240 K. Moreover, taking into account

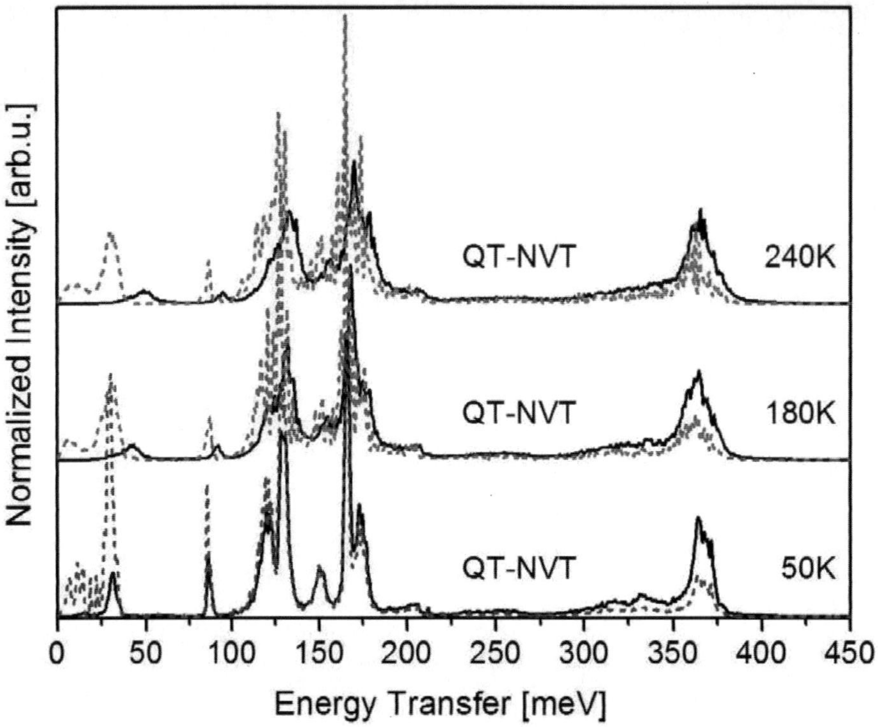

Figure 6. The variable-temperature proton-VDOS spectra calculated from QT BOMD simulations [25] The solid lines present the raw power spectra, while the dotted lines presents the NVE spectra deconvoluted from the NVT trajectory according to Rossi *et al.* [21]. Please note the frequency differences between the solid and dotted curves.

our remark on the possibility of accounting for the anharmonicity in the system by simply scaling the HLD-obtained values of the nuclear kinetic energy by a correction factor derived from the application of the virial theorem to a non-harmonic potential energy curve, we can further say that a $\frac{\langle E_K \rangle}{\langle E \rangle}$ ratio of 0.4859 would bring the HLD-obtained value into exact agreement with the NCS experiment. This in turn would imply a negative value of the anharmonicity parameter δ of -0.0288, which gives a positive quartic term in the potential. A strong disturbance of the nuclear dynamics and the unphysical upward VDOS shifts induced by the quantum thermostat (see Fig. 7) explain why the QT-NVT calculations overestimate the mean kinetic energy at the higher temperature conditions. While a questionable discrepancy of c.a. 10% between the $\frac{\langle E_K \rangle}{\langle E \rangle}$ values for the VDF and AVDOS data was noted, both sets of data are noticeably above the HLD reference and the experimental points.

In contrary, the results derived from the AVDOS analysis according to the NVE deconvolution proposed by Rossi et al. [21], brings the mean kinetic energy visibly below the reference data. In order to provide more robust benchmarks, the results were derived from both CASTEP and CP2K calculations. Both computational routes were found to be consistent, where the difference can be clearly attributed to the earlier-discussed differences in predicted VDOSs.

The obtained results provide us with a guess of a lower-limit of $\frac{\langle E_K \rangle}{\langle E \rangle}$ predicted computationally. Rigorous, direct calculations of nuclear momentum distributions (NMDs) require the use of open-path integral molecular dynamics formalism, which may lead to an unaffordable cost for real-chemistry problems [71, 72]. It motivates the search for less-accurate

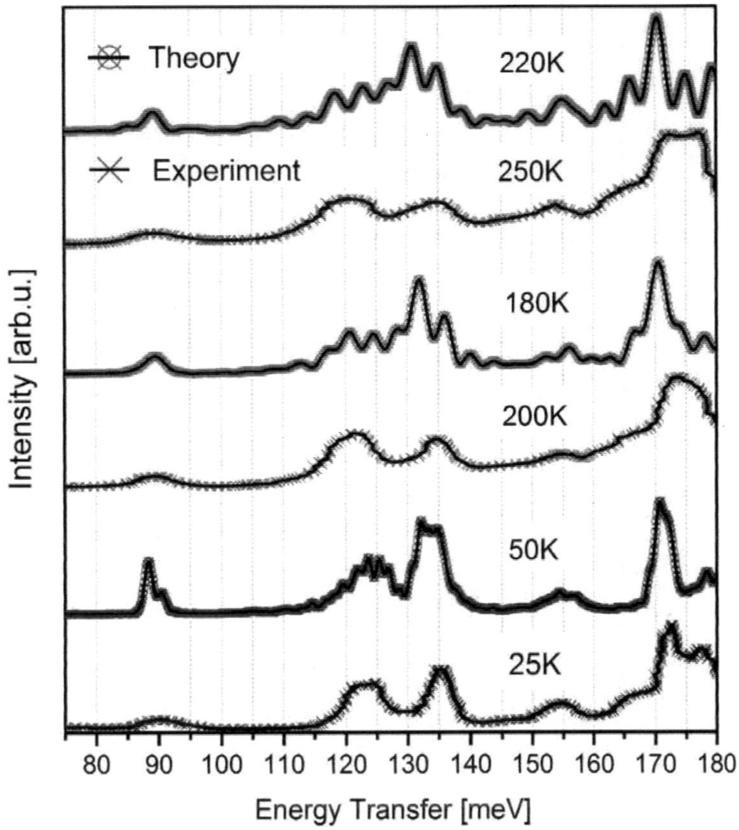

Figure 7. The temperature evolution of the INS spectrum of solid formic acid as measured with HRMECS spectrometer and predicted with QT NVT simulations (QT-GLE). The presented spectra were deconvoluted into the microcanonical ones [21,27]. The calculated energy transfers were semi-empirically scaled by the factor of 1.035.

but more efficient approaches, with the path integral generalized Langevin equation thermostat (PIGLET) as the primary computational route [73].

Table 2. Experimental values (VESUVIO) and first-principles predictions of the standard deviations of NMDs, σ_H [Å^{-1}] and the mean kinetic energies, <E_K>, [meV] of protons in solid FA. The results from Ref. [25] (DINS; HLD) are compared to the present results of simulations with a quantum colored-noise thermostat (QT; see the text for more details). The results were derived both from hydrogen-projected vibrational density of states (PDOS) and from a direct analysis of the velocity distribution function (VDF).

Temp. [K]	σ_H [Å^{-1}], (<E_K>[meV])				
	DINS	HLD [AVDOS]	QT-NVT [AVDOS]	QT-NVT [VDF]	QT-NVE (Deconvol.) [AVDOS]
240	4.86± 0.04 (147± 2)	4.93 (151.2)	5.17 (166.3)	4.97 (153.6)	4.47 (124.1)
180	4.76± 0.04 (141± 2)	4.92 (150.6)	5.16 (165.6)	4.97 (153.5)	4.41 (121.0)
LT-LIMIT	N/A	4.90 (149.4)	4.96 (153.1)	4.73 (139.2)	4.15 (107.0)

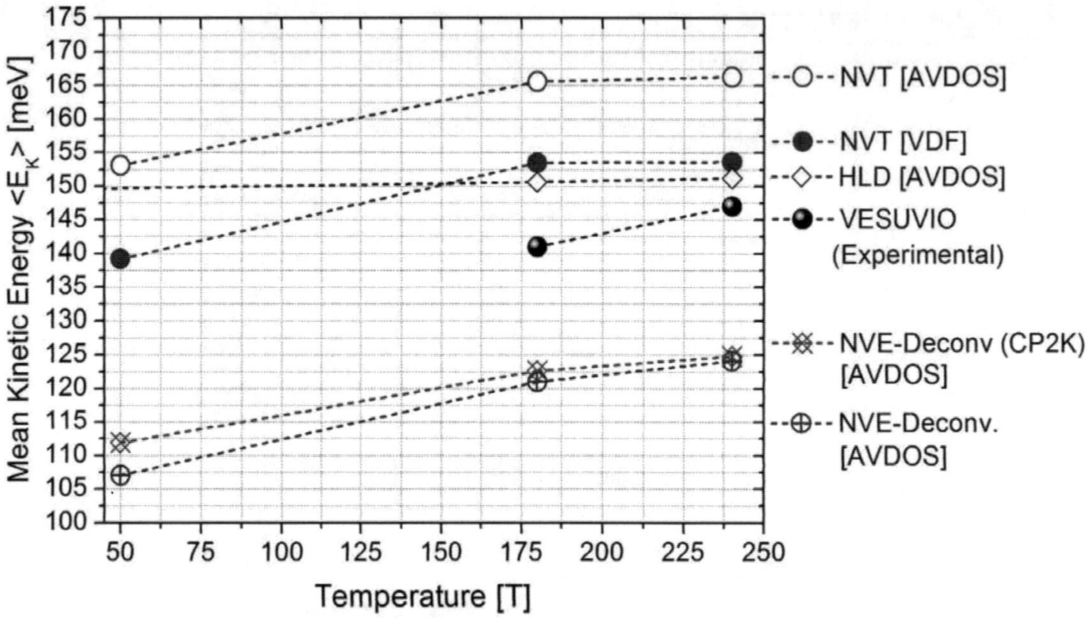

Figure 8. The mean kinetic energy, $\langle E_K \rangle$ [meV] for hydrogen atoms in formic acid as predicted by harmonic lattice dynamics (HLD) and anharmonic quantum-thermostatted simulations (NVT, [25] NVE-Deconv. [21]) and compared to experimental data recorded on VESUVIO spectrometer at 180 and 240 K. [25]

5. Conclusions

We have presented an extension of portable implementation of the quantum-colored noise thermostat towards condensed matter systems by means of ABIN code interfaced to CASTEP. Moreover, we have illustrated how the quantum-thermostated semi-classical simulations can be used in modeling of the INS and DINS response in considerably anharmonic hydrogen-bonded solids, with the case of solid FA.

The physics and chemistry of the crystalline FA is interesting for a multitude of reasons. First and foremost, it is a prominent example of an H-bonded system, where polymorphism had been postulated at a time when no sophisticated tools of computational chemistry and spectroscopy were available. Due to the rapid progress in both of these fields and the flurry of theoretical and experimental activity that followed, often a revision of original phase diagrams of solid molecular crystals is needed [77]. In this context, our previous [32], and present work on solid FA attempts to address the question of the postulated polymorphism by tackling it both experimentally and computationally. The overarching motivation here is that, as suggested by Cazorla and Boronat [77], the nuclear quantum delocalisation of the proton across the hydrogen bond may be easily mistaken for a superposition of two structures, especially when one uses a static probe with an infinite time window as diffraction. To this end, the result of this work seems to corroborate the result of our previous study, strongly suggesting that the nuclear momentum distribution of the proton is consistent with a picture in the direct space of spatial coordinates, in which the proton is quantum-delocalised across the hydrogen bond. In such a picture the distinction between the two formic acid crystal isomorphs, only differing by the position of the proton with respect to the donor and acceptor hydrogen, is lost. As a result, a hypothesis is plausible, in which one single formic acid phase with a single nuclear quantum proton state is present in the crystal, opening a route for extensive *ab initio* solid-state modeling and benchmarking.

In order to corroborate the hypothesis, a hierarchical series of computational approaches needs to be employed. One has to start from first-principles calculations of the model crystal structures, followed by calculations of the atom-projected vibrational densities of states. This can be further extended towards *ab initio* MD simulations with a standard thermostating scheme that does not treat the nuclear quantum effect-dictated properties in a special way (e.g. Nose-Hoover thermostat). Finally, employing the nuclear quantum-thermostat with an additional signal post-processing employing the deconvolution of the perturbed spectra in a way that can properly recover the unperturbed dynamic properties [21]. In doing this series of approximations, no convergence towards the experimental results, obtained from the neutron Compton scattering, is guaranteed from the outset, which largely motivated this work. In what we have established herein, the experimental NCS result is bracketed by the results obtained from the traditionally applied thermostated BOMD and the simulations with the nuclear quantum-thermostat with the deconvolution procedure. This result is important by itself as it shows that neither the route towards the optimal thermostating of the BOMD calculations nor the deconvolution procedure are completely optimised yet and further work is necessary to improve the agreement with the experiment. To this end, the NCS method is crucial as an ultimate benchmark of the *ab initio* results, as it does the comparison on atom-by-atom basis by testing the quality of the experimental data reproduction through the atom-projected VDOS, unlike any other neutron spectroscopic technique.

While NMDs continue to demonstrate its usefulness as benchmarking observables, another source of inspiration and motivation for our work reported here has been the fact that formic acid is an archetypic infinite hydrogen bond chain system and as such represent an ultimate challenge given by nature as far as the computational reproduction of the properties of hydrogen-bonded systems is concerned. Namely, contrary to the hydrogen-bonded systems, which crystalize as crystals of dimers, with benzoic acid and its derivatives given as the prime example, solid FA cannot be approximated computationally as a sum of smaller (dimeric) subsystems. Consequently, an *ab initio* calculation, at any level of theory and approximation, cannot be broken down into much less complicated lower-dimensional sub-problems. Most importantly in the context of our attempt to establish a hierarchical computational approach towards the high fidelity reproduction of the experimental nuclear momentum distributions, this structural feature of the hydrogen bond network in FA renders any Path Integral Molecular Dynamics calculations practically impossible given the current level of computing power of the high-performance-supercomputing. This apparent deficiency can be, however, turned into an opportunity by re-formulating the challenge. Namely, one can still formulate a quest towards an optimal reproduction of nuclear quantum effects by solely resorting to *ab initio* MD methods and place the centre of gravity of the challenge on the optimal thermostating and deconvolution scheme. To this end, we have found in this work that QT-GLE BOMD simulations can be considered as a particularly useful tool to support the modeling of the INS spectra of molecular materials at low temperatures. Such an approach can be found particularly valuable when combined with the microcanonical spectrum extraction according to Rossi et al. [21]. We have illustrated how the quantum thermostating tends to affect the vibrational density of states at elevated temperatures, leading to an unphysical bias in terms of frequencies. Such a dramatic disturbance can be, however, reduced to a great extent by employing *a posteriori* corrections [21].

On the whole, this work encourages the use of quantum thermostatting in the modeling of the low-temperature INS spectra in condensed matter systems. A natural extension of this methodology would employ more advanced and accurate approaches, e.g. thermostatted ring-polymer molecular dynamics (TRPMD) [74, 75]. A way of reducing the computational cost of TRPMD has been proposed by Kapil *et. al.* [78]. It boils down to the the use of multiple time steps in real and imaginary time in which one uses a cheap reference potential which captures well the short ranged part of the potential. The cheap reference is calculated on a large number of replicas and integrated with a small time step while the difference which contains the *ab initio* potential is calculated on a small number of replicas and integrated with a long time step.

Acknowledgement
The authors gratefully acknowledge the UK Science & Technology Facilities Council for financial support and access to the beam time and the research infrastructure of the ISIS Facility. This work was supported in part by PL-Grid Infrastructure and PROMETHEUS facility (Grant IDs: latticedynamics05; latticedynamics06). K. D. kindly thanks Mr Maciej Czuchry for his continuous technical support at PL-Grid. K.D. also acknowledges Ministry of Science and Higher Education of Poland and the Plenipotentiary of Poland to JINR, Dubna for supporting collaboration with the ISIS Molecular Spectroscopy Group. D.H. and P.S. were partially supported by Czech Science Foundation, project. number 18-16577S. We are grateful to the workshop organisers for the award for the poster presentation that led to this work.

References
[1] Marx D, Hutter J, *Ab Initio Molecular Dynamics: The Virtual Laboraory Approach*, 2010, Cambridge University Press, 2010.
[2] Dammak H, Chalopin Y, Laroche M, Hayoun M and Greffet J-J, 2009, *Phys Rev Lett* **103** 190601.
[3] Barrat J-L and Rodney D, 2011, *J Stat Phys* **144** 679.
[4] Bowman J, 1978, *J Chem Phys* **68** 608.
[5] Carter, S, Culik S and Bowman J, 1997, *J Chem Phys* **107** 10458.
[6] Bowman J, Christoffel K and Tobin F, 1979, *J Chem Phys* **83** 905.
[7] Christoffel K and Bowman J, 1982, *Chem Phys Lett* **85** 220.
[8] Thompson T and Truhlar D, 1980, *Chem Phys Lett* **75** 87.
[9] Wang X-G, Sibert E L and Martin J M, 2000, *J Chem Phys* **112** 1353.
[10] Barone V, 2005, *J Chem Phys* **122** 014108.
[11] Barone V, Bloino J, Guido C and Lipparini F, 2010, *Chem Phys Lett* **496** 157.
[12] Monserrat B, Drummond N D and Needs R J, 2013, *Phys Rev B* **87** 144302.
[13] Benoit D, PVSCF, source code available at http://pvscf.org.
[14] Gaigeot M P, 2010, *Phys Chem Chem Phys* **12** 3336.
[15] Cimas A, Maitre P, Ohanessian G and Gaigeot M P, 2009, *J Chem Theory Comput* **5** 2388.
[16] Williams R W, Schlucker S, Hudson B S, 2008, *Chem Phys* **343** 1.
[17] Van-Oanh N T, Falvo C, Calvo F, Lauvergnat D, Basire M, Gaigeot M P and Parneix P, 2012, *Phys Chem Chem Phys* **14** 2381.
[18] Ceriotti M, Bussi G and Parinello M, 2009, *Phys Rev Lett* **102** 020601.
[19] Ceriotti M, Bussi G and Parinello M, 2009, *Phys Rev Lett* **103** 030603.
[20] Ceriotti M, Bussi G and Parinello M, 2010, *J Chem Theory Comput* **6** 1170.
[21] Rossi M, Kapil V, Ceriotti M, 2018, *J Chem Phys* **148** 102301.
[22] Hollas D, Svoboda O, Oncak M, Slavicek P, ABIN, source code available at https://github.com/PHOTOX/ABIN
[23] Clark S J, Segall M D, Pickard C J, Hasnip P J, Probert M J, Refson K, Payne M C, 2005, *Z Kristallogr* **220** 567.
[24] Segall M D, Lindan P J D, Probert M J, Pickard C J, Hasnip P J, Clark S J, Payne M C, 2002, *J Phys Condens Matter* **14** 2717.
[25] Krzystyniak M, Drużbicki K, Romanelli G, Gutmann M. J., Rudić S, Imberti S, Fernandez-Alonso G, 2017, *Phys Chem Chem Phys* **19** 9064.
[26] Parker S, Fernandez-Alonso F, Ramirez-Cuesta A J, Tomkinson J, S. Rudić S, Pinna R, Gorini G, Castanon J F, 2014, *J Phys Conf Ser* **554** 012003.
[27] Loong C K, Trouw F, Iton L E, 1994, Hydrogen Bond Networks, edited by Bellissent-Funel M C and Dore J C (Kluwer, Netherlands) 231.
[28] Andreani C, Colognesi D, Mayers J, Reiter G F, Senesi R, 2005 *Adv Phys* **54** 377.
[29] Seel A G, Krzystyniak M, Fernandez-Alonso F, 2014, *J Phys Conf Ser* **571** 012006.
[30] Gorini G, Festa G, Andreani C, 2014, *J Phys Conf Ser* **571** 012005.
[31] Jackson S, Krzystyniak M, Seel A G, Gigg M, Richards S E, Fernandez-Alonso F, 2014, *J Phys Conf Ser* **571** 012009.
[32] Krzystyniak M, Drużbicki K, Fernandez-Alonso F, 2015, *Phys Chem Chem Phys* **46** 31287.

[33] Krzystyniak M, Richards S, Seel A G, Fernandez-Alonso F, 2013, *Phys Rev B* **88** 184304.
[34] Krzystyniak M, Fernandez-Alonso F, 2011, *Phys Rev B* **83** 134305.
[35] Krzystyniak M, Romanelli G, Fabian M, Gutmann M, Festa G, Arcidiacono L, Gigg M, Drużbicki K, Andreani C, Senesi R, Fernandez-Alonso F, 2018, *J Phys Conf Ser* **1021** 012026.
[36] Giannozzi P, Baroni S, Bonini N, Calandra M, Car R, Cavazzoni C, Ceresoli D, Chiarotti G L, Cococcioni M, Dabo I, Dal Corso A, Fabris S, Fratesi G, de Gironcoli S, Gebauer R, Gerstmann U, Gougoussis C, Kokalj A, Lazzeri M, Martin-Samos L, Marzari N, Mauri F, Mazzarello R, Paolini S, Pasquarello A, Paulatto L, Sbraccia C, Scandolo S, Sclauzero G, Seitsonen A P, Smogunov A, Umari P, Wentzcovitch R M, 2009, *J Phys Condens Matter* **21** 395502.
[37] Giannozzi P, Andreussi O, Brumme T, Bunau O, Buongiorno Nardelli M, Calandra M, Car R, Cavazzoni C, Ceresoli D, Cococcioni M, Colonna N, Carnimeo I, Dal Corso A, de Gironcoli S, Delugas P, DiStasio Jr R A, Ferretti A, Floris A, Fratesi G, Fugallo G, Gebauer R, Gerstmann U, Giustino F, Gorni T, Jia J, Kawamura M, Ko H-Y, Kokalj A, Kkbenli E, Lazzeri M, Marsili M, Marzari N, Mauri F, Nguyen N L, Nguyen H-V, Otero-de-la-Roza A, Paulatto L, Ponc S, Rocca D, Sabatini R, Santra B, Schlipf M, Seitsonen A P, Smogunov A, Timrov I, Thonhauser T, Umari P, Vast N, Wu X, Baroni, S, 2017, *J Phys Condens Matter* **29** 465901.
[38] Hutter J, Iannuzzi M, Schiffmann F, VandeVondele J, 2014, *Wiley Interdiscip Rev Comput Mol Sci* **4** 15.
[39] VandeVondele J, Krack M, Mohamed F, Parrinello M, Chassaing T, Hutter J, 2005, *Comp Phys Comm* 167 103.
[40] Perdew J, Burke K, Ernzerhof M, 1996, *Phys Rev Lett* **77** 3865.
[41] Tkatchenko A, Scheffler M, 2009, *Phys Rev Lett* **102** 073005.
[42] Grimme S, Antony J, Ehrlich S, Krieg H, 2010, *J Chem Phys* **132** 154104.
[43] Grimme S, Ehrich S, Goerigk L, 2011, *J Comput Chem* **32** 1456.
[44] Becke A D, Johnson E R, 2005, *J Chem Phys* **122** 154101.
[45] Ramer N J, Rappe A M, 1999, *Phys Rev B* **59** 12471.
[46] Tikhonov D, 2016 *J Chem Phys* **144** 174108.
[47] Colored-Noise Thermostats for Molecular Dynamics http://gle4md.org/
[48] Brehm M, Kirchner B, 2011, *J Chem Inf Model* **51** 2007.
[49] Kapil V, UNWIND, source code available at https://github.com/venkatkapil24/unwind
[50] Crystal Explorer 17.5 http://crystalexplorer.scb.uwa.edu.au
[51] Jayatilaka D, Grimwood D J, *Computational Science - ICCS 2003* **4** 142.
[52] Turner M J, Grabowsky S, Jayatilaka D, Spackman M A, 2014, *J Phys Chem Lett* **5** 4249.
[53] Turner M J, Thomas S P, Shi M W, Jayatilaka D, Spackman M A, 2015, *Chem Comm* **51** 3735.
[54] Albinati A, Rouse K D, Thomas M W, 1978, *Acta Crystallogr Sect B* **34** 2188.
[55] Riley K E, Platts J A, ez J, Hobza P, Hill J G, 2012, *J Phys Chem A* **116** 4159.
[56] Zelsmann H R, Bellon F, Marechal Y, Bullemer B, 1970, *Chem Phys Lett* **6** 513.
[57] Nahringbauer I, 1978, *Acta Cryst* **B34** 315.
[58] Grimme S, 2006, *J Comp Chem* **27** 1787.
[59] Tosoni S, Tuma C, Sauer J, Civalleri B, Ugliengo, 2007, *J Chem Phys* **127** 154102.
[60] Parker S F, Ramirez-Cuesta A J, Albers P W, Lennon D, 2014, *J Phys Conf Ser* **554** 012004.
[61] Parker S F, Lennon D, 2016, *J Phys Conf Ser* **746** 012066.
[62] Dymkowski K, Parker S F, Fernandez-Alonso F, Mukhopadhyay S, 2017, (submitted).
[63] http://docs.mantidproject.org/nightly/algorithms/Abins-v1.html
[64] Ramrez R, Lpez-Ciudad T, Kumar P, Marx D, 2004, *J Chem Phys* **121**3973.
[65] Lin C Y, George M W, Gill P M W, 2004, *Aust J Chem* **57** 365.
[66] Pestana L R, Mardirossian N, Head-Gordon M, Head-Gordon T, 2017, *Chem Sci* **8** 3554.
[67] Tikhonov D D, Sharapa D I, Schwabedissen J, Rybkin V V 2016, *Phys Chem Chem Phys* **18** 28325.
[68] Senesi R, Kolesnikov A I, Andreani C, 2014, *J Phys Conf Ser* **571** 012007.
[69] Andreani C, Krzystyniak M, Romanelli G, Senesi R, Fernandez-Alonso F, 2017, *Advances in Physics* **66** 1.
[70] Sivardiere J, *American Journal of Physics* **54**, 1100 (1986).
[71] Morrone J A, Lin L, Car R, 2009, *J Chem Phys* **130** 204511.
[72] Morrone J A, Srinivasan V, Sebastiani D, Car R, 2007, *J Chem Phys* **126** 234504.
[73] Ceriotti M, Parrinello M, Markland T A, Manolopoulos D E, 2010, *J Chem Phys* **133** 124104.
[74] Rossi M, Ceriotti M, Manolopoulos D E, 2014, *J Chem Phys* **140** 234116.
[75] Rossi M, Liu H, Paesani F, Bowman J, Ceriotti M, 2014, *J Chem Phys* **141** 181101.
[76] Ceriotti M, Manolopoulos D, 2012 *Phys Rev Lett* **109** 100604.
[77] Cazorla C, Boronat J, 2017 *Rev Mod Phys* **89** 035003.
[78] V. Kapil V, VandeVondele J, Ceriotti M, 2016 *The Journal of Chemical Physics* **144**, 054111.

IOP Conf. Series: Journal of Physics: Conf. Series **1055** (2018) 012004 doi:10.1088/1742-6596/1055/1/012004

Mass-selective neutron spectroscopy of glassy versus polycrystalline structures in binary mixtures of beryllium and zirconium

M Krzystyniak,[1,2] G Syrykh,[3] A Stolyarov,[4] R A Sadykov,[4] J Armstrong,[1] I da Silva,[1] G Romanelli,[1] and F Fernandez-Alonso[1,5]

[1] ISIS Facility, Rutherford Appleton Laboratory, Chilton, Didcot, Oxfordshire OX11 0QX, United Kingdom

[2] School of Science and Technology, Nottingham Trent University, Clifton Campus, Nottingham NG11 8NS, United Kingdom

[3] National Research Centre 'Kurchatov Institute', 123182, Moscow, Russia

[4] Institute for Nuclear Research,Russian Academy of Sciences, 117312, Moscow, Russia

[5] Department of Physics and Astronomy, University College London, Gower Street, London WC1E 6BT, United Kingdom

E-mail: matthew.krzystyniak@stfc.ac.uk

Abstract. Exploiting the unique tandem of VESUVIO and TOSCA inverted geometry spectrometers at the ISIS pulsed neutron and muon source in the United Kingdom, specifically the capability of VESUVIO to measure concurrently neutron diffraction and Compton spectroscopy, we have performed a global study of the structural and dynamical origins of disorder in Zr —Be metallic glasses. To this end, a polycrystalline $Zr_{30}Be_{70}$ and an amorphous $Zr_{40}Be_{60}$ systems were investigated in a wide range of temperatures ranging from 10 to 300K. For the first time, neutron diffraction has provided clues as to the structural composition of the polycrystalline $Zr_{30}Be_{70}$. The Rietveld refinement of the diffraction data has revealed that the polycrystalline system is made up of three distinct structural phases; a hexagonal phase, Be_2Zr, of the *P6/mmm* symmetry amounting to 87.11%, a second hexagonal phase, Be_5Zr, of the *P6/mmm* point group symmetry, amounting to 12.89%, as well as trace amounts of a third orthorhombic phase of unspecified stoichiometry. The overall sample stoichiometry, inferred from the dissection of the diffraction data, was in excellent agreement with the Compton results, both confirming the $Zr_{30}Be_{70}$ formulation. The analysis of INS data agreed very well with the theoretical results from the recursion method. The INS data were cross-validated by the nuclear momentum distributions of both Zr and Be, obtained from the analysis of the NCS data. Systematic differences between the crystalline and amorphous Zr —Be systems were identified in the whole temperature range and attributed to low-frequency mode softening, when going from crystalline to the amorphous phase.

1. Introduction

Despite the intense interest in metallic glasses for a variety of engineering applications, many details of their structure and dynamics remain a mystery. One of the iconic pieces of this unsolved puzzle is due to their phonon-related properties. Metallic glasses are at least two-component systems. From the point of view of lattice dynamics, two-component periodic systems should exhibit acoustic and optic vibrational modes. An acoustic dispersion relation has been observed in many metallic glasses [1, 2, 3, 4] by neutron and X-ray inelastic scattering methods. A model theoretically describing the dispersion relation of collective excitation in $Mg_{70}Zn_{30}$ with a combination of acoustic and optic modes has been made by Hafner [5]. Otomo [3] suggested that collective excitations in $Ni_{33}Zr_{67}$ originate from the combination of three 'optical' modes in the crystalline phase. Noticeably, in the same amorphous samples the dispersion of optical modes was not clearly observed.

Zr —Be amorphous alloys hold a special place among other metallic disordered systems. Because of the large difference in atom sizes and masses alloys can become amorphous in a wide range of compositions [6]. Zr —Be amorphous systems are also attractive in terms of industrial application. Physically vapor-deposited metallic Be has been used currently as a brazing filler metal [7]. ZrBe binary amorphous alloys were investigated as a new filler metal for joining zirconium alloy to supersede vapour-deposited beryllium [8].

One specific computational method of solid state physics is celebrated in the context of computational studies of metallic glasses, the recursion method [6]. The method relies on the presence of short-range order within a solid state matrix under investigation. In such case, the long-distance interactions can be neglected and the dynamical matrix of the whole system can be greatly reduced in dimensionality to a handful of force constants between neighbours and next-neighbours [9, 10]. Mathematically, the phonon modes are usually calculated in this method by tri-diagonalizing the dynamical matrix describing the forces in a basis of orthonormal operators [9]. It is the application of the recursion method to the calculation of generalised vibrational density of states (GVDOS) and atom-projected vibrational density of states (apVDOS) in amorphous $Zr_{1-x}Be_x$ systems with (x = 0.3 - 0.5) which is at the centre of the experimental and computational work by Bratkovsky *et al.* [6]. Further encouraging steps forward in the interpretation of the experimental and simulation results for Zr —Be amorphous systems were made by Syrykh *et al.* in the recent experiment on the IN4 spectrometer [11].

Investigation of modes in Zr —Be metallic glasses at low neutron momentum transfers fits well within the trends observed recently to investigate collective dynamics of fluids composed of disparate-mass atoms. For whole class of fluid mixtures, such as a rare-gas mixture [12], metallic alloy Li_4Pb [13], or liquid water [4, 12, 14], the emergence of two propagating 'kinetic modes' at intermediate wave vectors has been observed. In the investigations of Li_4Pb alloy [15] and $He_{65}Ne_{35}$ mixture [16] a non-acoustic origin of the high-frequency mode were suggested by a neutron experiment and MD study. Bryk *et al.* [17] showed that such an optic-like mode could vanish abruptly in the $Q \to 0$ limit. Computer simulation of the lattice dynamics of crystalline $Li_{22}Pb_5$ was carried out by Fernandez-Perea [18]. The calculated spectrum showed several bands corresponding to modes in which only the light atoms perform motions. The results support the idea put forward by Campa [19] regarding the possibility of detecting the presence of the 'kinetic modes' even if the fluid undergoes a freezing transition.

Motivated by the ongoing spectroscopic and structural research in ZrBe binary amorphous alloys and by their timely and relevant technological applications, we have embarked on a coherent research program employing a combination of neutron diffraction (ND), inelastic neutron scattering (INS) and neutron Compton scattering (NCS). Such an approach, demonstrated previously in the neutron studies on cesium hydrogen sulfate [20] and graphite intercalation compounds [21], has proven that, when applied to the same material, ND, INS and NCS span a uniquely large time and space window that enables global modelling and optimisation of structural and dynamical models of complex systems.

NCS has proven to be very successful in characterising systems where phonon dispersion curves are difficult to characterise or resolve or at high temperatures [22, 23]. This unique feature of the NCS technique is owed to the fact that, operating in the impulsive scattering regime, NCS measures total excitation curves of all motional degrees of freedom, through nuclear momentum distributions (NMDs), in a form of a collection of Doppler-broadened recoil peaks [22, 23]. The widths of the recoil peaks in momentum space are proportional to the first moments of atom-projected phonon density of states (apVDOS), no matter what the shape of the dispersion relation. This feature of NCS is of importance for measurements of phonon-related phenomena, like superconductivity, also exhibited by ZrBe binary systems [6], where the high-energy phonon modes are theoretically predicted but entirely confirmed experimentally. This NCS characteristic feature is in blatant contrast with structural probing techniques, like ND, where the Debye-Waller factor can be termed as the 'minus-first moment of the apVDOS'. To this end, NCS does not suffer from typical for ND intensity decrease as the temperature of the system under investigation is increased [22, 23]. Moreover, NCS is a local probe. Namely, due to the fact the measurement of momentum distributions yields wave functions in momentum space, the NCS technique is capable of directly measuring local effective potentials felt by nuclei in mass-resolved manner and thus to characterize the degree of anharmonicity or anisotropy of those potentials. It is this very sensitivity to the individual apVDOS and the local probe character that make the NCS a very robust benchmarking technique for *ab initio* predictions of NMDs, calculated both from molecular dynamics (MD) and density functional theory-based lattice dynamics (LD-DFT) [24, 25, 26, 21, 20, 22, 23, 27, 28].

Our first attempt to characterise ZrBe binary amorphous alloys concentrated on the $Zr_{40}Be_{60}$ system [29]. The work sought spectroscopic blueprints of phonon dispersion curves by jointly performing INS experiments at TOSCA instrument [30, 31, 32], and NCS and ND at VESUVIO instrument at Rutherford Appleton Laboratory (RAL) in the UK [33, 34]. The NCS-based characterisation of the dynamical behaviour of Zr an Be atoms has been performed, for the first time in a wide range of temperatures between 10 and 300 K. The characteristic nuclear quantum observables included the kinetic energy and standard deviations of nuclear momentum distributions. A comparison of the experimental values with the apVDOS spectra, refined by recursion method based on the data obtained with thermal neutron scattering, provided very satisfactory agreement. Also the experimental data were compared with the Debye model, which showed that the Debye temperature accounts better for the nuclear dynamics of Be nuclei than it is in the case of Zr. Interestingly, around the room temperature the nuclear dynamics of both isotopes was shown to be much less accurately accounted for by the Debye model. Moreover, the measurements at different temperatures provided an opportunity to evaluate the significance of anharmonicity in the dynamics of metallic glasses.

Continuing our line of research on the ZrBe binary amorphous alloys, this contribution reports on further efforts in multi-technique characterisation of this interesting and intriguing systems. In what follows, we will present results of joint application of ND, INS and NCS to contrast the amorphous and polycrystalline Zr —Be systems. Beside being motivated by our ongoing research programmes, both at the Kurchatov Institute and RAL, this work draws its inspiration from the seminal NCS work by Bermejo *et al.* [35] contrasting hydrogen and deuterium nuclear quantum effects in crystalline and glassy environments.

2. Results and discussion

2.1. Diffraction of amorphous and polycrystalline Zr —Be

In order to monitor the structure of the samples, two independent diffraction measurements were performed. The first, before the neutron experiment, was the X-Ray characterisation, performed at the diffractometer IPDS-II at the facilities of the Accelerator Center for Neutron Research at the 'Institute for Nuclear Research, Russian Academy of Sciences. The X-Ray diffractogramms of amorphous and polycrystalline Zr —Be samples are shown in Figure 1.

Following the X-Ray characterisation, concurrently with the NCS data, the ND data were recorded on the VESUVIO spectrometer [29]. The ND data recorded at T=10K together with the Al sample container diffraction data at T=10K are shown in Figure 2. Both techniques, XRD and ND, yield the same result: the amorphous sample (red traces) is characterised by

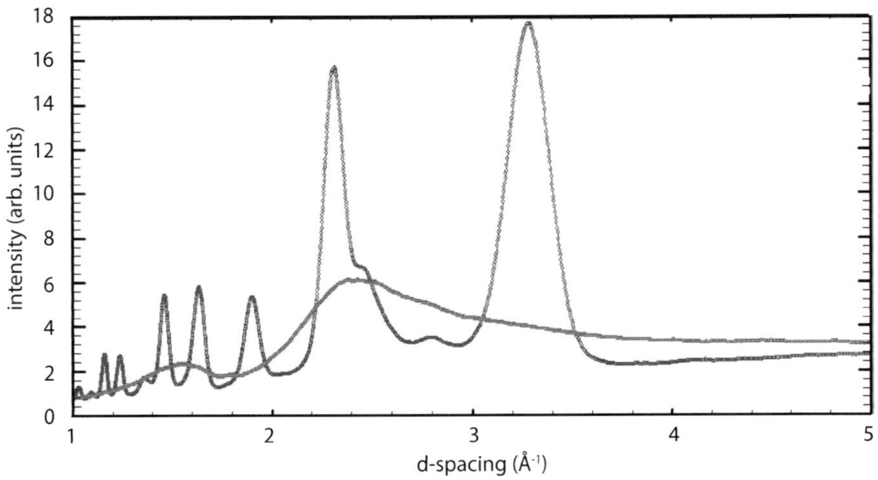

Figure 1. The X-Ray diffraction data of amorphous (red trace) and polycrystalline (black trace) Zr —Be samples measured on the diffractometer IPDS-II. See text for details.

diffuse, unresolved intensity in diffraction (the only Bragg peaks visible in the diffraction data come from the Al sample container), whereas the polycrystalline sample exhibits very well resolved diffraction peaks. The amorphous data exhibit large broad peak at the d-spacing of ca. 2.5 Å in very good accordance with the partial pair distribution functions, $G_{ab}(r)$, obtained by Bratkovsky *et al.* [6] using the structure model of amorphous Zr —Be alloys and more recent total scattering data obtained by Golovkova *et al.* [36]

Further refinement of ND VESUVIO data was performed in order to elucidate the structural composition of the polycrystalline Zr —Be system. To this end, structure refinement was performed using the TOPAS V 5.0 software [37] on the polycrystalline Zr —Be sample measured at T=10K. The result of this procedure is shown in Figure 3. Data were normalised, and the contribution to ND from the Al container subtracted prior to the analysis. Rietveld refinement, within agreement factors Rexp = 0.11% Rwp = 2.01% Rp = 4.46%, yielded three phases:

- The first hexagonal phase, Be_2Zr (ICSD code 58756), amounting to 87.11%, was refined, within the space group *P6/mmm*, with the following lattice parameters, $a= b = 3.800(3)$ [Å], $c = 3.226(2)$ [Å].
- The second hexagonal phase, Be_5Zr (ICSD code 58757), amounting to 12.89%, was refined, within the space group *P6/mmm*, with the following lattice parameters, $a= b = 4.564(3)$ [Å], $c = 3.477(2)$ [Å].
- The third phase of unspecified stoichiometry but contributing to the overall improvement of the refinement, of orthorhombic space group, with the following lattice parameters, a= 9.737(8) [Å], b = 4.577(3) [Å], and c = 3.072(3) [Å].

The results of the refinement are shown in Figures 4 and 5. Taking into account, the stoichiometric composition of both phases and their percentage contributions to the entire ND signal, obtained from refinement, the effective sample formulation yields $Be_{70}Zr_{30}$. Work is in progress, to calculate, based on the refined structure composition, the apVDOS and hence NMDs of Zr and Be within the LD-DFT method. This approach may prove superior to the recursion method, as the former is, in principle able to discern different contributions to the VDOS stemming from different structures present in the system under investigation, despite the same net sample stoichiometry. Moreover, LD-DFT is in principle capable of contrasting the

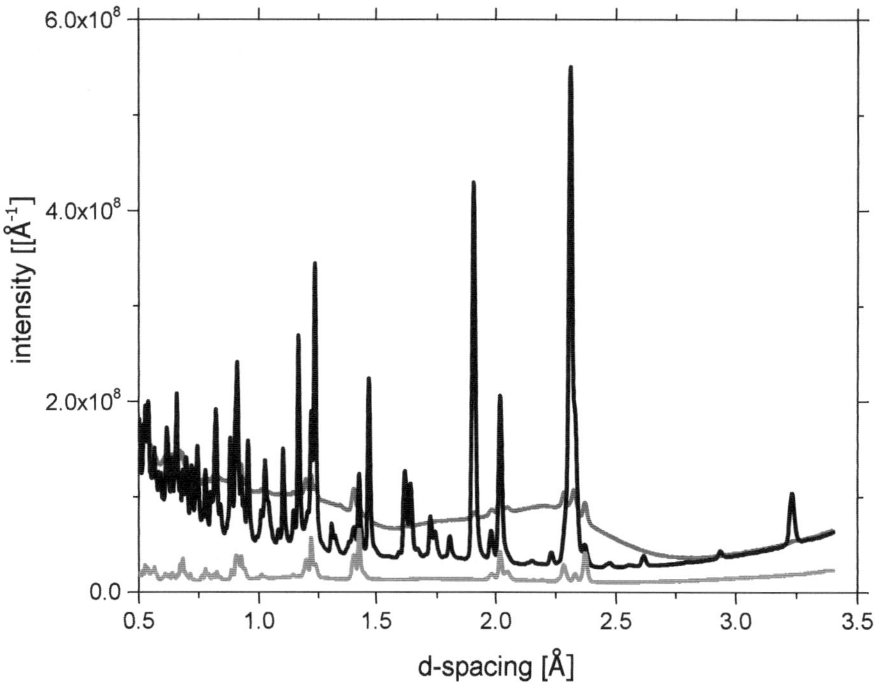

Figure 2. Neutron diffraction data of amorphous (red trace) and polycrystalline (black trace) Zr —Be samples recorded concurrently with NCS data at the VESUVIO spectrometer. Additionally, Al container diffraction pattern is shown as green trace. See text for details.

prediction for polycrystalline and amorphous phase, despite their identical formulation. The task envisaged here proves, however, very difficult due to the presence of the third phase and possible additional trace admixture of the amorphous phase.

2.2. INS and NCS

Inelastic and deep inelastic neutron scattering spectra were recorded on Tosca and VESUVIO spectrometers, respectively, both located at RAL [30, 31, 32, 33]. The details of the instrumental setup and the data reduction and analysis procedures have been extensively described elsewhere for both instruments. Thus, in what follows, we will provide a slightly alternative description aiming to emphasize the synergies, both at the instrumental and data analysis level, created by the strategy to collect and analyse INS and NCS data concurrently on the same system.

Also from the sample preparation point of view, both instruments present a favourable case as they can use the same sample container design, which can greatly facilitate the streamlining of multi-instrument global studies like the one presented here, especially in case of parametric studies as a function of external stimuli, like temperature. In the particular case of the experiments described here, both Zr —Be samples were measured as enclosed in a 6x6 cm^2 flat aluminum (Al) cells. The resulting total sample transmission was 92% for the polycrystalline and 87% in case of the amorphous sample, with the scattering power of the Al container identical and equal to 2% in both cases. The data used in the analysis described here were recorded at T = 10, 200 and 300K with the sample and container temperature stabilised using a dedicated helium closed-cycle refrigerator. With the average proton charge to the TS1 target of 165 μAhrs in case of both samples, the average time of both the NCS and INS experiments was 12 hours per sample per temperature.

IOP Conf. Series: Journal of Physics: Conf. Series **1055** (2018) 012004 doi:10.1088/1742-6596/1055/1/012004

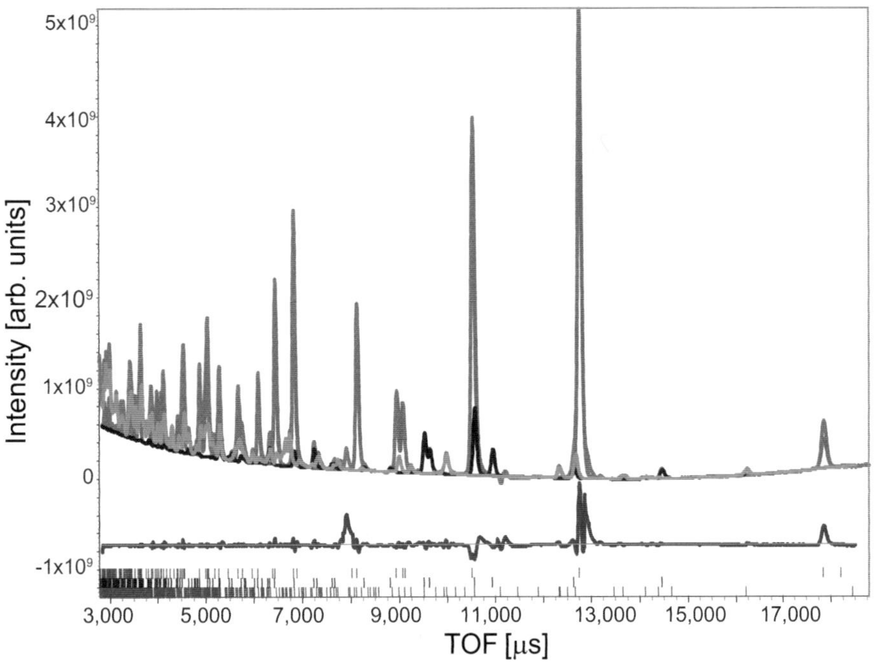

Figure 3. Neutron diffraction data of polycrystalline Zr —Be sample recorded at T=10K concurrently with NCS data at the VESUVIO spectrometer. Blue line - Be_2Zr phase, black line - Be_5Zr phase, green line unknown Be-Zr phase, red line - experimental ND data, violet line – residual. See text for details.

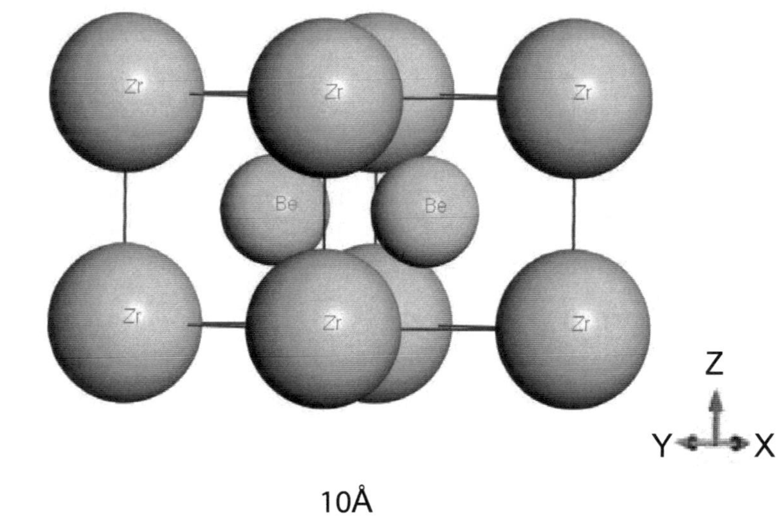

Figure 4. The unit cell of the first phase (ICSD code 58756), amounting to 87.11%, identified by Rietveld refinement of ND VESUVIO data. Eight Zr atoms contributing 1/8 and two Zr atoms make up the Be_2Zr unit formula. See text for details.

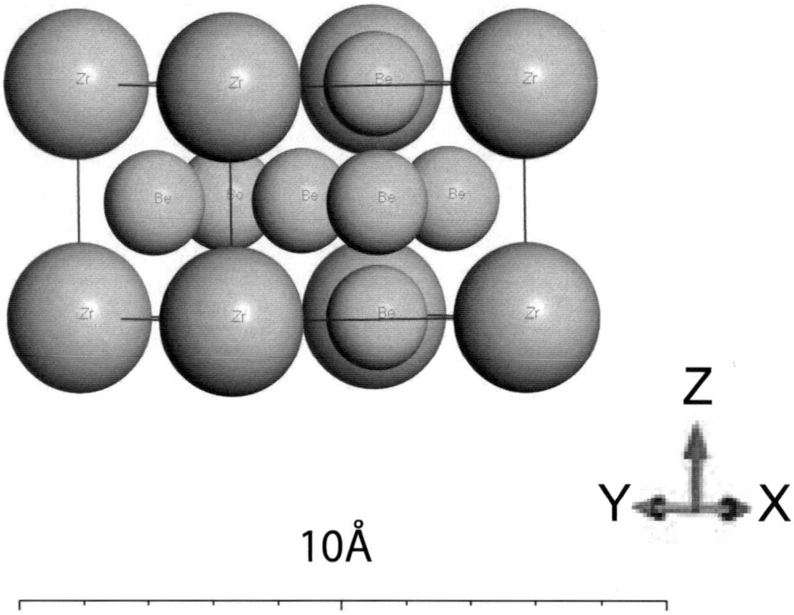

Figure 5. The second phase (ICSD code 58757), amounting to 12.89%, identified by Rietveld refinement of ND VESUVIO data. Eight Zr atoms contributing 1/8 and five Zr atoms make up the Be_5Zr unit formula. See text for details.

Tosca and VESUVIO are located at the Target Station 1 of the ISIS pulsed neutron and muon source at RAL. Both are inverted geometry instruments using a polychromatic broadband incident neutron beam viewed from sample position via water moderator. Moreover, both instruments use the same strategy to harness the full power of the ISIS source. Namely, as ISIS arranges the incident neutrons in pulses with the period between them being 20 ms (a sub-frame), both instruments are set up to receive 4 pulses in a single frame while the last pulse of the source is redirected towards the second target station [30]. Both Tosca and VESUVIO exploit the missing pulse to extend the energy transfer range of the instruments by measuring neutrons in the high time-of-flight (TOF) region. Tosca, measures the high TOF region neutrons to access the intense elastic line. On VESUVIO this high TOF region was previously extended to access the neutron energy range down to 1 meV quite successfully, extending the dynamics range of high resolution VESUVIO backscattering banks to access lattice spacing values in the range of ca. 7 Angstrom [38].

The INS spectra of amorphous and polycrystalline Zr —Be samples recorded at Tosca at T = 10K are shown in Figure 6. The recorded spectra agree very well with the atom-projected VDOS curves calculated by Bratkovsky *et al.* [6] by means of the recursion method, shown in the inset in Figure 6 as adopted from Ref. [6]), when one takes into account the sample stoichiometry and neutron neutron cross section weighting.

Figure 6 is the best starting point to briefly describe the NCS data analysis drawing from the analogies between the INS and NCS. Namely, in an INS spectrum both components of the atom-projected VDOS merge as a total, neutron-weighted (multiplied by the respective neutron

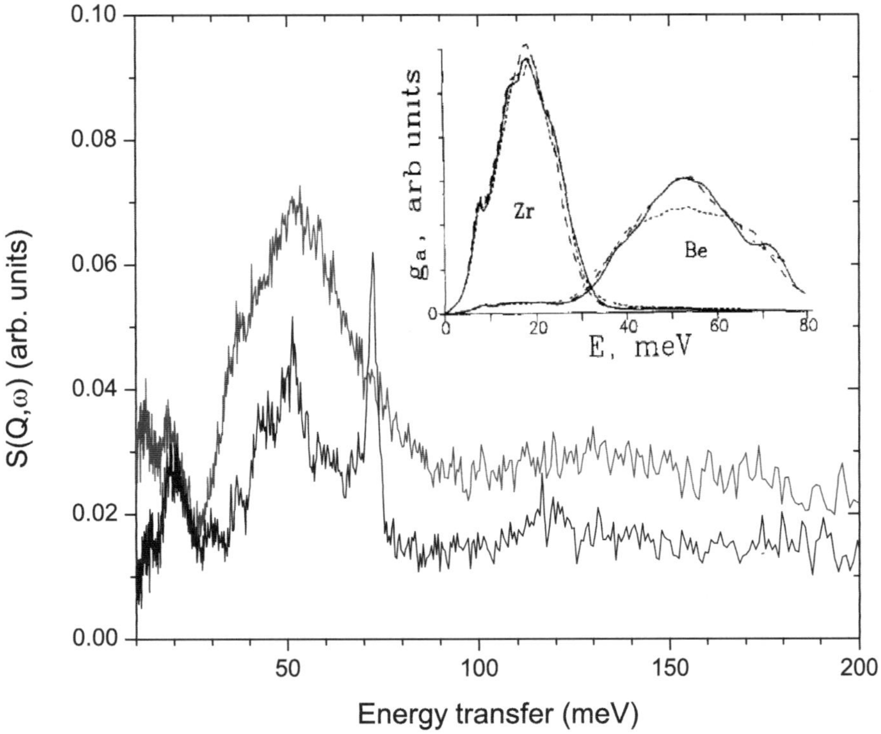

Figure 6. The INS spectra of amorphous (blue trace) and polycrystalline (black trace) Zr —Be samples recorded at Tosca at T = 10K. Inset: Partial densities of states, ga_{Zr}(E) and ga_{Be}(E), of Zr and Be respectively, calculated for a series of three amorphous systems $Zr_{1-c}Be_c$ with c = 0.3, 0.4 and 0.5. Figure adopted from Ref. [6]. The curves are calculated using the structure model for c = 0.3 (full line), 0.4 (dashed line) , and 0.5 (dotted lines). See text for details.

scattering cross-sections of Zr and Be) spectrum. This INS feature disables a detailed dissection of spectra into atom-projected contributions. Such a dissection could be in theory only performed from INS data in the lucky case of completely non-overlapping sets of INS modes stemming from two different nuclei. As one can see in Figure 6, even in case of such mass disparity as the one exhibited by Zr and Be in Zr —Be systems this task cannot be fully achieved. This clearly demonstrates the need for a method that does not suffer from 'coupling constant weighting', a method such as NCS. NCS, operating in the kinematic regime of up to hundreds of electron-Volt incident (termed as epithermal) neutron energies, and up to tens of $Å^{-1}$ of momentum transfer from neutrons to nuclei in the target sample, fulfils the assumptions of the Impulse Approximation (IA). [39, 24, 23]. In consequence, a billiard-ball type of scattering between the neutron and each nucleus in a sample leads to the creation of a TOF spectrum, consisting of a series of separate recoil peaks, each coming from the longitudinal momentum distribution of each nucleus in the sample. Moreover, the intensity of each NCS recoil peak scales with the product of its total bound scattering cross-section and the total number of atoms of given type in the incident neutron beam [39, 24, 23], thus rendering the NCS method completely free from the 'neutron weighting problem'.

The second, very important in the context of the presented joint INS and NCS data and the global modelling approach to Zr —Be systems under investigation, NCS feature is that the individual recoil peak shapes, when transformed from TOF to longitudinal momentum transfer domains to form nuclear momentum distributions, NMDs, are, to a good approximation

Gaussian in shape with second moments given by the first moments of the apVDOS [39, 24, 23]. This very useful feature of the NCS method provides a direct link between the experiment and *ab initio* modelling. Concretely, *ab initio* prediction of an NMD width, σ, can be calculated, for instance, by using the LD-DFT approach within harmonic description of nuclear motions under the Born-Oppenheimer approximation [40, 26] as:

$$\sigma_{Mn}(\widehat{q})^2 = \frac{M_n}{N_q \bar{\hbar}} \sum_{q \in 1BZ} \sum_{\lambda=1}^{N_\lambda} \times (e_n(\lambda, q) \cdot \widehat{q})^2 \frac{\omega(\lambda, q)}{2} \coth\left(\frac{\omega(\lambda, q)}{2k_B T}\right), \tag{1}$$

where $\omega(\lambda, q)$ are phonon frequencies, $e_n(\lambda, q)$ are polarization vectors and T is the sample temperature. The double summation in Eq. 1 runs over all N_q vectors in the first Brillouin zone, and all phonon branches N_λ.

From Eq. 1, the spherically averaged value of the second moment of the nuclear momentum distribution for a given nucleus n, $\sigma_{Mn}(\widehat{q})^2$, can be calculated as [41]

$$\sigma_n^2 = \frac{1}{3}\left[\sigma_n(\widehat{x})^2 + \sigma_n(\widehat{y})^2 + \sigma_n(\widehat{z})^2\right] \tag{2}$$

Interestingly, in the context of the atom-projected VDOS data shown above, yet another method of the NMD widths calculation has risen to popularity, the recursion method [9]. In a nutshell, recursion method amounts to recursively by tri-diagonalizing the dynamical matrix describing the forces in a basis of orthonormal operators [9]. This method can lead to substantial acceleration of computational effort, as compared to LD-DFT. By exploiting the system symmetry, or lack of long-range order, the dynamical matrix can be greatly simplified and described by a handful of force constants.

In what follows, we will complement already published data on the amorphous $Zr_{40}Be_{60}$ system [29] with new data collected on the $Zr_{40}Be_{60}$ system which is polycrystalline. The NCS data treatment protocols are described elsewhere in great detail [24, 23, 25, 26, 42]. Here, we will only briefly mention those aspects of the particular data reduction that were crucial for extraction of the NMD widths and nuclear kinetic energies in case of the specific systems under investigation. Specifically, for the NCS data analysis on both Zr —Be systems data recorded in backscattering were exclusively chosen due to much better mass-separation, much needed in case of three disparate nuclear masses, of 9 (Be), 27 (Al), and 91 (Zr) atomic mass units. The NCS backscattering data analysis can be summarised in the following steps:

- Raw backscattering TOF data were fitted sequentially detector by detector using Equation:

$$C_\theta(t) = A'[\frac{E_0 I(E_0)}{Q}]_t \times \sum_{n=1}^{N} I_n M_n J_n[y_n(t)] \otimes R_n[y_n(t)] \tag{3}$$

  where A' is a mass–independent experimental constant and the mass–independent factor $[\frac{E_0 I(E_0)}{Q}]_t$ depends on the incident neutron spectrum, $I[E_0(t)]$, the initial neutron energy, $E_0(t)$, and the momentum transfer $Q(t)$, all explicit functions of t, $J_M(y_M(t))$, are NMDs of nuclei in the sample in the IA limit with integrated peak intensities I_M proportional to the scattering densities and total (bound) neutron-scattering cross sections, $I_M = AN_M 4\pi b_M^2$ where $4\pi b_M^2$ is a total (coherent plus incoherent) bound neutron scattering cross section.
- the obtained values of the fitting coefficients, I_M and σ_M were used as input parameters to calculate the sample-dependent back scattering corrections due to multiple scattering (MS) from the sample and container;
- after subtraction of the MS contributions, a cumulative sum of weighted TOF spectra was formed with spectra normalised by the total TOF integral and summed in three groups of

backscattering detectors: 3-46, 47-90, and 91-134. Each sum of normalised detector data was fitted with a respective sum of TOF-normalised fitting curves, each member of the sum described by Eq. 3, i.e., $\sum_{i=1}^{N} \frac{C(\theta_i,t)}{\int C(\theta_i,t)dt}$. Underlying NMDs for all nuclei were assumed to be the sums of a pure Gaussian with standard deviations, σ and an additional term, proportional to a Gaussian multiplied by a third-order Hermite polynomial, accounting for the departures from the IA approximation, termed as Final States Effects (FSEs), with FSE magnitude k constrained within the harmonic approximation to $k = \frac{\sqrt{2}}{12}\sigma$. This method, referred to as CAAD, has been already successfully tested in treatment NCS data in Mass-selective Neutron Spectroscopy (MANSE) beyond the proton [24, 23, 43].

- The fitting parameters for each cumulative sum were the 3 values of NMD widths and 3 values of NMD relative intensities of Be, Al, and Zr. No stoichiometric fixing was employed in order to check the obtained sample stoichiometry against the diffraction data, collected concurrently on VESUVIO. In what follows, we will present the results of the CAAD analysis for the first group of detectors as the NCS data collected from detector 3-46 were found to exhibit the best signal-to-noise and background ratios and best mass-separation due to relatively high scattering angle values.

The first result coming from the CAAD fitting, is shown in Figure 7. The first observation is that, due to different total scattering power, greater in case of the amorphous sample, the share in the total CAAD curve, of the Al container recoil peak is smaller. However, in case of both samples the scattering power levels are good, thus resulting in overall good signal-to-noise ratios and enabling precision NMD fitting. This result demonstrates why the CAAD method was chosen for data analysis in this particular case. Namely, the method has the advantage of retaining the mass-resolution contrast in the data for better visualisation but capitalises on the cumulative nature of the signals of nearby detectors to maximise the signal to noise ratio and reduce otherwise high degree of data sparsity. The CAAD method proves robust for the determination of both NMD widths and relative peak intensities of both Be and Zr. To demonstrate this, we first mention that the relative intensities of the polycrystalline sample were, as averaged over all three temperatures of measurement and detectors 3-46, 0.23 ± 0.01 for Zr and 0.54 +/- 0.01 for Be. This, gives the effective sample stoichiometry of $Be_{70}Zr_{30}$, to within 2% accuracy identical to the one obtained from the ND studies, reported above. The values for the amorphous Zr —Be system are 0.34 ± 0.01 for Zr and 0.56 +/- 0.01 for Be, yielding the stoichiometry $Be_{60}Zr_{40}$ to within 2% accuracy equal to the values reported in our previous work [29].

The results can be better summarised in form of a table (Table 1) comparing the values of the the kinetic energies of Be and Zr, in both Zr —Be systems. The values reported are contrasted with theoretical predictions, from the Debye and recursion methods. Importantly in this context, the predictions of the VDOS and nuclear kinetic energy from the recursion method depend only on the short range order. It is thus expected that, as far as the local order dictating the high-frequency modes is concerned, the recursion method provides reliable predictions for both polycrystalline and amorphous systems. The most striking feature, present consistently at all temperatures is that the polycrystalline sample exhibits significantly (above the value of one standard deviation) higher value of the kinetic energy for both Zr and Be nuclei. This result is in line with the result of pioneering work by Bermejo *et al.* [35], where differences of ca. 6 meV in kinetic energy were observed for hydrogen (H) and deuterium (D) between polycrystalline and amorphous glasses in the temperature range from 10 to 110K. Following the authors of this seminal work [35], we will mention here that the low temperature regime of both studies, (i.e., between 10 and 200K), minimises the chances of anharmonic effects to manifest themselves and the response of the 'lattice' is dominated by one-phonon terms. In this spirit, the low temperature results show already another two important trends. Firstly, the experimental results for the

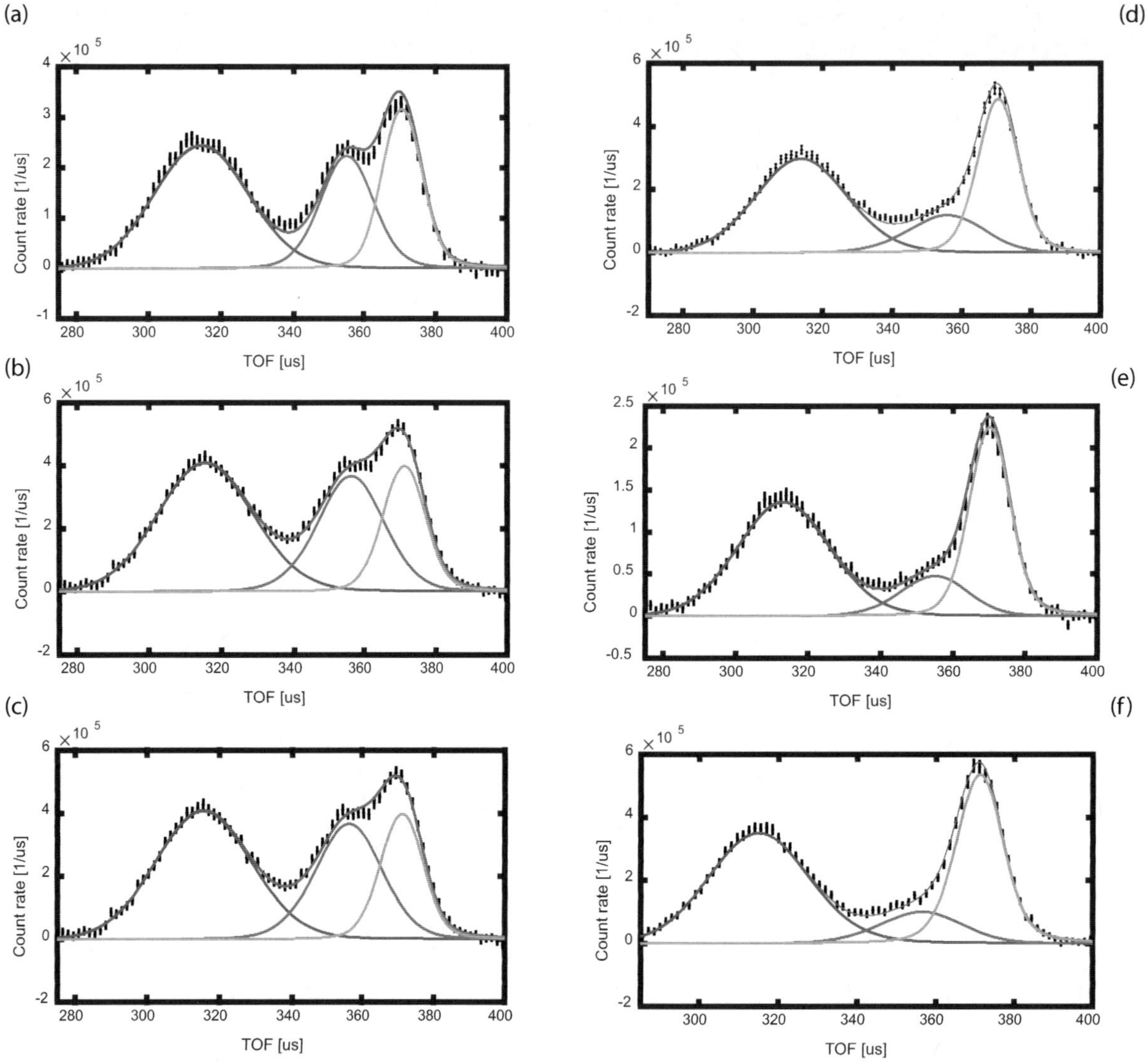

Figure 7. MANSE analysis by CAAD method of the Zr —Be systems for the first group of detectors, 3-46, in backscattering. Left panel: polycrystalline, right panel: amorphous samples. Temperature of the measurement 10K (a,d), 200K (b,e), and 300K (c,f). See text for details.

polycrystalline sample show more convergence towards the theoretical values obtained from the recursion model, thus taking into account albeit short range but still local order. The differences between the polycrystalline and amorphous samples in terms of their kinetic energies are smaller than it was the case for H and D in glass samples investigated by Bermejo et al [35] due to the fact that both Be and Zr are, in the sense of their Zero Point Energy values (ZPEs) much more classical in nature than it is the case for H and D. This means that their momentum distributions, even in the chemically bound state, resemble more those of systems of classical non-interacting gas particle, the celebrated Maxwell-Boltzman distribution [22, 23]. In this spirit, a yardstick of the degree of nuclear momentum distribution modification due to chemical binding and ordering was proposed in NCS work on cesium hydrogen sulfide. [20] The variable introduced there, the quantum energy excess, QE, was defined as ratio of experimental to classical kinetic energy

$E_{cl} = \frac{3}{2}kT$, where k is the Boltzman constant and T the temperature [20]. The values of E_{cl} are mass independent and yield 1.29, 25.8 and 38.8 meV at T=10K, 200K, and 300K, respectively. The QE values range from as much as 23.26 for polycrystalline Be at 10 K to 1.16 for amorphous Zr at 300K (see Table 2). The general trend is that the QE values for the polycrystalline system are systematically higher than those for the amorphous one, both for Be and Zr (see Table 2). In the spirit of the QE, we can say that both Zr and Be exhibit a larger degree of binding and ordering in the polycrystalline system, independently on the temperature and degree of system anharmonicity.

Table 1. The value of kinetic energies of Zr and Be in amorphous and polycrystalline Zr —Be system (in brackets). All values of the kinetic energy are in meV and the values of temperature in K. See text for details.

T [K]	*Theory* E_{Zr}	*Experiment* E_{Zr}	*Debye* E_{Zr}	*Theory* E_{Be}	*Experiment* E_{Be}	*Debye* E_{Be}
10	18	15 ± 1(16 ± 1)	14.0	28	26 ± 1(30 ±1)	28.4
200	36	32 ± 1.5(38.9 ± 1.3)	28.4	49	47 ± 1.5(51.9 ± 1.3)	50.1
300	46	45 ± 2(47.1 ± 1.5)	40.0	67	62 ± 2(64.6 ± 1.3)	58.8

Table 2. The values of quantum energy excess, QE, of Zr and Be in amorphous and polycrystalline Zr —Be system (in brackets). See text for details.

T [K]	QE_{Zr}	QE_{Be}
10	11.63 ± 0.78 (12.40 ± 0.77)	20.16 ± 0.78 (23.26 ± 0.78)
200	1.24 ± 0.06 (1.51 ± 0.05)	1.82 ± 0.06 (2.01 ± 0.05)
300	1.16 ± 0.05 (1.21 ± 0.04)	1.60 ± 0.05 (1.66 ± 0.03)

The fact that the kinetic energies of the polycrystalline Zr —Be system are systematically higher than the values for its amorphous counterpart has one more very important consequence. Namely, as mentioned in the beginning of this section, the second moments of the NMDs can be termed as first moments of the atom-projected VDOS. Thus, the presence of a previously unaccounted for mode, like an optical mode, associated with high frequency vibrations, will be signalled in NCS by the increase of the NMD width as the first moment of the VDOS will shift to the higher frequencies. Glasses and polycrystalline samples present an interesting case in this respect. As reported more than two decades ago by Bermejo et al [35], glasses had been already then subject to heated debate in the literature as to the origin of the low-frequency modes, known to be responsible for most of the thermal and transport anomalies in disordered matter. Highly localised soft modes had been at that time made mainly responsible for this phenomenon, with the soft modes postulated to be either already existing in the crystalline state, or alternatively, to be already present in the glasses. Drawing the motivation from the very same study of Bermejo [35], we can seek the origins of different dynamical response of amorphous and polycrystalline Zr —Be systems to dynamics anomalies in glass formation, visible in the energy-momentum window of the NCS method. In doing so, we will account for the difference between the glassy and crystalline environments by approximating the overall NMD profiles of Zr and Be as a convolution of the local and lattice modes, assuming the local modes to give numerically the same contribution to the kinetic energies of both nuclei. Work is in progress to account for the lattice modes of the crystalline phase found in the ND refinement described above. As this will require some computational effort as far as LD-DFT or MD is concerned, here we will thus only give a quick assessment using the experimental results. In the following, we will only use the 10K values to avoid possible complications arising from the anharmonicities in the both systems. As can be seen from Table 1, the differences in kinetic energies for Zr are ca. 1 meV at 10K. The corresponding values for Be are ca 4 meV at 10K. These values lie outside the energy regime investigated recently by INS spectroscopy using IN-4 and BRISP spectrometers, where opposite direction trends of phonon dispersion lines were discovered for low (ca 14 meV) and high (ca 45 meV) energy modes. The high-energy modes were thus shown to be largely independent from type of disordered system composed of Zr and Be [44]. The low frequency modes, however, may be exhibiting softening behaviour and their dispersion curve may differ from system to system, depending on the nature of the local order or disorder of the glassy matrix. Thus, on the whole, it seems to be more plausible that the low-frequency mode band centered at ca 15 meV softens when going from crystalline to glassy environment in the Zr —Be systems under investigation.

2.3. Summary and outlook

In conclusion, exploiting the the unique tandem of VESUVIO and Tosca inverted geometry spectrometers, in case of VESUVIO augmented with the capability to perform high resolution diffraction, has led to a number of interesting results shedding more light on the possible origins of disorder in Zr —Be binary mixtures by comparing ND, INS and NCS data recorded for the polycrystalline $Zr_{30}Be_{70}$ and an amorphous $Zr_{40}Be_{60}$ systems.

Firstly, the phase assignment of the polycrystalline Zr —Be system led to the prediction of sample stoichiometry, thus corroborating the NCS results. The unique combination of INS, NCS and ND has lead to the elucidation of the structural and dynamical features responsible. Secondly, the INS data were cross-validated by the NMDs of both Zr and Be, obtained from NCS. Finally and most importantly, the crystalline Zr —Be system showed systematically higher values of kinetic energies of Zr and Be, as compared to its amorphous counterpart. A plausible explanation of this trend, present at all temperatures from ambient down to 10 K, is due to low-frequency vibrational mode softening, when going from crystalline to the amorphous phase.

On the theoretical front, the INS and NCS analysis provided a robust benchmarking for the theoretical results obtained from the recursion method. To this end, the small deficiency of the recursion method to account for the long range order in the crystalline system was exposed by the NCS, emphasizing the need for *ab initio* calculation. To this end, the results of the diffraction analysis have provided structural information for the ongoing further structure elucidation and *ab initio* nuclear dynamics work by employing DFT-based lattice dynamics and molecular dynamics.

On the whole, the work has provided clues as to hitherto unexplored aspects of origins of structural and dynamical differences between the order and glassy environments, thus paving the way for future global studies with more challenging, structurally complicated materials of technological relevance.

3. Acknowledgments

G.Syrykh also thanks Russian Scientific Fund (grant 16-12-10065) for partial support. We are grateful to Dr Svemir Rudić for help with the experimental measurements on TOSCA instrument and his careful reading of the manuscript and many insightful comments and suggestions. The authors gratefully acknowledge the UK Science and Technology Facilities Council for financial support and the ISIS Facility for access to beamtime.

[1] Suck J B *et al.* 1983 *Phys. Rev. Lett.* **50** 49
[2] Calvo M *et al.* 1993 *Journal of Non-Crystalline Solids* **156-158** 53
[3] Otomo T, Arai M, Suck J B and Bennington S 2002 *Journal of Non-Crystalline Solids* **312-314** 599 – 602
[4] Calvo M *et al.* 2006 *Phys. Rev. Let.* **96** 135501
[5] Hafner J 1983 *J. Phys. C.* **16** 5773
[6] Bratkovsky A M *et al.* 1993 *J. of Non-Cryst. Solids* **156158** 72
[7] Page R D 1976 Canadian power reactor fuel Tech. Rep. AECL-5609 Atomic Energy of Canada
[8] Park C H *et al.* 1998 *Journal of Nuclear Materials* **254** 34
[9] Haydock R 180 *Solid State Phys.* **35** 215
[10] Jiang Q, Xia H, Zhang Z and Tian D 1992 *Chem. Phys. Lett.* **192** 93
[11] Syrykh G *et al.* 2008 *J. Phys. Condens. Matter* **20** 104241
[12] Montfrooij W *et al.* 1989 *Phys. Rev. Lett.* **63** 544
[13] Bosse J *et al.* 1986 *Phys. Rev. Lett.* **57** 3277
[14] Schram R P C 1991 *Phys. Rev. A* **44** 8062
[15] Alvarez M *et al.* 1998 *Phys. Rev. Lett.* **80** 2141
[16] Enciso E *et al.* 1995 *Phys. Rev. Lett.* **74** 4233
[17] T B and I M 2002 *J. Phys.: Condens. Matter* **14** L445
[18] Fernandez-Perea R *et al.* 1999 *Phys. Rev. E* **59** 3212
[19] Campa A and Cohen E G D 1990 *Phys. Rev. A* **41**(10) 5451–5463
[20] Krzystyniak M, Druzbicki K and Fernandez-Alonso F 2015 *Phys. Chem. Chem. Phys.* **17**(46) 31287–31296
[21] Krzystyniak M *et al.* 2011 *Faraday Discuss.* **151**(0) 171–197
[22] Andreani C, Senesi R, Krzystyniak M, Romanelli G and Fernandez-Alonso F 2017 *Neutron Scattering - Applications in Biology, Chemistry, and Materials Science* (*Experimental Methods in the Physical Sciences* vol 49) ed Fernandez-Alonso F and Price D L (Academic Press) pp 403 – 457
[23] Andreani C *et al.* 2017 *Adv. Phys.* **66** 1–73
[24] Andreani C *et al.* 2005 *Adv. Phys.* **54** 377–469
[25] Krzystyniak M 2010 *J. Chem. Phys.* **133** 144505
[26] Krzystyniak M and Fernandez-Alonso F 2011 *Phys. Rev. B* **83**(13) 134305
[27] Krzystyniak M *et al.* 2017 *Phys. Chem. Chem. Phys.* **19** 9064–9074
[28] Druzbicki K *et al.* 2018 Nuclear dynamics and phase polymorphism in solid formic acid: quantum termostated born oppenheimer molecular dynamics study. submitted
[29] Syrykh G F *et al.* 2017 *JETP Letters* **105** 591–594
[30] Pinna R S *et al.* 2017 *Nucl. Instr. Meth. A* **870** 79
[31] Pinna R S *et al.* 2015 *EPJ Web of Conferences* **83** 03013
[32] Parker S *et al.* 2014 *J. phys.: Conf. Ser.* **554** 012003
[33] Pietropaolo A and Senesi R 2011 *Physics Reports* **508** 45
[34] `http://www.isis.stfc.ac.uk/instruments/vesuvio/`
[35] Bermejo F J *et al.* 1994 *Physics Letters A* **189** 333 – 339 ISSN 0375-9601
[36] Golovkova E A, Surkov A V and Syrykh G F 2015 *Physics of the Solid State* **57** 266–269
[37] Coelho A A 2018 *Journal of Applied Crystallography* **51** 210–218
[38] Romanelli G *et al.* 2017 *Measurement Science and Technology* **28** 095501
[39] Fernandez-Alonso F and Price D L 2013 *Neutron Scattering Fundamentals* (*Experimental Methods in the Physical Sciences* vol 44) ed Fernandez-Alonso F and Price D L (Academic Press) pp 1 – 136
[40] Clark S J *et al.* 2005 *Z. Kristallogr.* **220** 567
[41] Colognesi D, Degiorgi E and Pace E 2001 *Physica B* **293** 317
[42] Mayers J and Abdul-Redah T 2004 *J. Phys.: Cond. Matter* **16** 4811
[43] Krzystyniak M *et al.* 2014 *Journal of Physics: Conference Series* **571** 012002
[44] Syrykh G, Orecchini A, de Francesco A, Laloni A and Stoliarov A 2017 *JETP Letters* **105** 488–491

Neutron resonance capture analysis and chemometric tools: an integrated approach

M Nardini[1,2], C Andreani[1,3,5,6], R Senesi[1,3,5,6], A Scherillo[7], F D'Agostino[8], L Romano[8], C Scatigno[1,4], and G Festa[1,3,5]

[1] Università degli Studi di Roma "Tor Vergata", Centro NAST, Via della Ricerca Scientifica 1, Rome, 00133 Italy
[2] Physics Institute, Fondazione Policlinico Universitario A. Gemelli, Catholic University of the Sacred Heart, Largo Francesco Vito 1, Rome, 00168 Italy
[3] Università degli Studi di Roma "Tor Vergata", Dipartimento di Fisica, Via della Ricerca Scientifica 1, Rome, 00133 Italy
[4] Università degli Studi di Roma "Tor Vergata", Dipartimento di Scienze e Tecnologie Chimiche, Via della Ricerca Scientifica 1, Rome, 00133 Italy
[5] Centro Fermi - Museo Storico della Fisica, Centro Studi e Ricerche "Enrico Fermi", Piazza del Viminale 1, Rome, 00184 Italy
[6] CNR-IPCF Sezione di Messina, Viale Ferdinando Stagno d'Alcontres 37, Messina, 98158 Italy
[7] ISIS Facility, Rutherford Appleton Laboratory, Chilton, Didcot, Oxfordshire OX11 OQX, UK
[8] Sapienza - University of Rome, Italian Institute of Oriental Studies, Rome, Italy

E-mail: `roberto.senesi@uniroma2.it`

Abstract. A joint approach based on the Neutron Resonance Capture Analysis (NRCA) integrated by chemometric tools namely Principal Component Analysis (PCA) is used to determine the semi-quantitative isotopic composition of a large set of Sumerian pottery. Here, we present NRCA results from the experimental campaign carried out on the INES beamline at the ISIS pulsed neutron and muon source (Rutherford Appleton Laboratory, UK). The potteries come from the archaeological site still under excavation of Abu Tbeirah, a 3rd millennium BCE site located in Southern Iraq, in the S-E periphery of the city of Nasiriyah. NRCA allows to determine the presence and the relative isotopic amounts, while the Principal Component Analysis distinguishes the elements/isotopes linked to the raw materials from the decay products due to contaminants that affect the potteries under investigation.

1. Introduction

Neutron Resonance Capture Analysis (NRCA) is an experimental technique whereby a neutron with energies in excess of few eV and up to hundreds of keV is resonantly absorbed by a nucleus, and a prompt-gamma cascade is emitted. The energy of the resonance is characteristic of the element and isotope that produce the gamma, and can be used to identify the chemical composition of the investigated sample by the comparison with (n,γ) cross section libraries [1]. Moreover, the intensity of the resonance is proportional to the number of absorbing centres via the (n,γ) cross section. NRCA was developed at the GELINA facility in Belgium, a pulsed neutron source with a linear electron accelerator [2], and at the ISIS pulsed neutron and muon source in the UK [3, 4], where the intense flux at electron-volt (epithermal) [5] neutron energies

IOP Conf. Series: Journal of Physics: Conf. Series **1055** (2018) 012005 doi:10.1088/1742-6596/1055/1/012005

is available. Experiments have been performed on the Italian Neutron Experimental Station (INES) [6] over the last decades [7, 8, 9], and more recently on the VESUVIO spectrometer at ISIS [10], whose NRCA capabilities have been discussed in a separate contribution to this volume as an application to materials-science investigations [11]. Here, we present a joint semi-quantitative approach that integrates NRCA measurements with statistical tools via a multivariate analysis namely Principal Component Analysis (PCA) to reveal possible correlations between the isotopes for the investigated samples, bringing to the fore the main properties of the pottery data set. From a chemometric point of view, the primary aim of this analysis is determining the most significant reference basis representing the experimental data and minimizing the noise. The output of the statistical approach is described through the Principal Components (PCs) that allow to define the parameters' variance due to the different correlations among the dataset. These techniques allow to identify member of clusters and to study the variance and correlations along the data [12].

2. Experimental set-up

The entire set of 36 pottery samples found at the site of Abu Tbeirah [13] was investigated through the NRCA technique at the Italian Neutron Experimental Station (INES), ISIS pulsed neutron and muon source in the UK. Artefacts were placed within the instrument tank, with the primary flight path between the sample and the TS1 water moderator $L_1 = 22.804$ m. Prompt-gamma radiation was detected using an Yttrium Aluminium Perovskite crystal $YAlO_3$ coupled with a photomultiplier and positioned above the tank at a distance $L_2 \simeq 1.0$ m from the sample position. As the gamma radiation travels at the speed of light, the time needed for a photon to travel from the sample to the detector can be neglected compared to the time a neutron takes to move from the moderator to the sample position. Therefore, time-of-flight (ToF) measurements were performed, and the energies of incident neutrons are directly related to ToF values via the *a priori* knowledge of L_1, using the relation:

$$E = \frac{1}{2} m_n \frac{L_1^2}{ToF^2} \tag{1}$$

with E the energy of the incident neutron, and m_n its mass. Measured spectra show the sharp resonances, and an environmental background of photons from the moderator, generated within the blockhouse, and within the neutron beam stop. The simpler analysis of NRCA spectra is based on the indexing of the observed resonances by comparing the centre of each measured peak to tabulated libraries [1], up to hundreds of keV. Some examples of (n,γ) cross sections for chlorine, iron, sodium, and manganese are shown in Figure 1.

3. Results and discussion

NRCA spectra and pictures of two potteries are shown in figure 2. The indexing of the observed resonances results in the recognition of potassium, chlorine, iron, manganese and sodium as the principal signals in the NRCA spectra.

It could be noted that sample 22 is characterized by an high signal from the ^{35}Cl respect to sample 23, both sample show a small peak for ^{39}K and sharp and close peaks of ^{23}Na and ^{56}Fe. A broad peak from ^{55}Mn is present in both spectra, this broad shape depends on ^{55}Mn (n, γ) cross section, see figure 1 (e). A semi-quantitative determination of the relative amounts of these isotopes/elements are then attempted based on the normalisation procedure:

$$\rho_i = \sum_j \frac{I_{r_j,i}}{d \cdot \sigma_i(t_{r_j,i})} \tag{2}$$

where $I_{r_j,i}$ is the peak amplitude, from the detector output, of the j-th resonance of i-th element, $\sigma_i(t_{r_j,i})$ is the value of (n,γ) cross section calculated at time $t_{r_j,i}$ where one can find the

IOP Conf. Series: Journal of Physics: Conf. Series **1055** (2018) 012005 doi:10.1088/1742-6596/1055/1/012005

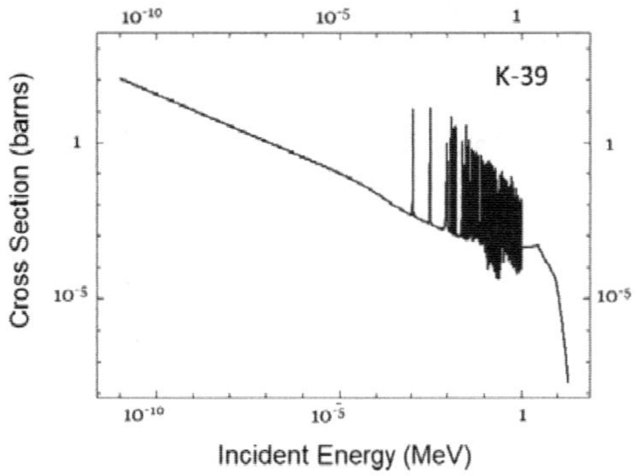

(a) ^{39}K *(n,γ) cross section.*

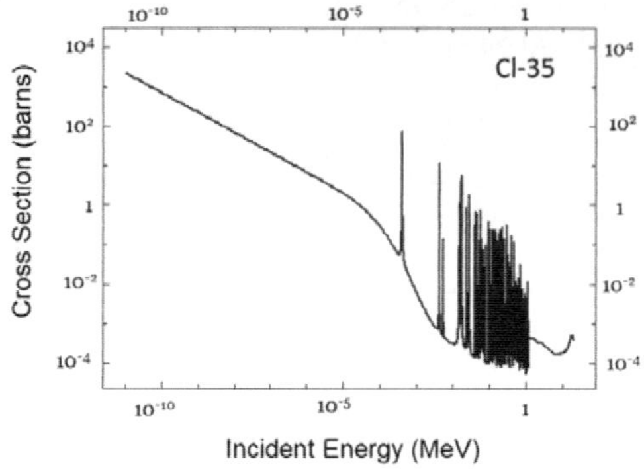

(b) ^{35}Cl *(n,γ) cross section.*

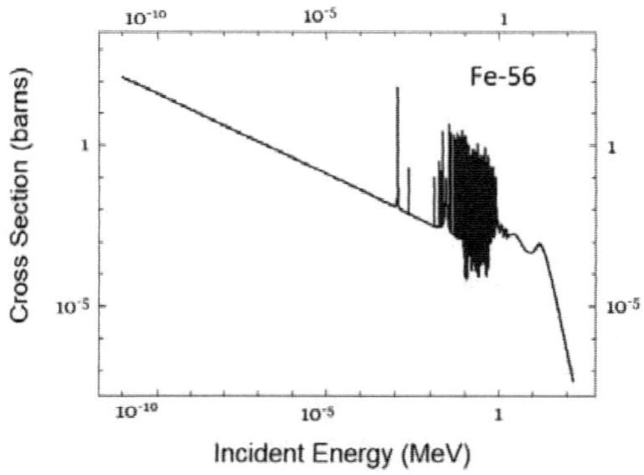

(c) ^{56}Fe *(n,γ) cross section.*

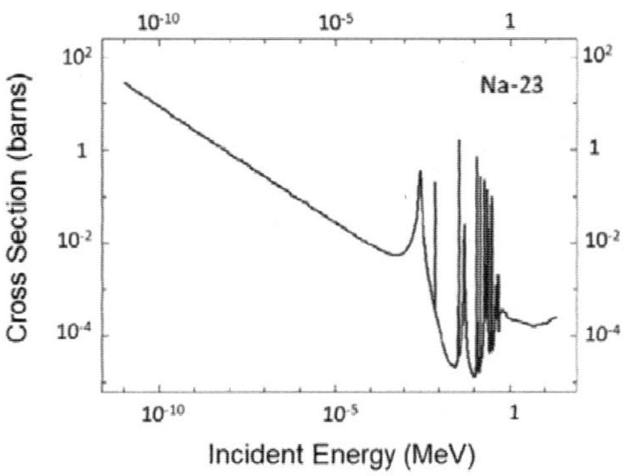

(d) ^{23}Na *(n,γ) cross section.*

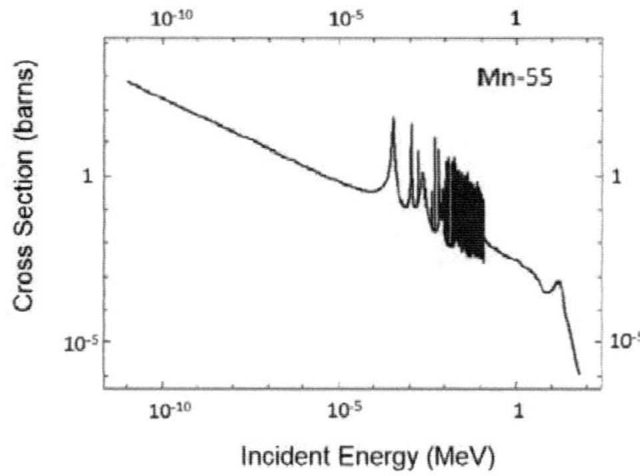

(e) ^{55}Mn *(n,γ) cross section.*

Figure 1: Cross sections for the (n,γ) process of ^{35}Cl, ^{56}Fe, ^{23}Na, and ^{55}Mn elements.

maximum of the resonance peak r_j for the i-th element, and d is the value of sample thickness. The resulting values for the relative concentrations ρ_i (in arbitrary units) of the most abundant isotopes are listed in Table 1. These values are obtained through a normalization process to be able to compare results obtained from the same set of samples. Errors in table are calculated using error propagation theory from equation 2.

Data are treated through a chemometric tool namely PCA. The dataset is organized by a

(a) *Sample 22.*

(b) *Sample 23.*

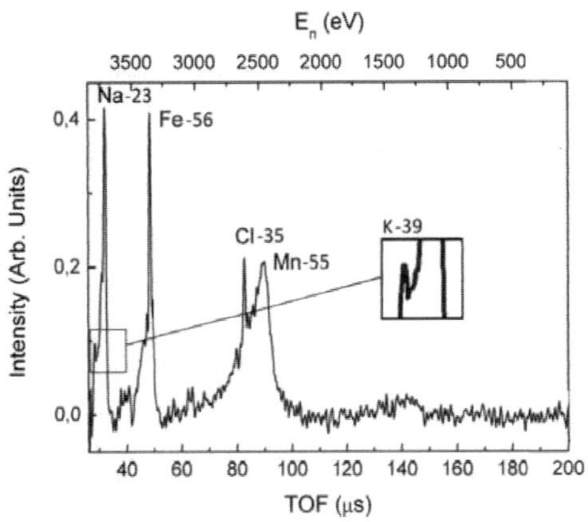

(c) *NRCA spectrum of sample 22.*

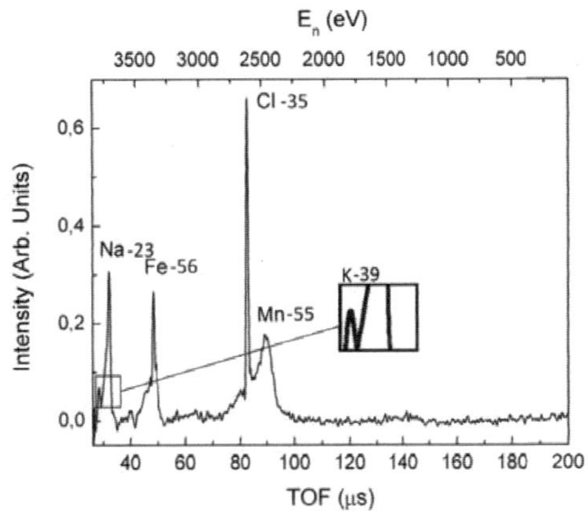

(d) *NRCA spectrum of sample 23.*

Figure 2: Pictures and Neutron Resonance Capture Analysis spectra with isotopes assignement of pottery samples.

matrix with 36×10 entries, samples and variable respectively. The algorithm used is the singular value decomposition (SVD), with the cross-validation method, and mean centre data as a model input. The software used is the Unscrambler package - version 10 (chemometric software package — Camo, Woodbridge, NJ, USA) [14]. PCs Loading plot of the relative amount of the isotopic composition on the entire set of samples is showed in figure 3; it shows the correlation among the detected isotopes/elements.

Two principal components (PCs) are extracted, representing 97% of the total variance (PC1: 89% and PC2: 8%). The first Loadings plot (Fig. 4a) shows two groups: ^{35}Cl and ^{23}Na, characterized by a positive value of PC1 and PC2 (that means that are markers of the whole set of data) and ^{56}Fe, ^{39}K and ^{55}Mn grouped in the center of PCs with a small variability, indicating an homogeneity of the element distributions on the analyzed data. ^{35}Cl and ^{23}Na are clearly linked to the decaying products such as sodium chloride contamination, while the second group is linked to the primary minerals (present in the raw material) and secondary minerals (products of thermally induced reactions), as well as the firing minerals (formed after the production, during their use and mainly their burial, as a result of either transformation of metastable firing minerals or infiltration of any solution) [15]. Actually, ^{56}Fe could be related

IOP Conf. Series: Journal of Physics: Conf. Series **1055** (2018) 012005 doi:10.1088/1742-6596/1055/1/012005

Table 1: Normalized peak intensities of the whole set of 36 samples resulting from the Neutron resonance capture analysis spectra.

Sample	ρ_{K-39}	ρ_{Na-23}	ρ_{Fe-56}	ρ_{Cl-35}	ρ_{Mn-55}
1	0.025 ± 0.006	1.8 ± 0.4	0.009 ± 0.002	1.1 ± 0.2	0.0030 ± 0.0008
2	0.013 ± 0.004	1.3 ± 0.3	0.007 ± 0.001	0.33 ± 0.07	0.0028 ± 0.0007
3	0.016 ± 0.003	1.1 ± 0.2	0.006 ± 0.001	0.44 ± 0.07	0.0026 ± 0.0005
4	0.012 ± 0.001	0.7 ± 0.1	0.006 ± 0.000	0.38 ± 0.03	0.0019 ± 0.0002
5	0.023 ± 0.005	1.3 ± 0.3	0.007 ± 0.001	0.33 ± 0.07	0.0031 ± 0.0007
6	–	1.0 ± 0.2	0.005 ± 0.001	0.6 ± 0.1	0.0030 ± 0.0008
7	0.020 ± 0.004	1.2 ± 0.2	0.007 ± 0.001	0.7 ± 0.1	0.0032 ± 0.0006
8 B	0.031 ± 0.007	1.9 ± 0.4	0.010 ± 0.002	0.26 ± 0.07	0.0036 ± 0.0009
8 W	0.018 ± 0.004	1.2 ± 0.2	0.010 ± 0.002	0.5 ± 0.1	0.0044 ± 0.0009
9	0.027 ± 0.006	1.3 ± 0.3	0.007 ± 0.001	0.5 ± 0.1	0.0031 ± 0.0007
10	0.014 ± 0.002	0.7 ± 0.1	0.005 ± 0.001	0.63 ± 0.07	0.0017 ± 0.0003
11	0.019 ± 0.003	1.3 ± 0.2	0.008 ± 0.001	0.66 ± 0.09	0.0029 ± 0.0005
12	–	1.9 ± 0.5	0.012 ± 0.003	1.3 ± 0.3	0.004 ± 0.001
13	0.017 ± 0.004	1.0 ± 0.2	0.005 ± 0.001	0.9 ± 0.1	0.0020 ± 0.0005
14	0.033 ± 0.009	1.9 ± 0.5	0.010 ± 0.002	1.1 ± 0.3	0.004 ± 0.001
15	0.021 ± 0.007	1.3 ± 0.4	0.007 ± 0.002	0.3 ± 0.1	0.004 ± 0.001
17	0.020 ± 0.005	1.5 ± 0.3	0.007 ± 0.001	1.2 ± 0.2	0.0023 ± 0.0006
18 L	0.016 ± 0.003	1.3 ± 0.2	0.007 ± 0.001	0.27 ± 0.06	0.0041 ± 0.0007
18 S	0.022 ± 0.005	1.1 ± 0.2	0.006 ± 0.001	0.27 ± 0.06	0.0031 ± 0.0007
19 L	0.010 ± 0.004	0.7 ± 0.2	0.005 ± 0.001	0.19 ± 0.06	0.0021 ± 0.0007
19 S	0.012 ± 0.004	0.6 ± 0.2	0.004 ± 0.001	0.19 ± 0.06	0.0018 ± 0.0006
20	–	1.2 ± 0.2	0.007 ± 0.001	0.24 ± 0.06	0.0034 ± 0.0007
21 L	0.013 ± 0.004	1.0 ± 0.2	0.005 ± 0.001	0.39 ± 0.08	0.005 ± 0.001
21 S	–	1.0 ± 0.2	0.005 ± 0.001	0.27 ± 0.06	0.0028 ± 0.0007
22	0.007 ± 0.001	0.8 ± 0.1	0.006 ± 0.000	0.21 ± 0.02	0.0023 ± 0.0002
23	0.009 ± 0.002	1.1 ± 0.2	0.006 ± 0.001	0.27 ± 0.05	0.0034 ± 0.0005
24	0.013 ± 0.004	1.3 ± 0.3	0.008 ± 0.001	0.33 ± 0.07	0.0028 ± 0.0007
25	–	0.9 ± 0.2	0.004 ± 0.001	0.12 ± 0.04	0.0023 ± 0.0006
26 L	0.005 ± 0.002	1.1 ± 0.2	0.007 ± 0.001	0.27 ± 0.06	0.0023 ± 0.0006
26 S	–	0.9 ± 0.3	0.006 ± 0.002	0.14 ± 0.06	0.0029 ± 0.0009
27	–	1.0 ± 0.2	0.006 ± 0.001	0.21 ± 0.06	0.0027 ± 0.0008
28	–	1.0 ± 0.2	0.008 ± 0.002	–	0.0036 ± 0.0009
29	–	0.6 ± 0.1	0.005 ± 0.000	0.32 ± 0.02	0.0020 ± 0.0002
30	–	0.5 ± 0.1	0.003 ± 0.000	0.21 ± 0.02	0.0010 ± 0.0002
31	–	0.9 ± 0.2	0.006 ± 0.001	0.59 ± 0.09	0.0028 ± 0.0005
32 L	–	0.6 ± 0.1	0.006 ± 0.001	0.18 ± 0.03	0.0023 ± 0.0004
32 S	–	0.6 ± 0.1	0.005 ± 0.001	0.13 ± 0.04	0.0018 ± 0.0005
33	–	0.8 ± 0.2	0.005 ± 0.001	0.55 ± 0.09	0.0025 ± 0.0005
34	–	1.0 ± 0.2	0.006 ± 0.001	0.9 ± 0.2	0.0026 ± 0.0006
35	–	0.8 ± 0.2	0.004 ± 0.001	0.50 ± 0.09	0.0023 ± 0.0005
36	–	0.9 ± 0.2	0.004 ± 0.001	0.6 ± 0.1	0.0020 ± 0.0005
37	0.008 ± 0.002	0.5 ± 0.1	0.006 ± 0.001	0.13 ± 0.03	0.0025 ± 0.0004

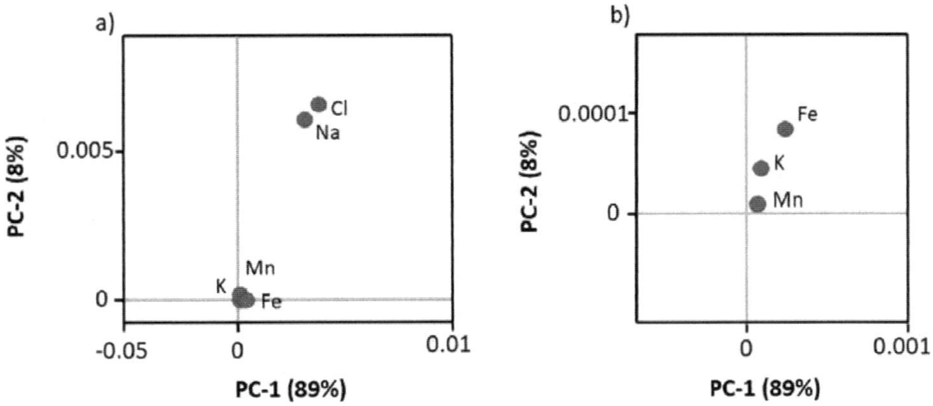

Figure 3: Principal Component Analysis (PCA) of the NRCA measurements on the ceramics performing in the entire ceramic set (36 samples). a) PCA Loading plot of the relative amount of the isotopic composition on the entire set of samples; b) Zoom of the PCA Loading plot in the origin region.

to the hematite or inorganic impurities (aluminoferrites and ferrites) responsible of the reddish hue of potteries [16].

4. Outlook and conclusions

A new integrated approach based on Neutron Resonance Capture Analysis (NRCA) and chemometric tools namely Principal Component Analysis (PCA) is used to determine the semi-quantitative elemental/isotopic composition of a large set of Sumerian potteries from the Abu Tbeirah archaeological site. The NRCA technique shows that the main isotopes present in the entire set of samples are ^{39}K, ^{56}Fe, ^{55}Mn, ^{23}Na and ^{35}Cl. The PCA technique allows the determination of isotope correlations, pointing out the main primary or firing minerals (such as ^{39}K, ^{56}Fe, ^{55}Mn) and the decay products such as sodium chloride (^{23}Na and ^{35}Cl) as pollutant, into two separate groups. The closed group around the center of the PCs belonging to raw materials, give us clues about the homogeneity of the pottery set and on the similar manufacturing processes. Actually, a groundwater aquifer, affected by probable saline intrusion results into characteristic fingerprints of the geo-hydrological surrounding.

Acknowledgements

This work was supported within the CNR-STFC Agreement 2014–2020 (No. 3420 2014–2020) concerning collaboration in scientific research at ISIS Spallation Neutron Source, and partially supported within the SMART Campus Project (Regione Lazio 2015-2017). We would like to thank Dr G Romanelli for useful discussions.

References

[1] Evaluated nuclear data file URL `https://www-nds.iaea.org/exfor/endf.htm`
[2] Postma H and Schillebeeckx P 2005 *Journal of radioanalytical and nuclear chemistry* **265** 297–302
[3] http://wwwisisstfcacuk Last accessed on February 2018
[4] Pietropaolo A, Gorini G, Festa G, Reali E, Grazzi F and Schooneveld E M 2010 *Applied spectroscopy* **64** 1068–1071

[5] Gorini G, Festa G and Andreani C 2014 Epithermal neutron instrumentation at isis *Journal of Physics: Conference Series* vol 571 (IOP Publishing) p 012005
[6] Pietropaolo A, Festa G, Grazzi F, Barzagli E, Scherillo A, Schooneveld E and Civita F 2011 *EPL (Europhysics Letters)* **95** 48007
[7] Perelli Cippo E, Borella A, Gorini G, Kockelmann W, Pietropaolo A, Postma H, Rhodes N J, Schillebeeckx P, Schooneveld E M, Tardocchi M, Wynants R and ANCIENT CHARM collaboration 2010 **623** 693–698
[8] Andreani C, Aliotta F, Arcidiacono L, Borla M, Di Martino D, Facchetti F, Ferraris E, Festa G, Gorini G, Kockelmann W *et al.* 2017 *Journal of Analytical Atomic Spectrometry* **32** 1342–1347
[9] Festa G, Pietropaolo A, Grazzi F, Sutton L, Scherillo A, Bognetti L, Bini A, Barzagli E, Schooneveld E and Andreani C 2013 *Measurement Science and Technology* **24** 095601
[10] Andreani C, Krzystyniak M, Romanelli G, Senesi R and Fernandez-Alonso F 2017 *Advances in Physics* **66** 1–73
[11] Romanelli G, Krzystyniak M and Fernandez-Alonso F 2018 *Journal of Physics Conference Series*
[12] Scatigno C, Prieto-Taboada N, García-Florentino C, de Vallejuelo S F O, Maguregui M and Madariaga J M 2018 *Environmental Science and Pollution Research* **25** 6285–6299
[13] D'Agostino F, Kadhem A, Romano L, Vidale M and Angelozzi M 2011 *Rivista degli studi orientali* 17–34
[14] Unscrambler x software URL `http://www.camo.com/rt/Products/Unscrambler/unscrambler.html`
[15] De Benedetto G, Laviano R, Sabbatini L and Zambonin P 2002 *Journal of Cultural Heritage* **3** 177–186
[16] De Bonis A, Cultrone G, Grifa C, Langella A, Leone A P, Mercurio M and Morra V 2017 *Ceramics International* **43** 8065–8074

IOP Conf. Series: Journal of Physics: Conf. Series **1055** (2018) 012006 doi:10.1088/1742-6596/1055/1/012006

Nitrogen doping and the performance of superconducting radio-frequency niobium cavities: insights from neutron diffraction and neutron Compton scattering

M Krzystyniak [1,2], M J Gutmann[1], G Romanelli[1], Y Trenikhina[3], A Romanenko[3], and F Fernandez-Alonso[1,4]

[1] ISIS Facility, Rutherford Appleton Laboratory, Chilton, Didcot, Oxfordshire OX11 0QX, United Kingdom
[2] School of Science and Technology, Nottingham Trent University, Clifton Campus, Nottingham NG11 8NS, United Kingdom
[3] Fermi National Accelerator Laboratory, Batavia, IL 60510-5011, USA
,
[4] Department of Physics and Astronomy, University College London, Gower Street, London WC1E 6BT, United Kingdom

E-mail: matthew.krzystyniak@stfc.ac.uk

Abstract. Neutron Compton scattering and neutron diffraction have been applied to investigate the influence of nitrogen doping of niobium on the performance of superconducting radio-frequency niobium cavities. To this end, a comparative study of the neutronic response of two samples has been performed. An electro-polished and nitrogen-doped niobium sample was compared with a standard, a niobium sample that has only undergone the electro-polishing procedure. The first piece of information, provided by neutron diffraction, is that additional conditioning of the electro-polished cavity material, through doping with nitrogen, leads to a systematically larger niobium lattice expansion, which provides an upper conservative limit of nitrogen concentration consistent with values reported in the literature. Furthermore, neutron Compton scattering shows a broadening of the niobium momentum distribution in the nitrogen-doped sample, as compared to the standard, thus indicating an increased degree of ordering and binding of niobium in the metal lattice. On the whole, these observations suggest that nitrogen-doping leads to some degree of lattice ordering, most likely due to increased hydrogen trapping, in agreement with previous results using surface spectroscopy.

1. Introduction

Modern particle accelerators greatly rely on state-of-the-art superconducting low-dissipation niobium (Nb) radio-frequency cavities (SRFCs). Understanding the structure-performance relation of SRFCs is thus of crucial importance. One of the most important findings in this respect has been that SRFC performance is compromised by the presence of hydrogen (H) near the cavity surface [1,2]. Depending on the level of H concentration, two scenarios of the quality-factor (Q) degradation are known: (i) H Q-disease which degrades cavity performance even at low accelerating gradients; and (ii) High-Field Q-slope (HFQS) which is caused by the strong increase in residual resistance at high accelerating fields. H can be present in SRFCs both as an interstitial in the Nb lattice and in the form of non-superconducting ϵ and β Nb hydride phases, as evidenced by Scanning Electron Nano-area Diffraction (SEND) [2]. It has been shown that several micrometer-sized Nb hydride inclusions, which are not superconducting and form in Nb near the SRFC surface upon cavity cool down, are responsible for Q-disease [1].

Several mitigation strategies for Q-factor degradation in SCRFs have been identified so far. Temperature-dependent TEM studies have demonstrated precipitation of Nb nanohydrides within the penetration depth at cryogenic temperatures in HFQS-limited electropolished (EP) cavities [2]. 800 °C bake of Nb cavities was empirically found to eliminate Q-disease and HFQS-free cavities which underwent 120 °C bake showed significantly reduced hydride formation [1,2]. A reduction of Nb-hydride precipitation appears to be caused by the introduction of stable H-vacancy complexes during 120 °C bake [3,4]. Recently, the Linac Coherent Light Source-II Project at Stanford accepted a novel nitrogen doping (N-doping) technology [5] for the production of 280 cavities. Discovered at Fermilab, N-doping systematically improves Q up to a factor of three [2]. A combination of annealing in a partial pressure of N_2 gas and subsequent EP of the Nb cavities leads to several hundred ppm of N on the Nb surface. Being extremely efficient at medium fields, N-doped cavities are limited by the quench at high accelerating fields.

Current research efforts by the SRFC community are concentrated on finding the origin of the limitation of the N-doped cavity quench, seeking to understand how N-doping and Nb sample baking cause an increase in performance. In an attempt to address these efforts, this work reports on investigations of nitrogen-doping-induced changes to structure and dynamics of conditioned niobium samples mimicking SRFC cavities. In this work, we have extended the traditional SCRFC research programme, chiefly consisting of surface probes, by employing a combination of neutron Compton scattering (NCS) and neutron diffraction (ND).

The advocated approach seems very timely, as electron and X-ray diffraction-based methodologies have already been successfully applied in the characterisation of SCRFs [2] and ND is a natural extension of this line of research. NCS can also be a source of valuable information about the nuclear dynamics that can be linked to local material properties, especially at the interface between ordered and disordered environments [19,23,24]. In this latter context, NCS provides a unique capability as it can be used to measure nuclear kinetic energies and other nuclear quantum effects (NQEs) in a mass-resolved manner [19,23,24]. Moreover, NCS can in principle access NQEs at any temperature, in a manner unaffected by the Debye-Waller factor [19,23,24]. NCS can also be used in a wide variety of samples and sample environments, with samples of different shapes, sizes, densities, textures and transparencies to radiation, thus beating other neutron-based materials characterisation techniques like reflectometry or small-angle neutron scattering, where sample thickness and surface roughness may render experiments unfeasible [24].

The proposition to employ NCS as a new technique for cavity material characterisation is also very timely in the light of recent modelling efforts repoerted in the SCRFs-related literature where Zero Point Energy (ZPE) effects have been shown as a very relevant ingredient, crucial for the understanding of the energetic, thermodynamic and electric properties of SCRFs [7]. More importantly, however, NCS together with ND continues to be relatively unexploited yet powerful combination of techniques as applied on the VESUVIO+ spectrometer at the ISIS Facility in the UK [10, 11, 19, 22–24].

2. Methodology and Results

In what follows, the methodology and results of NCS and ND experiments on two Nb samples mimicking two different types of Nb SCRFs will be described. To this end, two Nb samples, representing two different cavity-material treatment protocols, were prepared from the same Nb sheets that cavities are made out of and investigated at temperatures of 80 K and 300 K. The samples were prepared using the same chemical processes and the same furnaces that are used for the actual cavities: an EP sample (hereinafter denoted as Nb1) and a sample that has been subjected to N-doping (hereinafter denoted as Nb4).

In essence, an NCS spectrum measured by a particular detector (at a given scattering angle θ) and at a given time-of-flight (TOF) value, t, is a sum of neutron Compton recoil peaks centred at mass-dependent TOF values, (see Fig. 1, lower inset). Mathematically [10,12,14–16,18,19,23,24],

$$C(\theta, t) = A\frac{E_0 I[E_0]}{q}\left[\sum_M c_M J_M(x_M) \otimes R_M(x_M)\right], \tag{1}$$

where, M is the mass of a given nuclide, $(M = Al, Nb)$, A is a proportionality constant, E_0 and $I[E_0]$ are the incident neutron energy and intensity, respectively, and q the neutron momentum transfer to the sample. Moreover, $J_M(x_M)$ is the unit-area-normalised longitudinal momentum distribution (NMD) of a nucleus of mass M, $R_M(x_M)$ is the mass-dependent resolution function, the symbol, $\otimes$, implies the convolution product between the aforementioned functions, and c_M are coefficients proportional to the product of the number density and the total bound neutron cross section of a given nuclide [19]. The experimental curve for an isolated recoil peak for a given mass M (see Fig. 1, main), expressed in the harmonic and isotropic lattice limit (HL) is of the form $J_M(x_M) \otimes R_M(x_M)$ with

$$J_M(x_M) = \frac{\exp(-x_M^2)}{\sqrt{2\pi\sigma_M^2}}[1 - \frac{\frac{\sqrt{2}}{12}\sigma_M}{q}H_3(x_M)] \tag{2}$$

where $x_M = (y_M)/(\sigma_M\sqrt{2})$, y_M is the longitudinal nuclear momentum for a given nuclide of mass M, and σ_M is the standard deviation of $J_M(x_M)$ [19, 23, 24]. The second term in Eq. (2) accounts for deviations from the IA, also known as final states effects (FSEs), expressed here in the HL limit [14, 16] .

In this work, NCS data were measured at a total of 132 detectors over the angular range of 130 to 170 degrees using a double-difference configuration offering improved resolution in back scattering [8, 19]. The samples were loaded into aluminium (Al) containers of 0.1 mm wall thickness and placed inside a dedicated closed-cycle helium refrigerator. Each sample was a plain Nb slab of cross-sectional area 2x3 cm^2 and 3 mm thickness. Sample Nb1 was prepared to contain no traces of nitrogen (N). Sample Nb4, subject to N doping, was expected to contain sub-percent N concentrations.

The NCS data reduction and fitting was performed according to standard procedures, detailed elsewhere [10, 12, 14–16, 18, 23, 24]. The procedure applied to raw TOF spectra, recorded by each backscattering detector, consisted of a pre-fit of the raw data, the calculation of sample and container-dependent multiple scattering (MS) signal, the subtraction of the MS signal and the MS-corrected container signal, the subsequent focusing of corrected data into the longitudinal momentum space of Nb, and finally the fit of the focused data to obtain the longitudinal Nb momentum distribution curves, from which the values of the standard deviations of Nb longitudinal NMDs were extracted.

Focused and normalised-to-unit-area NMDs of Nb1 and Nb4, recorded at 80 K, are shown in Fig. 1. It can be clearly seen that both the maximum value and the position of the maximum of the NMD is lower for Nb4 compared to Nb1. To guide the eye, an inset has been added and placed on the left hand side within the main figure. Green dashed lines and a green arrow indicate this trend. The observed shift of the NMD maximum, y_{max}, towards lower (more negative) y_{Nb} values, given the mathematical from of the NMD in the HL limit [Eq. (2)], signifies an increase of the NMD standard deviations. A rough estimate, based on the position and the direction of the green arrow, gives a shift of y_{max} *ca.* -2.5 Å^{-1} between samples Nb1 and Nb4. It is in fact possible to estimate the standard deviation (STD) of an NMD within the harmonic lattice limit (HL) solely based on the position of the NMD peak maximum. To this end, another contribution to this volume series has been prepared [25]. In a nutshell, the method uses the fact that, within the HL and in the presence of FSEs, Eq. (2) can be differentiated with respect to the x_M variable to find the maximum of $J_M(x_M)$. This maximum proves to be a function of σ_M which is a solution of a rather complicated multiply-nested radical equation [25]. However, in addition to the exact analytic solution one can obtain an approximate solution and use the fact that an FSE correction only slightly shifts the position that the NMD maximum would have otherwise had in the IA limit, $y_{max} = 0$. Thus, consequently $y_{max}^2 \ll (\sigma_M^2 + s_M^2)$, where s is the standard deviation of the mass-dependent instrument resolution function, $R_M(x_M)$. One then can obtain an approximate solution to be refined through an iterative process with the zeroth iteration solution given by [25]:

$$y_{max}^{(0)} = -\frac{\sigma^4}{q(\sigma_M^2 + s_M^2)} \tag{3}$$

Equation (3) can be used to estimate the STD change of NMDs when going from sample Nb1 to Nb4. However, due to the fact that, as can be seen in Figure 1, the NMD of Nb1 has its maximum very close the the IA limit, *i. e.*, $y_{max} = 0$, using Eq. (3) to estimate the NMD STD for the Nb1 sample would lead to a very unprecise estimation. Thus, a different strategy was adopted. Using the combined knowledge of the backscattering kinematics for $M = 93$ amu and the instrument resolution function $R_M(x_M)$ characteristics published elsewhere (see, *e.g.* [23]) we obtain two important clues to this end. Firstly, the values of the momentum-transfer, q for the backscattering detectors under investigation vary only very slightly between 90 and 94 Å^{-1} and we can assume, for the sake of this estimation, an average value $q = 92$ Å^{-1}. Secondly, the STD of the total instrument resolution function for $M = 93$ amu also does not vary markedly across the whole range of the backscattering detectors with an average of $s = 29 \pm 1$ Å^{-1}. Furthermore, expecting a difference between NMD STDs of Nb1 and Nb4 to be very small, one can plug in $\sigma + d\sigma$ into Equation (3) and then Taylor-expand treating $d\sigma$ as a small expansion parameter while keeping q and s constant. This leads to the following expression:

$$\Delta y_{max}^{(0)} = y_{max}^{(0)}(\sigma + d\sigma) - y_{max}^{(0)}(\sigma) \simeq \frac{d\sigma\left(\frac{4\sigma^3}{s^2+\sigma^2} - \frac{2\sigma^5}{(s^2+\sigma^2)^2}\right)}{q} \tag{4}$$

Substituting $q = 92$ Å^{-1}, $s = 29$ Å^{-1}, and $\Delta y_{max}^{(0)} = -2.5$ Å^{-1} first into Equation (3) and then, using the obtained solution, to Equation (4), one obtains $\sigma = 17$ Å^{-1} and $d\sigma = 7$ Å^{-1} for the Nb4 sample. The obtained value of $d\sigma$ shows the right trend visible in the NCS data. Its numerical value is, however, rather too high. This is a simple consequence of the fact that the method to estimate STD of an NMD from the NMD peak position maximum relies inherently on the sensitivity of the peak position to both q and σ, a sensitivity which is the largest for smaller masses (*e.g.* protons).

Another robust method of the estimation of the STDs of NMDs come to mind in the context of the above calculation, the difference method [26]. In this method, the difference of two

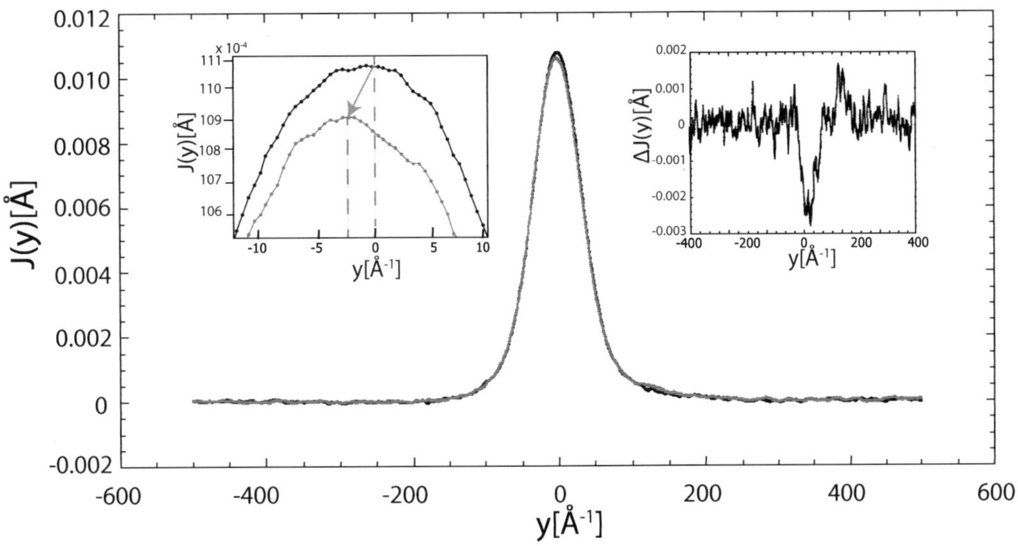

Figure 1. Main: Focused and normalised-to-unit-area $J(y_{Nb})$ data, recorded for Nb1 (black trace) and Nb4 (red trace) at 80 K. Left inset: close-up around the $J(y_{Nb})$ maximum. Right inset: difference between $J(y_{Nb4})$ and $J(y_{Nb1})$. See text for details.

NMDs is fitted with the differenced NMD shapes modelled either as simple Gaussian functions for harmonic and isotropic local potential, or and an analytical Gauss-Hermite expansion for an anharmonic or anisotropic potential. In case of Gaussian NMDs, whenever the difference δ between two NMDs, characterised by two different values of σ, σ_2 and σ_1, is small compared to their average, σ, the difference of NMDs can be expressed as a Taylor expansion in powers of δ/σ [26]. Unlike the method relying on FSEs and peak positions, described above, the NMD differencing method was demonstrated to work well for heavy masses, with copper and aluminium samples, measured at ambient and low temperatures, used as worked examples in the original work introducing the method [26]. Again here, a quick look on the difference between $J(y_{Nb4})$ and $J(y_{Nb1})$, shown as an inset on the right hand side of the main frame of Figure 1, shows that the quality of the difference signal is rather poor with the shoulders characterised by low signal-to-noise ratio and spurious spikes and other unresolved signals that would have not allowed for high precision estimation of the difference of NMD widths. However, similarly as in the case of the peak-maximum method described above, also here, the general trend can be inferred from the shape of the difference signal. This shape is characteristic for a difference between an NMD characterised by a larger width (here, sample Nb4) and the one with a narrower NMD (here, sample Nb1).

Encouraged by the qualitative insights yielded by both the NMD peak-maximum and the NMD-difference method, the fitting of focused Nb NMDs, recorded at T=80 K and T=300 K, was performed using Eq. (2). The fitting results confirm the qualitative insights pointing at systematic broadening of the NMD profile of the Nb4 sample, as compared to the Nb1 sample. The results of the fitting procedure, including the values of the STDs of NMDs for Nb1 and Nb4, are listed in Table 1.

Following the NCS analysis, we proceeded to analyze the ND data concurrently recorded using the high resolution ($\frac{\Delta\lambda}{\lambda} = 0.003$) backscattering VESUVIO+ detector bank [11]. The data were first normalised by point-wise division with Vanadium (V) data separately recorded

Table 1. Nb NMD widths together with 1-STD errors (in units of Å^{-1}). See text for details.

Sample	T = 80 K	T = 300 K
Nb1	11.6 ± 0.1	18.2 ± 0.1
Nb4	11.8 ± 0.1	18.5 ± 0.1

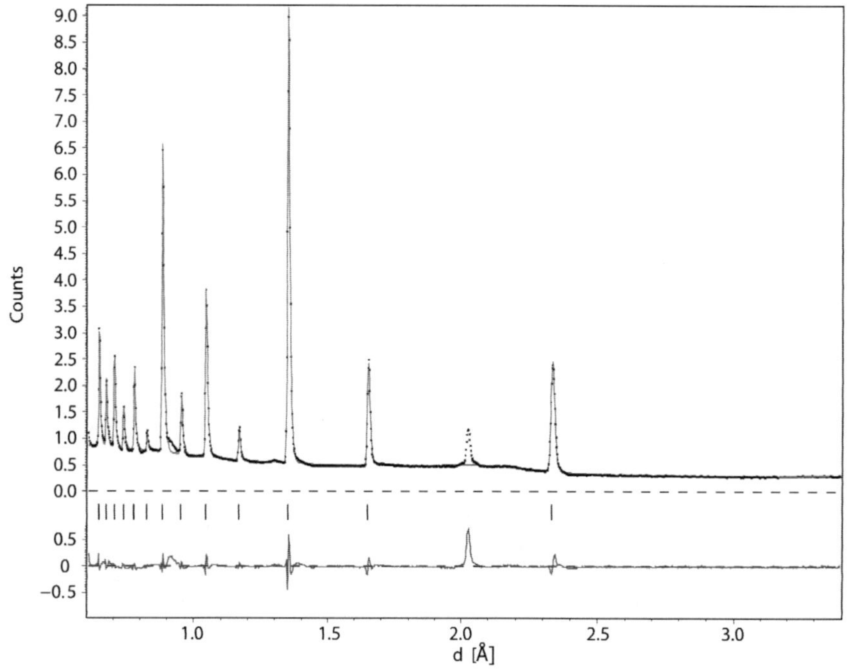

Figure 2. Nb4 ND data recorded on VESUVIO+ at 300 K. Black markers – ND data, red line – Le Bail-refinement of the data, blue line – residuals.

on the same instrument. Before division, V data were smoothed to exclude regions with Bragg peaks by fitting a fifth-order Chebyshev polynomial. Finally, Le Bail-refinement was performed using the JANA2006 program [27] with a background simulated by a fifth-order Chebyshev polynomial. An example of the result of this procedure is shown in Fig. 2 for the Nb4 sample at T=300 K. Figure 3 shows the ND data and fit for the Nb1 sample at T=300 K.

Table 2. Nb lattice constants together with 1-STD errors (in units of Å) from the Le Bail-refinement of the ND data recorded for samples Nb1 and Nb4 at T=80 K and T=300 K. See text for details.

Sample	T = 80 K	T = 300 K
Nb1	3.2899 ± 0.0002	3.2954 ± 0.0001
Nb4	3.2923 ± 0.0002	3.2978 ± 0.0002

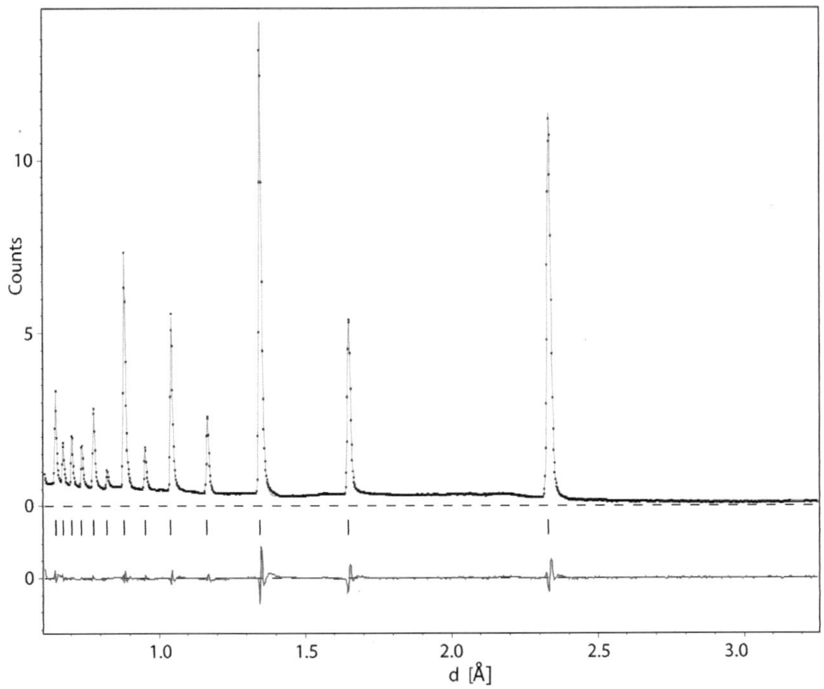

Figure 3. Nb1 ND data recorded on VESUVIO+ at 300 K. Black markers – ND data, red line – Le Bail-refinement of the data, blue line – residuals.

The Le Bail fits, both at T=80 K and T=300 K (see Table 2) yield Nb4 lattice spacing values that are systematically, by ca. 0.002 Å (ca. 10 STDs) higher than the Nb1 lattice spacing values. This result, together with the result of the NCS analysis described above, constitutes, to the best of authors' knowledge, the first demonstration of a rather unprecedented capability of Vesuvio+ to pick up very subtle structural and dynamical signatures of sample conditioning-induced local disorder for nuclei with such a high atomic mass (93 amu). In what follows, we will present a qualitative interpretation of this striking result.

3. Discussion

It is generally believed that the presence of adsorbed gasses in the Nb lattice leads to a systematic increase in lattice spacing. Diffuse X-ray scattering data of H trapping at N sites in the Nb lattice in the temperature range of 50–450 K by Metzger showed that, for a 1.1% N concentration, the lattice parameter change due to the presence of H, $\Delta a_H = a(NbN_{0.011}H_{0.008}) - a(NbN_{0.011})$, increases with increasing temperature, whereas the analogous figure for N, $\Delta a_N = a(NbN_{0.011}) - a(Nb)$, shows a less pronounced increase (see Fig. 5 in Ref. [28]). The values of Δa_H, inferred from this X-ray work, are ca. 0.0011 Å and ca. 0.0014 Å, at T=80 K and T=300 K respectively, whereas the values of Δa_N are ca. 0.0045 Å and ca. 0.0047 Å. Moreover, Nb lattice constants have been shown to depend linearly on H and N concentration at T < 300 K [36, 37]. Additionally, one can assume that: (i) at sub-percent H and N concentrations lattice expansion due to the uptake of N is independent on the expansion due to the uptake of H; (ii) both Nb1 and Nb4 contain the same amount of H due to the presence of the protective Nb_2O_5 surface layer which remains intact during sample conditioning; and (iii) Nb1 does not contain any traceable amounts of N. Based on these assumptions and on the values of the lattice constants listed in Table 2 one can obtain an upper conservative limit of the N concentration in Nb4 of $0.0024/0.0045 \cdot 0.011 \simeq 0.6$ %, a value ca. 40% less than the value obtained by Metzger [28]. Considering the fact that the above limits obtained from ND are bulk values, this result is consistent with the maximum N concentrations measured using the SIMS technique over the first few microns of the surface of the N-doped Nb cavity samples (10^{20} N atoms per cm^3, corresponding to an average stoichiometry of $NbN_{0.0018}$ [33]).

The ND results presented above are further corroborated by the NCS results by noting that Nb4 NMD widths are systematically and consistently (at both T=80 K and T=300 K) higher than Nb1 widths. In general, increased NMD widths measured in a crystalline sample signal indicates a surplus of states in the atom-projected vibrational density of states (apVDOS) due to tighter binding of a given atom in the lattice [19, 23, 24] which may be the consequence of lattice ordering, *e.g.*, when going from a glassy to a crystalline state [23, 24].

Using lattice theory when analysing X-ray diffuse scattering distribution in the Nb-H-N system, Dosch *et al* [31] showed that any H site which contributes more than 0.05 Å to the displacement of the two Nb neighbours of N in the direction of the crystallographic *c*-axis must be excluded from being a possible trap site for H. Only the nearest Nb neighbours of H experience a displacement of this order of magnitude [32]. Inelastic neutron scattering experiments in the Nb-N-H system show that the local mode energies of H in the untrapped and trapped states remain unchanged [30], thus leading to the conclusion that H remains in tetrahedral sites while being trapped at a remote enough position from the N atom to leave the local potential undisturbed.

Based on these observations, a model was built in which H atoms are placed so far apart from the N atoms so that no single Nb atom becomes a nearest neighbour for both of them [28]. In such a scenario, the dynamical coupling of Nb to H and N, the changes of the local potential felt by H, N, and Nb, and finally the changes in apVDOS are too small to manifest themselves markedly in Nb NMDs. Moreover, using the Dosch model [31] and the fact that the Debye temperature of Nb is constant in the temperature region 80–300 K [29], one can predict that untrapped protons present in the Nb lattice should systematically increase the displacement of the two Nb neighbours by at least 0.05 Å, on top of the temperature effect associated with the Boltzmann statistics of vibrational states. Using the pure-Nb Debye temperature of 284 K and assuming ca. 0.025 Å displacement surplus per Nb atom in the lattice, one obtains a decrease of the Nb NMD width of the order of 0.2 $Å^{-1}$. This prediction is in a very good agreement with the trend visible in Table 1: the Nb NMD width in Nb1, at both 80 K and 300 K, is lower by 0.2 $Å^{-1}$ from the Nb NMD widths in Nb4. This observation signals lattice-induced NMD changes and seems to point towards an increased fraction of untrapped protons in Nb1, as compared to Nb4 in the temperature range 80–300 K.

On the whole, N doping and baking of the Nb in Nb4 seems to lead to some degree of Nb lattice ordering induced by increased H trapping (as compared to Nb1). This conclusion is in line with the result of Pfeiffer postulating H trapping by interstitial N atoms where only N-H pairs (and no higher agglomerates) are formed as the temperature is lowered [6]. However, two scenarios are still possible within this model. If the N concentration exceeds the H concentration in the Nb lattice, the variation of the temperature does nothing but to reversibly create and cleave N-H pairs in dynamics equilibrium [28]. In this scenario, for any choice of reasonable parameters, all the H atoms will be trapped below 120 K and the percentage of trapped H atoms decreases with increasing temperature with the final trap concentration at 300 K still amounting to about 40-60% [35]. Conversely, if the H concentration is higher than the N concentration, some H atoms might still not be trapped below 120 K and the final trap concentration may be much lower than 40% at room temperature. The presented analysis points towards the latter scenario.

4. Summary and Outlook

In summary, the analysis of ND and NCS results in electro-polished and N-doped Nb samples, prepared to exactly mimic two different types of state-of-the-art SRFCs, seems to be consistent with previous electron microscopy results indicating the coexistence of nano or micro domains of at least two H-containing Nb phases embedded in the Nb lattice [1–4]. Moreover, the results presented here are in line with previous surface spectroscopic work suggesting that N is present as an interstitial in the Nb lattice, possibly binding or trapping H [2,6]. The NCS analysis presented in this work shows that N doping and baking of the Nb sample leads to some degree of Nb-lattice ordering induced by increased H trapping. Due to trapping, the H-induced SRF resistivity increase is reduced and the occurrence of precipitation is shifted to higher H concentrations thus greatly improving SRFC performance. However, the N concentration in the Nb lattice may still be insufficient to trap all H atoms below 120 K.

Further NCS and ND research should be carried out along two clearly visible lines of work. Firstly, more data should be acquired at cryogenic temperatures to shed more light on the N doping-induced changes in cavity performance at temperatures relevant of the operation of these devices. Secondly, more systematic studies need to be performed on various SRF cavity cut-outs in order to prune out sample-specific trends from those induced by material conditioning.

On the methodological front, further *ab initio* materials modelling-augmented studies are highly desired. To this end, single-crystal ND and NCS measurements would provide a clear cut assessment of the role played by anisotropy of material properties. Moreover, single-crystal ND would enable an appraisal of the thermal diffuse scattering and single-crystal NCS would open the possibility to assess the degree of NMD anisotropy.

Traditionally, NCS has been used as a local probe of nuclear dynamics in condensed matter systems and molecules [23]. Capitalising on more than three decades of evolution of NCS as a technique, this work demonstrates, for the first time, an attempt to extend the realm of NCS applicability to simple devices, superconducting radio-frequency cavities. Moreover, this work paves the way for more use of ND and NCS in future studies of the performance of technologically relevant devices. The synergies between ND and NCS cannot be more emphasized as in the case of their concurrent application to the same system on the same neutron beamline. Further research using this paradigm in mind should, in this particular case, lead to SRFC performance optimisation and the improvement of SRFC surface-processing technologies.

5. Acknowledgments

The authors gratefully acknowledge the UK Science and Technology Facilities Council for financial support and the ISIS Facility for access to beamtime.

[1] F. Barkov, A. Romanenko, Y. Trenikhina, and A. Grassellino, 2013 *J. Appl. Phys.* **114**, 164904.
[2] Y. Trenikhina, A. Romanenko, J. Kwon, J.-M. Zuo, and J. F. Zasadzinski, 2015 *J. Appl. Phys.* **117**, 154507.
[3] A. Romanenko, C.J. Edwardson, P.G. Coleman, and P.J. Simpson, 2013 *Appl. Phys. Lett.* **10**, 232601.
[4] B. Visentin, M.F. Bathe, V. Moineau, and P. Desgardin, 2010 *Phys. Rev. ST Accel. Beams* **13**, 052002.
[5] A. Grassellino, A. Romanenko, D. Sergatskov, O. Melnychuk, Y. Trenikhina, A. Crawford, A. Rowe, M. Wong, T. Khabiboulline and F. Barkov, 2013 *Supercond. Sci. Technol.* **26**, 102001.
[6] J. J. Rush, T. J. Udovic, N. F. Berk, D. Richter and A. Magerl, 1999 *Europhys. Lett.* **48**, 187.
[7] D. C. Ford, L. D. Cooley, D. N. Seidman, 2013 *Supercond. Sci. Technol.* **26**, 095002.
[8] T. Abdul-Redah, M. Krzystyniak, C. A. Chatzidimitriou-Dreismann, 2005 *Phys. Rev. B.*, **72**, 052202.
[9] M. Krzystyniak et al, in preparation.
[10] M. Krzystyniak, M. A. Adams, A. Lovell, N. T. Skipper, S. M. Bennington, J. Mayers and F. Fernandez-Alonso, 2011 *Faraday Discuss.* **151**, 171.
[11] G. Romanelli et al, 2017 *Meas. Sci. Technol.*, **28**, 9.
[12] S. Jackson, M. Krzystyniak, A. G. Seel, M. Gigg, S. E. Richards and F. Fernandez-Alonso, 2014 *J. Phys. Conf. Ser.*, **571**, 012009.
[13] M. Krzystyniak, A. Seel, S. E. Richards, M. J. Gutmann and F.Fernandez-Alonso, 2014 *J. Phys. Conf. Ser.*, **571**, 012002.
[14] M. Krzystyniak, 2010 *J. Chem. Phys.*, **133**, 403.
[15] M. Krzystyniak, K. Drużbicki and F. Fernandez-ALonso, 2015 *Phys. Chem. Chem. Phys.*, **46**, 31287.
[16] M. Krzystyniak, S. Richards, A. G. Seel and F. Fernandez-Alonso, 2013 *Phys. Rev. B.*, **88**, 184304.

[17] M. Krzystyniak and F. Fernandez-Alonso, 2011 *Phys. Rev. B*, **83**, 134305.
[18] M. Krzystyniak, A. G. Seel, S. Richards, M. J. Gutmann and F. Fernandez-Alonso, 2014 *J. Phys. Conf. Ser.*, **571**, 012002.
[19] C. Andreani, D. Colognesi, J. Mayers, G. F. Reiter and R. Senesi, 2005 *Adv. Phys.*, **54**, 377.
[20] A. G. Seel, M. Krzystyniak and F. Fernandez-Alonso, 2014 *J. Phys. Conf. Ser.*, **571**, 012006.
[21] G. Gorini, G. Festa and C. Andreani, 2014 *J. Phys. Conf. Ser.*, **571**, 012005.
[22] M. Krzystyniak, K. Druzbicki, G. Romanelli, M. J. Gutmann, S. Rudic, S. Imberti and F. Fernandez-Alonso, 2017 *Phys. Chem. Chem. Phys.*, **19**, 9064.
[23] C. Andreani, M. Krzystyniak, G. Romanelli, R. Senesi and F. Fernandez-Alonso, 2017 *Adv. Phys.*, **66**, 1.
[24] C. Andreani, R. Senesi, M. Krzystyniak, G. Romanelli and F. Fernandez-Alonso in D. L. Price and F. Fernandez-Alonso Eds., 2017 *Neutron Scattering Fundamentals* vol 49 (Academic Press)
[25] M. Krzystyniak et al., J. Phys. Conf. Series., submitted (2018).
[26] G. Romanelli, M. Krzystyniak, 2016 *Nuclear Instruments and Methods in Physics Research A* **819**, 84.
[27] V. Petricek, M. Dusek and L. Palatinus, 2014 *Z. Kristallogr*, **229**, 245.
[28] T. H. Metzger, U. Schubert and J. Peisl, 1985 *J. Phys. F: Met. Phys.*, **15**, 779.
[29] R. I. Sharp, 1969 *J. Phys. C.*, **2**, 306.
[30] A. Magerl et al., 1983 *Phys. Rev. B* **27**, 927.
[31] H. Dosch, U. Schubert, H. Metzger and J. Peisl, 1984 *J. Phys. F: Met. Phys.*, **14**, 2467.
[32] H. Metzger, H. Behr and J. Peisl, 1982 *Solid State Commun.*, **40**, 789.
[33] D. Gonella, Ph. D. Thesis, Cornell University (2016).
[34] J. Peisl, 1976 *J. Physique Coll.*, **12**, 47.
[35] G. Pfeiffer and H. Wipf, 1976 *J. Phys. F: Met. Phys.*, **6**, 167.
[36] H. Pfeiffer and J. Peisl, 1977 *Phys. Lett.*, **60A**, 363.
[37] H. Zabel and J. Peisl, 1979 *J. Phys. F: Metal Phys.*, **9**, 1461.

IOP Conf. Series: Journal of Physics: Conf. Series **1055** (2018) 012007 doi:10.1088/1742-6596/1055/1/012007

Fractal dimension as a scaling law for nuclear quantum effects: a neutron Compton scattering study on carbon allotropes

J Armstrong[1], M Krzystyniak[1,2], G Romanelli[1], S F Parker[1], K Drużbicki[3,4] and F Fernandez-Alonso[1,5]

[1] ISIS Facility, Rutherford Appleton Laboratory, Chilton, Didcot, Oxfordshire OX11 0QX, United Kingdom

[2] School of Science and Technology, Nottingham Trent University, Clifton Campus, Nottingham NG11 8NS, United Kingdom

[3] Faculty of Physics, Adam Mickiewicz University, Umultowska St. 85, 61-614, Poznan, Poland

[4] Frank Laboratory of Neutron Physics, Joint Institute for Nuclear Research, Dubna, JoliotCurie 6, 141980, Russia

[5] Department of Physics and Astronomy, University College London, Gower Street, London WC1E 6BT, United Kingdom

E-mail: `matthew.krzystyniak@stfc.ac.uk`

Abstract. In this work, we tackle the problem of the sensitivity of neutron Compton scattering, measured through the widths of nuclear momentum distributions, to the degree of complexity and ordering of the structural motif characterising the surrounding environment felt by a particular nucleus (carbon). In doing so, we replace the usual concept of the bond strength categorised in terms of its thermodynamical or electronic properties with a novel observable inspired by the language of mathematical topology, the Hausdorff-Besicovitch fractal dimension. We derive a relatively straightforward connection between the fractal dimension of a given system under consideration and the nuclear kinetic energy. To achieve this, we modify the concept of the energy equipartition theorem for solid-state systems composed of carbon atoms where the atom-ordering topology does not follow a simple two or three-dimensional order, but rather atoms are placed along curves in space that have an intermediate dimension related to the varying amounts of information they contain. A series of results from past neutron Compton scattering studies, as well as new results on Buckminsterfullerene (C_{60}), correlate with the topological measures of surface roughness and bending, as categorised quantitatively by the fractal dimension of the system. Namely, for the same formal chemical binding motif (sp^2 C) and with decreasing system dimensionality from nearly 3 towards 1, the quantum nature of the system becomes more pronounced. The simple scaling law developed in this work allows for relatively simple assessment of the nuclear "quantumness" of a given system with potentially important ramifications in the *ab initio* modelling of nuclear quantum effects in condensed matter.

> One must say- instead of points, straight lines and planes - tables, chairs and beer mugs
>
> David Hilbert.

1. Introduction

From the point of view of a materials chemist, Neutron Compton scattering (NCS) can be termed as "nuclear quantum thermometry" of condensed matter systems and molecules [1–3]. As a neutron spectroscopic technique, NCS owes its unique character to its capability to measure, in a mass-resolved manner, nuclear momentum distributions (NMDs). NMDs are, in turn, the most basic quantum observables that can be related to nuclear quantum effects (NQEs) [1–3].

In fact, NMDs of light and medium-weight atomic species are dominated by the zero-point-energy contribution to the nuclear kinetic energies [1–3]. Mathematically, the second moments of NMDs are proportional to the first-moments of atom-projected vibrational densities of states (apVDOS). Consequently, the NMD widths, from the *ab initio* modelling methodology point-of-view, can be computed by optimising the geometry of a given system of atoms or molecules, e.g. in a crystal, and computing the apVDOS. This relatively simple mathematical procedure has profound physical ramifications. Namely, by virtue of the closure relation for the quantum harmonic oscillator (QHE), this route of calculation of the NMDs can be termed as being due to "trivial nuclear quantum effects". This name is fully deserved as NMDs turn out to be in this case completely computable by resorting to electronic structure theory dictating the properties of the forces entering the dynamical matrix, used further to compute the apVDOS [4]. Furthermore, by comparing the NMD width predicted using apVDOS, obtained from *ab initio* calculation performed at a given temperature, with its counterpart form an NCS experiment one can assess the degree in which "trivial nuclear quantum effects" determine NMD of a given nucleus. This computational route has got very important ramifications for heavy-weight nuclei, whose NMD widths are almost entirely dictated by their respective apVDOS. Thus, the "non-trivial nuclear quantum effects", such as nuclear wave function coherence and nuclear quantum tunnelling, contribute to NMDs of heavy-weight nuclei in such a minuscule manner that their contribution can be neglected with impunity. Conversely, for light-to-medium weight both "trivial" and "non-trivial nuclear quantum effects" may simultaneously shape the NMDs and the interplay between them is, in general, dictated by the system temperature and nuclear mass.

Are then the NQEs exhibited by heavy-weight nuclei completely obscured from the view of the NCS? The answer to this question can only be given after a systematic search of available heavy-weight NMD data for condensed-matter systems has been performed and correlations between NQE-related properties and NCS observables are established. Such a systematic review of NCS has recently been published [1, 3]. Despite the fact that some degree of correlation has been observed between the formal electric charge of the nuclei in lattice sites, their electronegativity, the character of bonding (covalent, ionic, metallic, or long-range dispersive) and the widths of NMDs, the full picture is still largely missing.

One of the main questions still remaining to be addressed is how the degree of complexity and ordering of the structural motif characterising the surrounding environment felt by a particular nucleus correlates with its NMD shape. For instance, it is not clear what level of sensitivity the technique has to length scales characterising phenomena such as dynamical or static disorder, surface roughness or the dimensionality of a condensed-matter system. The sensitivity of NCS to first-order structural phase changes is generally believed to be very high [1, 3], as the local environment of the atom is changed during the transition. In the case of dynamical disorder, however, there may be no change in the average local environment, or it may involve a continuous change with temperature. It is also not clear what is the critical length-scale attributed to a dynamic or static distortion in a condensed-matter system or imperfection of the crystal lattice that can lead to a visible change of the nuclear kinetic energy with reference to a perfectly ordered system. It is also not clear how this critical length-scale depends on the dimensionality of the system, its temperature, the nuclear mass and the chemical binding motif dominating the surrounding of a given nucleus. These open questions have largely motivated the work presented here.

As the correlations mentioned above may be obscured by a number of different, often cross-correlated factors, the optimal experimental protocol should seek to establish links between the topological features of a family of systems exhibiting the same chemical binding motif. The chemistry of carbon offers a great opportunity in this respect as one can easily identify a large family of carbonaceous systems, all of which will be characterised by the sp^2 hybridisation. Amongst carbonaceous systems, Buckminsterfullerene (C_{60}) represents a particularly interesting case. It is characterised by dynamical disorder in a wide range of temperatures and it undergoes two phase transitions. At high temperature (>260 K), C_{60} forms a face-cantered-cubic structure, where the C_{60} molecules are able to perform isotropic free rotation about their lattice sites. Below 260 K, the system undergoes a first-order phase transition to a simple cubic structure with orientational anisotropy. Below 90 K, all rotational motion ceases to exist, and the system becomes trapped in a given orientation.

In addition to C_{60} serving as a prototypical system for studying dynamical disorder, it also represents a system of geometric curiosity from a nuclear kinetic energy perspective. C_{60} is quasi-2D quantum system in the sense that the C_{60} molecules themselves can be best thought of as folded 2D sheets. This means that the shape of carbon NMDs in solid C_{60} should be simultaneously dictated by the vibrational properties of planar carbon systems and by the librations and translations associated with the entire C_{60} molecules. In this sense, C_{60} falls into the hierarchy somewhere between a 2D-system like graphene and a 3D-system like graphite.

In what follows, we compare a series of relatively simple carbon-based compounds, bulk carbon, two-dimensional carbon sheets, and C_{60} with the aim to place them on a common scale characterising the degree of the correlation between the complexity of the structural motif characterising the surrounding environment felt the carbon and its kinetic energy. To this end, we will attempt to use a geometrical scale, well-suited for nanoscale systems, the fractal dimension.

2. Methodology

2.1. Ab initio methods

As the *ab initio* methods, whose results were used in this work to calculate carbon apVDOS and carbon NMDs in both low and high-temperature phases of solid C_{60}, are very well known and described in great detail elsewhere, here only a brief account of the methodology employed will be given.

In the solid state from the sublimation temperature, 800 K, down to 260 K, C_{60} undergoes rapid rotation (10 ps at room temperature) and the structure can be approximated as a face centred cubic (*fcc*) lattice of spheres, space group $Fm3\bar{m}$ with one molecule in the primitive cell [5]. Below 260 K, an ordered simple cubic (*sc*) phase is obtained, space group $Pa\bar{3}$ with four molecules in the primitive cell [6]. For the low-temperature ordered phase the *ab initio* work, reported in the literature, essentially exactly reproduces the structure from neutron diffraction work [6]. However, as noted by Parker *et al.* [6], in order to simulate the high-temperature disordered C_{60} phase, a computationally intractable periodic-DFT calculation using a very large unit cell consisting of many disordered C_{60} molecules would be required. It is, however, possible to use an ordered *fcc* phase with space group $Fm3\bar{m}$ that can be obtained by aligning the three twofold axes along the Cartesian x, y and z-axes [7]. In order to prevent an unphysical expansion of the lattice, due to van der Waals-nature of this solid C_{60} phase, the the *ab initio* geometry optimisation needs to be carried out at the experimental lattice parameter of 14.160 Å [6].

Thus, as an input for our spectroscopic calculations reported in this work we have used the output generated from the existing results of the *ab initio* work by Parker *et al.* who performed periodic DFT calculations of both crystalline structures using the plane-wave pseudo-potential method implemented in the CASTEP code [6]. Exchange and correlation were approximated using the PBE functional, and a norm-conserving pseudopotential for carbon was generated using the kinetic-energy optimised method with a plane-wave cut-off energy of 750 eV. The equilibrium structure was obtained by Broyden-Fletcher-Goldfarb-Shanno (BFGS) geometry optimization algorithm with the residual-force-convergence within 0.0015 eV Å^{-1}. In light of practically dispersion-less phonon spectra, observed both at the spectroscopic and computational level, phonon frequencies for this study were obtained by diagonalisation of dynamical matrices computed using density-functional perturbation theory at zero wave-vector (Γ-point-LD calculation) [6].

Interestingly, in the context of the topological analysis presented in this work, yet another method of apVDOS calculation is worth mentioning, namely the recursion method [8]. The recursion method is based on a theorem stating that the projected density of states is stable with respect to changes in distant parts of the system and that it can be calculated recursively by tri-diagonalising the dynamical matrix describing the forces in a basis of orthonormal operators [8]. Mathematically, the diagonal elements of the dynamical matrix, D, are solved for using recursive equations [8]. In case of the solid C_{60}, D depends on the geometry of the C_{60} molecule, the mass of the carbon nucleus, and four force constants used to model the four bonds formed between a given carbon and its nearest neighbours. The recursion method has been demonstrated to satisfactorily reproduce the apVDOS of carbon in molecular C_{60} using two of the four force constants to model the stretching of the pentagonal and hexagonal bonds, respectively, and the remaining two force constants to model the bending of the pentagonal and hexagonal bonds, respectively.

2.2. From ab initio calculations to NMD observables

Predictions of the widths of the NMDs for nuclei in molecular crystals can be calculated using two computational routes:

- the lattice dynamics (LD) computational route, whereby one performs *ab initio* calculation [usually within the density functional theory (DFT) under the Born-Oppenheimer approximation] consisting of geometry optimisation, followed by solving the eigenproblem of the dynamical matrix of the whole crystal and calculation of phonon-dispersion curves for both intra and intermolecular modes [3, 9];
- a mode-decoupling route, whereby one neglects the rotation-vibration interactions and any interplay between translational and internal motion which enables the decomposition of the nuclear kinetic energy into three separate contributions: molecular translations and librations (referred to as external contributions), and internal molecular vibrations [10, 11] which can be accounted for using two different types *ab initio* calculations: (i) a single-molecule geometry optimisation followed by vibrational analysis, or (ii) a dispersion-less Γ-point LD calculation.

As presented in more detail in Ref. [9], the first route amounts to the calculation of the second moment of an NMD for a nucleus of mass, M, $\sigma_M(r)^2$, that can then be decomposed into three Cartesian directions in space, for which a unit vector $r = (x, y, z)$ can be defined, according to:

$$\sigma_M(r)^2 = \frac{M^2}{N_q \hbar} \sum_{q \in 1BZ} \sum_{\lambda=1}^{N_\lambda} [e(\lambda, q) \cdot r]^2 \frac{\hbar\omega}{2} \coth\left(\frac{\hbar\omega}{2k_B T}\right), \quad (1)$$

where $\omega(\lambda, q)$ are phonon frequencies, $e(\lambda, q)$ are polarization vectors and T is the sample temperature, and k_B the Boltzmann constant. The double summation in Eq. (1) runs over all N_q vectors in the first Brillouin zone, and all phonon branches N_λ.

In the second route the computation of $\sigma_M(r)^2$ is performed under the assumptions that (1) molecular librations can be dealt with classically, (2) vibrational displacements of nuclei from their equilibrium positions are small if compared with their average distances from the centre of mass, and (3) internal vibrations can be considered as purely harmonic and treated quantum mechanically [10,11]. The translational contribution to the kinetic energy of a nucleus of a mass M reads:

$$\langle E_k^{tra} \rangle = \frac{1}{2}\frac{M}{M_T} M_T v_{CM}^2, \tag{2}$$

where v_{CM}^2 is the molecule center-of-mass velocity and M_T is the total mass of the molecule. Making use of the classical equipartition theorem, we obtain $\langle E_k^{tra} \rangle = \frac{3}{2}\frac{M}{M_T} k_B T$.

In case of librations in the solid, the contribution to the kinetic energy $\langle E_k^{lib} \rangle$ of a nucleus of a mass M in a molecule can be expressed, in the coordinate system coinciding with the molecular principal inertia axes, in the following way:

$$\langle E_k^{lib} \rangle = \frac{1}{2} M k_B T \sum_{i,j=1}^{3} r_i R_{ij} r_j, \tag{3}$$

where the tensor R_{ij} is closely linked with the Sachs–Teller (S-T) mass tensor [10,11] for M, which often appears in the classical treatment of molecular rotations:

$$M_{ij}^{-1} = R_{ij} + M_T^{-1}\delta_{ij}. \tag{4}$$

The contribution to $\sigma_M^2(r)$ from Eqs. 2 and 3 can then be defined as:

$$\sigma_{S-T}^2(r) = \frac{M^2 k_B T}{\hbar^2} \left(\frac{1}{M_T} + \sum_{i,j=1}^{3} r_i R_{ij} r_j \right). \tag{5}$$

The internal vibrational contribution to the kinetic energy, $\langle E_k^{vib} \rangle$ can be worked out decomposing the nuclear displacements, u_M into normal modes q_λ by means of the amplitude vectors C_M^λ:

$$u_M = \sum_{\lambda=1}^{N_\lambda} C_M^\lambda q_\lambda. \tag{6}$$

The variance $\sigma_M(r)^2$ for r parallel to the direction of the nuclear momentum p can be calculated using the following relation [10,11]:

$$\sigma_{intra}^2(r) = \frac{M^2}{\hbar^2} \sum_{\lambda=1}^{N} (C_M^\lambda)^2 \sum_{j=1}^{N_\lambda} \frac{1}{N_\lambda} \frac{\hbar\omega_\lambda^j}{4} \coth\left(\frac{\hbar\omega_\lambda^j}{2k_B T}\right), \tag{7}$$

where $\hbar\omega_\lambda^j$ is the harmonic excitation energy of the j-th component of the λ-th mode, and N_λ is the number of components of the same mode [10,11].

The multivariate Gaussian vibrational contribution, $\sigma_M^2(r)$ can be chosen so as to coincide with the molecular principal inertia axes, and combining Eqs. 5 and 7 one has [10,11]:

$$\sigma_M^2(r) = \sigma_{S-T}^2(r) + \sigma_{intra}^2(r), \tag{8}$$

from which the spherically averaged value of the second moment of the nuclear momentum distribution for a given nucleus $\sigma_M(r)^2$ can be calculated [12]:

$$\sigma_M^2 = \frac{1}{3}\left[\sigma_M(x)^2 + \sigma_M(y)^2 + \sigma_M(z)^2\right] \tag{9}$$

2.3. Nuclear momentum distributions from NCS

Detailed accounts of the NCS method, both in terms of its instrumental and mathematical foundations, can be found elsewhere [1–3]. Here, we will only emphasize the aspects of the method that, in our opinion, guide the reader to a better understanding of the results of the concurrent measurements using inelastic (INS) and NCS spectroscopy presented in this work. In an nutshell, the kinematics of NCS leads to the description of the scattering in terms of the Impulse Approximation (IA), whereby a billiard-ball type of collision of neutrons with individual nuclei in the sample leads to instantaneous excitation of all nuclear degrees of freedom. Moreover, the energy and momentum conservation, invoked by the IA, leads to a description in which each binary collision of a neutron with an individual nucleus produces a separate recoil peak in the time-of-flight spectrum [1–3].

It is this mass selectivity that underlies the very uniqueness of NCS as a spectroscopic method. For a given fixed sample stoichiometry, the intensity of the NCS recoil peak of each nuclear isotope present in the sample scales *only* as the product of the total bound scattering cross-section and the relative quantity of this particular isotope [1–3]. A very important consequence of this scaling is that the shape of the NMD of each nuclear isotope is *not* affected by the scattering cross sections and the relative quantities of *other* isotopes from the same sample. Thus, there is no mixing of neutronic responses of different isotopes as NCS, unlike INS, is free from the complication of "neutron-weighting". This "neutron weighting" implies that, unlike in NCS, in INS in order to reproduce an experimental spectrum one must calculate a linear combination of apVDOS curves, predicted for each individual isotope by an *ab initio* calculation, weighted by the neutron scattering cross-sections of each isotope. This important distinction between INS and NCS has many consequences. The one that is most explored in the NCS literature is that the second moment of the NMD for a given isotope is *only* proportional to the first moment of its apVDOS [1–3].

In the time-of-flight (TOF) domain, t, and for a given fixed scattering angle, θ, the NCS spectrum, $C_\theta(t)$, is described as:

$$C_\theta(t) = A' \left[\frac{E_0 I(E_0)}{Q}\right]_t \times \sum_{n=1}^{N} I_n M_n J_n[y_n(t)] \otimes R_n[y_n(t)]. \tag{10}$$

Mathematically, the mass selectivity of the NCS method can be expressed by the fact that the whole expression, $C_\theta(t)$, can be further decomposed into subexpressions describing spectral contributions of recoil peaks of individual isotopes, $J_n[y_n(t)] \otimes R_n[y_n(t)]$. This decomposition is facilitated by another unique feature of NCS, West scaling [1–3]. In a nutshell, West scaling is a consequence of the conservation of momentum and energy during a binary collision of a neutron with a nucleus of mass M, and it reduces the shape of the curve describing the recoil peak of M to a peak function, mathematically expressed as $(M_N/\hbar^2 Q(t)) J_N(y_{M_{N(t)}}) \otimes R_n[y_n(t)]$, which is centered at the recoil energy $E_r = \hbar^2 Q^2/2M$ with $\hbar = 2.0446$ (meV amu)$^{1/2}$ Å [2, 13–19]. It is this very function $J(y(t))$ in this expression which is a peak function expressed in a domain of the West scaling variable $y(t) = (M_N/\hbar^2 Q(t)) \left(E_0(t) - E_1 - {}^{-}\hbar Q(t)^2/2M\right)$, where E_1 is the final neutron energy which is fixed in an inverted-geometry instrument (thus t-independent). Physically, $J_N(y_{M_N})$ is the so called NCS profile [13,14] and formally corresponds to the longitudinal NMD of the target nucleus along y_{M_N}. Instrumentally, each recoil peak is represented by the convolution of the NMD of a given nucleus with the nucleus-specific instrument resolution function, which can also be expressed in the domain of the West scaling variable of a given mass M [1–3].

The fact that the width of the Vesuvio resolution function increases rapidly with increasing mass of the recoiling nucleus [20] renders fitting anisotropic NMD models to experimental NCS data, recorded for nuclei with masses higher than few atomic mass units, practically infeasible.

In consequence, carbon NMDs are usually fitted by spatially averaged, Gaussian distributions, despite the fact that the *ab initio* modelling predicts usually anisotropic carbon NMDs. In this case, $J_{IA}(y)$ for a given mass M in the IA limit depends only on the magnitude of y but not on $\hat{Q}$, described in the IA limit, by a purely Gaussian term only:

$$J_{IA}(y) = \frac{1}{\sigma\sqrt{2\pi}} \exp\left(\frac{-y^2}{2\sigma^2}\right), \tag{11}$$

where σ, the standard deviation (STD) of the NMD, with the present choice of units, is expressed in units of $Å^{-1}$.

For finite values of Q, corrections to the IA are known as "Final-State Effects" (FSEs) [2,19] and must be applied. Usually, the longitudinal NMD in the presence of the FSEs corrections is expressed as a series of the form:

$$J(y) = J_{IA}(y) + J_{FSE}(y) = J_{IA}(y) - \frac{M\langle\nabla^2 V\rangle}{36\hbar^2 q}\frac{d^3}{dy^3}J_{IA}(y) + ..., \tag{12}$$

where $J_{IA}(y)$ is the IA result given by Eq. (11). $\langle\nabla^2 V\rangle$ is the mean value of the Laplacian of the potential energy of the nucleus expressed in meV $Å^{-2}$ (cf. Refs. [19, 21]).

An explicit connection between the FSE magnitude k and σ can be established in the harmonic approximation, $k = \sigma\frac{\sqrt{2}}{12}$ [22–24] with the FSE term, given by the Eq. 12, expressed using the third order Hermite polynomial, $H_3\left(\frac{y}{\sigma\sqrt{2}}\right)$:

$$J(y) = \frac{1}{\sigma\sqrt{2\pi}} \exp\left(\frac{-y^2}{2\sigma^2}\right)\left[1 - \frac{\sigma\frac{\sqrt{2}}{12}}{Q}H_3\left(\frac{y}{\sigma\sqrt{2}}\right)\right]. \tag{13}$$

3. INS and NCS results for solid C_{60}

3.1. Carbon vibrational density of states and INS

As a starting point for our later discussion we show the results of the application of the recursion method for the calculation of the carbon apVDOS in an isolated C_{60} molecule taken from the existing in the literature work [8]. The apVDOS calculated in this manner (see Figure 1) was reported to have a reasonable qualitative agreement with the experiments considering that the Raman and infrared spectra are related to the VDOS through matrix elements whose exact nature is imprecisely known [8].

Following the brief account of the results of the recursion method, we give a short account of the high resolution inelastic neutron scattering (INS) at the inverted geometry neutron spectrometer TOSCA at ISIS spanning energy transfers up to 4000 cm^{-1} [25–27]. The spectral resolution of TOSCA is similar to that of infrared and Raman techniques, approximately 2% of the energy in question. The measurements were performed on 5 g of C_{60} as bought from Sigma Aldrich using a flat aluminum cell of cross-sectional area 4x4.8 cm^2. The sample was studied at a temperature of 10 K using a dedicated closed-cycle helium refrigerator, so as to minimize the effect of the Debye-Waller factor, thus allowing us to fully distinguish the individual peaks. Data were reduced using the Mantid Software [28]. The INS spectrum of low-temperature phase of C_{60} recorded at TOSCA is shown in Figure 2. The experimental spectrum is compared in Figure 2 with the synthetic spectrum generated from the output of the *ab initio* periodic-DFT LD calculation performed for the low-temperature phase of C_{60} by Parker *et al.* [6], briefly described in section 2.1.

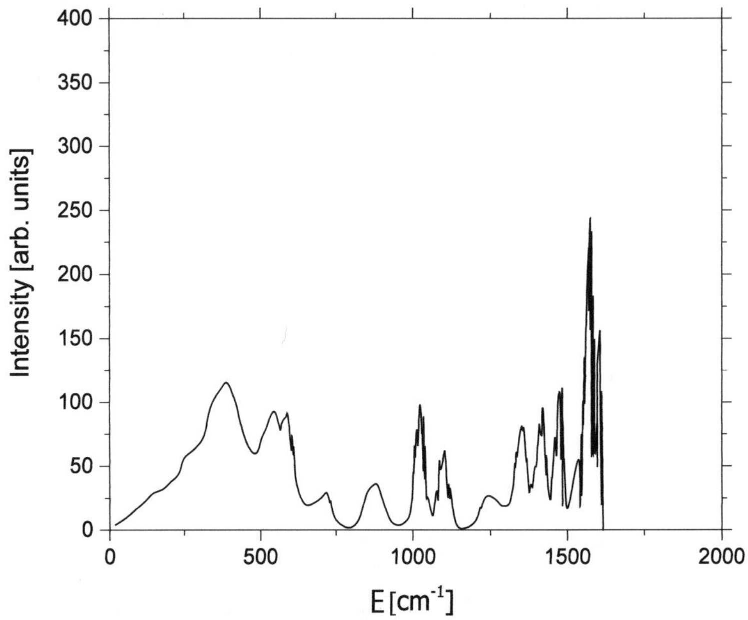

Figure 1. The result of the application of the recursion method to the solution of the dynamical matrix equation for the C_{60} molecule, taken from Ref. [8]. See text for further details.

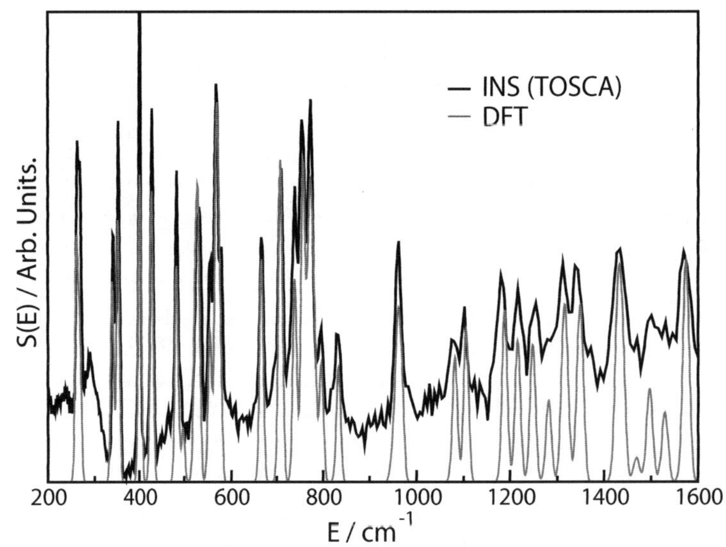

Figure 2. The INS spectrum of C_{60}, recorded at the inverted geometry neutron spectrometer TOSCA, at the ISIS Pulsed Neutron and Muon Source, Rutherford Appleton Laboratory, United Kingdom (black trace) compared with the synthetic spectrum generated from the output of the the *ab initio* periodic-DFT LD calculation by Parker *et al.* [6] (read trace). See text for further details.

IOP Conf. Series: Journal of Physics: Conf. Series **1055** (2018) 012007 doi:10.1088/1742-6596/1055/1/012007

3.2. Nuclear kinetic energy of carbon in solid C_{60} from NCS

All NCS measurements were performed on the latest incarnation of the indirect-geometry NCS spectrometer VESUVIO located at the ISIS Pulsed Neutron and Muon Source, United Kingdom [20,29,30]. The NCS forward scattering data reduction performed in this work follows the literature protocol [9]. In short, the protocol amounted to the sequential, detector-by-detector, pre-fit of the data, the calculation of sample-dependent signal corrections (multiple scattering and gamma background), the focusing of corrected data in the West scaling space of carbon, and finally the fit of the focused data to obtain the longitudinal carbon NMDs, from which the values of the second moments and kinetic energies of carbon were calculated.

For the VESUVIO experiment on C_{60}, the sample was a pellet of C_{60} in powder form. The specimens were enclosed in a 6x6 cm^2 flat aluminum (Al) cell resulting in a 92% transmission level of the sample in the Al container, and 2% transmission level of the empty Al container, respectively. The data used in the analysis were recorded in the temperature range 5–300 K, with the sample temperature stabilised using a dedicated helium closed-cycle refrigerator.

In the final NCS data analysis step, a maximum-likelihood-based model selection was performed, based on the Bayesian approach implemented in the Fabada algorithm in the MantidPlot software [31]. Examples of the results of the application of this procedure for C_{60} are shown in Figure 3. Inspired by our previous NCS work [32] two NMD models were compared: (i) a model in which the NMD is described by a purely Gaussian shape, and (ii) an NMD model described mathematically by a sum of a Gaussian function and a Gaussian multiplied by a fourth-order Hermite polynomial accounting for the extra peakedness (excess kurtosis) of the NMD (not shown here). The Bayesian analysis preferred the purely Gaussian NMD model for NCS data recorded at every temperature between 5 K and 270 K. The probability-distribution-function (PDF) histograms of standard deviations of purely Gaussian NMDs, σ, calculated using the Fabada algorithm for carbon NCS data recorded at 5 different temperatures, are shown in Fig. 3. The inset of Fig. 3 shows carbon NMD recorded at 5 K (blue trace) together with the fitted Gussian NMD curve (red dashed trace). The values of the standard deviations of purely Gaussian NMDs, σ, calculated using the Fabada algorithm, are listed in Table 1 and denoted as "Experiment". These values are compared in Table 1 with lattice dynamics (LD) DFT predictions, calculated using Eq. (1) from the output of the DFT-based phonon calculation [6] and denoted as "DFT", and the values obtained from the recursive method, denoted as "Recursive".

Table 1. The values of the standard deviations of purely Gaussian NMDs, σ, calculated using the Fabada algorithm for carbon NCS data recorded at 5 different temperatures, denoted as "Experiment", are compared with LD DFT predictions, calculated using Eq. (1) and denoted as "DFT", and the values obtained from the recursive method, denoted as "Recursive". All values of NMD widths are given in $Å^{-1}$ and the values of temperature in K. See text for further details.

T [K]	5	80	100	250	270
DFT	12.63	12.65	12.66	12.85	12.91
Experiment	12.8 ± 0.2	12.9 ± 0.2	13.0 ± 0.3	13.4 ± 0.3	13.4 ± 0.2
Recursive	12.07	12.08	12.10	12.46	12.54

In order to assess the relative contributions of vibrational, translational and librational contributions to carbon kinetic energies in solid C_{60}, in what follows we will resort to the model

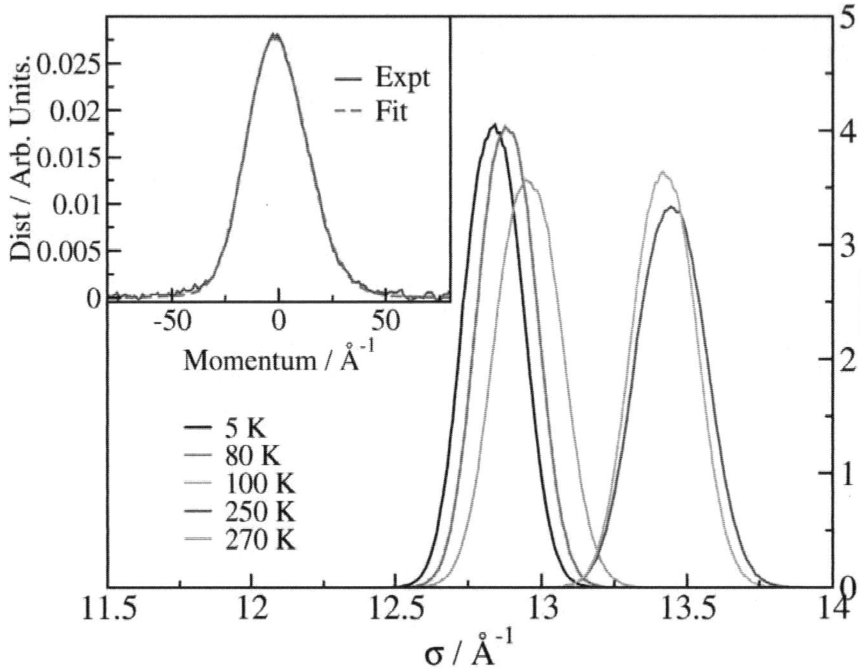

Figure 3. Inset: carbon NMD recorded at 5 K (blue trace) together with the fitted Gussian NMD curve (red dashed trace). Main: PDFs of standard deviations of carbon NMDs, σ, calculated using the Fabada algorithm for carbon NCS data recorded at 5 different temperatures. See text for further details.

of the nuclear kinetic energy for a nucleus in a molecular crystal, introduced by Colognesi et al. [10, 11]. It is worth noting that the model can be employed in two independent routes:

- In the first route, all components of the S-T tensor, R_{ij}, are computed based on the optimised geometry of the molecule obtained from the *ab initio* calculation, followed by the calculation of the rotationally averaged value $\langle\sigma^2_{S-T}\rangle$ from Eq. 5;
- In the second route, the rotationally averaged contribution due to internal vibrations, $\langle\sigma^2_{intra}\rangle$ from Eq. 7, is subtracted from the experimental value of the total second moment obtained from fitting the NMDs using the isotropic Gaussian model, denoted as σ^2, yielding $\sigma^2 - \langle{\sigma_{intra}}^2\rangle$.

Physically, the first route amounts to a classical rigid-body calculation for uncorrelated translational and hindered rotational motions of entire C_{60} molecules. Contrary to the first route, the second one does not assume any uncorrelated rigid body rotation and translation. It assumes, however, the lack of correlation between the intramolecular vibration of carbon nuclei in C_{60} molecules and other modes contributing to the total nuclear kinetic energy of carbon. In the case of the solid C_{60} under investigation, the intramolecular vibration contribution to carbon kinetic energy was obtained based on existing results from a periodic-DFT LD calculation in the CASTEP code obtained, in light of practically dispersion-less phonon spectra, by diagonalisation of dynamical matrices at zero wave-vector (Γ-point-LD calculation) [6].

The STD values of NMDs, resulting from of both routes of calculation performed for the C_{60} as a function of temperature, denoted as $\sqrt{\langle{\sigma_{S-T}}^2\rangle}$ and $\sqrt{\sigma^2 - \langle{\sigma_{intra}}^2\rangle}$, respectively, are plotted in Fig. 4.

The first striking feature evident in Figure 4 is that the prediction for the sum of the

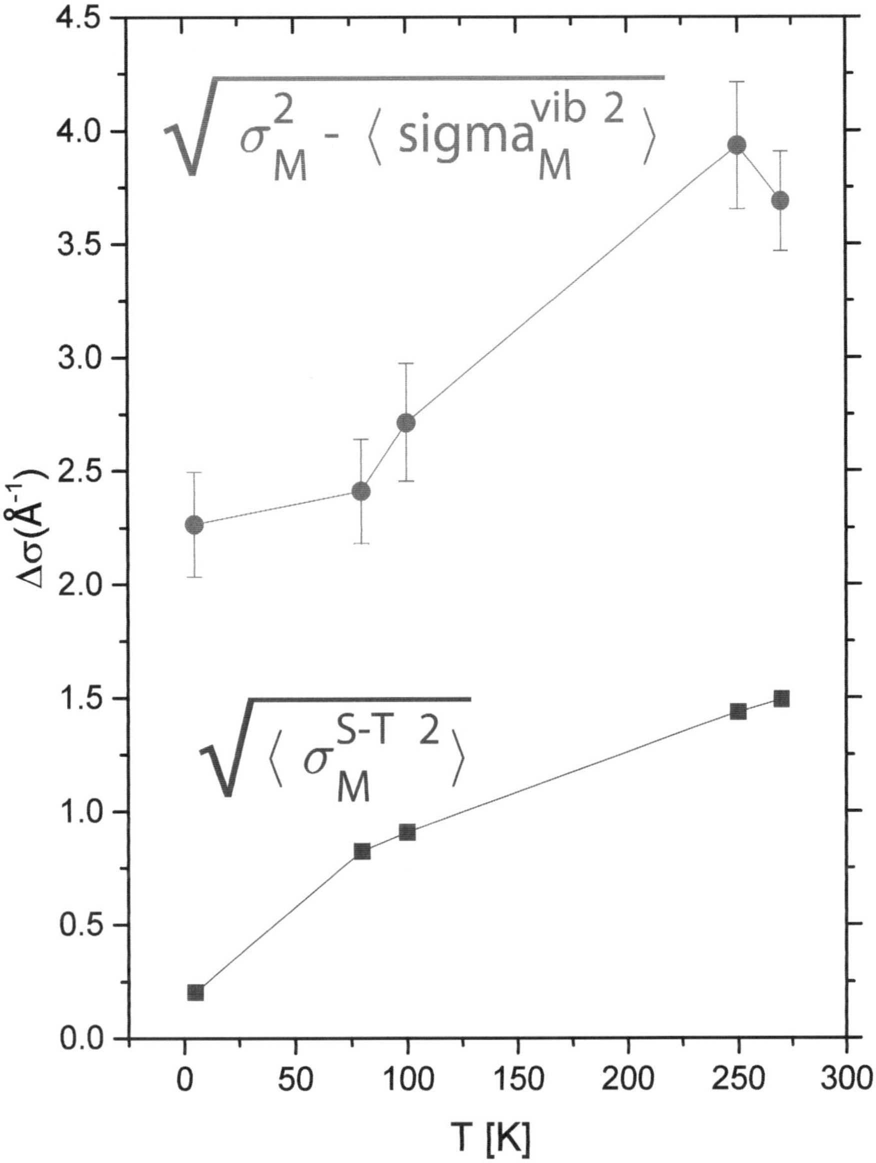

Figure 4. Comparison of the results of two routes to calculate the prediction of the sum of translational and rotational contributions to the STD of the NMD of a carbon nucleus in solid C_{60}: first route – blue trace, second route–red trace. See text for further details.

translation and hindered rotation contributions to the STD of carbon NMD in C_{60}, calculated based on the rigid-body model using the S-T tensor, is systematically lower than the prediction based on the difference between the experimental STD value and the rotationally averaged vibrational contribution, obtained using *ab initio* periodic-DFT approach. In other words, the experiment seems to account for more kinetic energy stored in translations and hindered rotations than the ST model predicts. This significant discrepancy between the results of both routes of the calculation becomes more intuitive in the light of the following remark. The way in which the rigid-body prediction for the nuclear kinetic energy of carbon due to translation and hindered rotation of entire C_{60} molecules is calculated is reminiscent of the Translation-Libration-

Screw-rotation (TLS) model of rigid-body harmonic anisotropic displacement parameters (ADPs), introduced in crystallography by Schomaker and Trueblood [33]. The TLS method has a very well known deficiency. Namely, whereas it is successfully used to predict ADPs of heavy atoms, it is known to fail for the calculation light-atom ADPs based on the extrapolation of heavy-atom ADPs using the librational and translational motion tensors of the same molecule [34,35]. The reason for this failure is that light-atom ADPs depend more on medium-to-high frequency modes. Thus, more sophisticated schemes, including the TLS+ONIOM method, are necessary, as they combine the rigid TLS model with *ab initio* computed estimates of the contribution to ADPs from internal degrees of freedom [35].

However, the LD-inspired approach is likely to fail in the case of lightweight atoms as a result of the classical nature of the treatment of the nuclear degrees of freedom in the *ab initio* simulation, which ignores "non-trivial nuclear quantum effects" [33]. In the end, only path-integral molecular dynamics (PIMD) simulations are capable of capturing these effects, especially at cryogenic temperatures [36]. Moreover, any rigid-body-inspired method, augmented with classical *ab initio* calculations for the vibrational contribution due to internal degrees of freedom in a molecular crystal, will render itself much more useless with respect to the calculation of the shape of an NMDs than it is in the case of the ADP prediction using the TLS. The reason for this relatively larger inaccuracy in the case of the NMD calculation is that, from the point of view of LD, the ADPs can be termed as "minus-first moments of apVDOS", whereas the widths of NMDs can be termed as "plus-first moments of apVDOS" [3,37]. In consequence, the widths of NMDs are much more sensitive to medium-to-high frequency modes than the values of ADPs. Also, NQEs, including ZPE effects, may constitute another source of systematic errors for the LD-based predictions of NMD widths of lightweight nuclei, similarly as it is in the case of ADP predictions. NQEs have been shown to be important already at room temperature for nuclei as heavy as sulfur [38]. For oxygen, a nucleus one-third heavier than carbon, even at 750 K, the NQE contributions render its kinetic energy ca. 10% higher than the classical mean kinetic energy per atom [39]. Despite the relative importance of NQEs, even for medium-weight nuclei, the fully-pledged PIMD-based modelling of both ADPs and NMDs renders itself very often prohibitively expensive due to the fact that it requires the use of open-path PIMD [40,41]. Even less-accurate but more efficient methodology, based on the path integral generalized Langevin equation thermostat (PIGLET) [42], is beyond the scope of the present study.

However, despite the lack of a PIMD-based prediction, the qualitative appraisal of the dynamical behaviour of C_{60} molecules in the solid phase below room temperature, based on results shown in Figure 4, is still possible due to the fact that the NQE contribution to the carbon NMD is expected to be approximately constant over the entire range of temperatures and equal to the value in the limit of zero temperature. This zero-temperature-limit contribution can be assessed by noting that in the limit of zero temperature all rotation and translation of entire C_{60} molecules should cease. Moreover, in the limit of zero temperature, the ZPE contribution to the vibrational kinetic energy should be in principle accounted for by the LD-based calculation. Thus, the value of $\Delta\sigma$ obtained in the limit of $T \rightarrow 0$ K, equal to 2.25 ± 0.25 Å^{-1}, should represent an upper conservative bound of NQE contributions to the carbon NMD width. This assessment is further corroborated by the observation that $\Delta\sigma$ is, within one STD, constant between 0 K and 90 K (see red trace and data points in Figure 4).

Encouraged by the argumentation presented in the above paragraph and motivated by the assessment described in the literature [5], one can attempt to perform an appraisal of the structural behaviour in solid C_{60} between 90 K and 260 K in terms of the role played solely by reorientational motion of entire C_{60} molecules via hops between two energetically similar configurations corresponding, however, to one local and one global potential energy surface (PES) minimum. The first obvious remark to start with is that a rigid-body calculation, unlike a molecular dynamics-based one, does not explore the potential energy surface (PES) and thus the NMD widths obtained using the S-T tensor cannot be linked to any PES minima. However, the picture widely accepted in the literature [5] is that at high temperatures, T > 260 K, C_{60} molecules are able to perform nearly isotropic free rotation about their lattice sites. Below 260 K, C_{60} rotations are restricted to hops between the two favourable orientational positions. With increasing temperature, the hopping frequency increases while also shifting the population of molecules in the global PES minimum to the local one. David *et al.* [5] put forward a model of uniaxial reorientation of entire C_{60} molecules principally about the (111) direction via 60° hops between two energetically similar configurations that involve the facing of either a pentagon or a hexagon opposite a carbon-carbon double bond of a neighbouring C_{60} molecule (designated P and H configurations, respectively). They considered a more appealing description of the reorientation between these two configurations that involves smaller, ca. 36°, hops about the (110) directions that are straddled by carbon-carbon double bonds [5]. In this model, any anticlockwise rotation of ca. -36° about [101] returns the molecule to a new P configuration and any clockwise rotation of ca. 36° along [011] results in a new H configuration. The resultant pseudo-random sequence of uniaxial reorientations has a quasi-isotropic nature when averaged over time, whereby all 60 degenerate P and 60 degenerate H configurations may be accessed [5]. Finally, below 90 K all rotational motion ceases to exist, and the system becomes trapped in a given orientational ordering. Interestingly, there is a kink in the residual value of the carbon NMD width at 90 K, clearly seen in Fig. 4, followed by a linear trend continuing until 260 K, the temperature at which C_{60} molecules are able to perform isotropic free rotation about their lattice sites. This kink might signal an onset of the quasi-isotropic hopping motion, described in the model by David [5], that cannot be accounted for within the sample rigid-body S-T-based calculation.

A molecular dynamics simulation of the C_{60} crystal lattice in the face-centred-cubic structure under progressively raised pressure was performed by Baameur *et al.* [43], to illustrate the limits of the crude pseudo-atom approach and the validity of the potential in the rigid molecule, when seeking for possible phase transitions. Baameur achieved a good description of C_{60} crystal properties using the Cheng and Klein potential in the low-pressure region at 5 K. Furthermore, the carbon apVDOS was worked out from the Fourier transform of the Velocity Auto Correlation Function (VACF). The obtained apVDOS had a peak at 15 cm^{-1} (1.86 meV). Converting this value to into STDs of carbon NMD, calculated within the isotropic Gaussian NMD model, yields $\Delta\sigma$ equal to 1.94 Å^{-1} at 5 K, in very good agreement with values plotted in Figure 4. Thus, one can put the upper bound on the total amount of the nuclear kinetic energy of carbon due to molecular reorientation motion of entire C_{60} molecules via hops between two energetically similar configurations at 260 K by assuming that the whole difference between $\Delta\sigma$ at 260 K and 90 K is due to this type of hopping motion. This assumption yields the kinetic energy of the hopping motion of *ca.* 6 meV at 260 K. Classical constant-pressure molecular-dynamics simulations using a pairwise-additive atom-atom Lennard-Jones (LJ) potential have been demonstrated to provide an excellent account of the carbon apVDOS, whereas the quality of the reproduction of the librational modes, including the one of the high-temperature rotator phase of both solid C_{60} and solid C_{70}, was demonstrated to be less satisfactory [44, 45]. The librational phonon density of states for a triclinic lattice of solid C_{70} at 100 K, calculated using classical molecular dynamics (CMD) approach with a LJ potential [44], peaks at the kinetic energy of *ca.* 11 cm^{-1} (1.36 meV), thus corresponding for isotropic Gaussian NMD model, to $\Delta\sigma$ equal to 1.66 Å^{-1} above the zero-temperature-limit value of 1.94 Å^{-1}, attributed to translational motion only. Applying a classical rigid-body ST-inspired correction to account for the difference in size of C_{60} and C_{70} molecules, one comes up with $\Delta\sigma$ equal to $\sqrt{\frac{60}{70}}1.66 = 1.54$ Å^{-1}. Thus, the total carbon NMD width at 100 K should yield $\sqrt{1.66^2 + 1.94^2} = 2.48$ Å^{-1}, which is in very good agreement with the $\Delta\sigma$ calculated from the NCS experiment. On the whole, the classical MD calculations using the LJ potential for the solid C_{60} yield satisfactory agreement with experiment. The values of librational energies lie between 1.3 and 2.5 meV, compared to the observed range of 1.8 to 4.0 meV [45–47]. The bond-charge model proposed by Spirk *et al.* [48] yields phonon-dispersion curves with the highest energy phonon of 5.3 meV, as compared to the highest value observed experimentally of 6.8 meV, with the calculated librational energies between 0.9 and 2.2 meV. This is also seen in the calculation of the carbon apVDOS at 100 K, which shows a librational peak at ca. 13 cm^{-1} (1.6 meV), as compared to the experimental value of 2.7 meV [48].

On the whole and quite surprisingly, there have not been not many studies in the literature of a detailed characterisation of individual contributions to the nuclear kinetic energy of carbon in solid fullerides as a function of temperature. NCS offers a novel route in this respect. The procedure proposed here involving the dissection of the total carbon kinetic energy into contributions due to translation, libration, and vibration, leads to results, shown in Figure 4, that account very well for the distinction between the translational and librational modes of solid C_{60} between 5 K and 260 K. The obtained results correctly reproduce the librational contribution and its temperature dependence due to the reorientational motion of entire C_{60} molecules via hops between PES minima between 90 K and 260 K. However, due to the fact, that in many studies, the theoretical predictions of the reorientation motion are based on MD simulations results, no phonon-dispersion relations for translations and librations can be computed. Thus, more detailed LD-based simulations are necessary to provide the benchmarking against the results of INS and NCS experiments. Phonon dispersion curves and the nuclear kinetic energies obtained from such studies could be, in principle, capable of discerning different hopping scenarios. Thus a combination of an INS and NCS experiments, performed over a wide range of temperatures, and contrasted with the results of *ab initio* predictions, could lead to a better understanding of the role of the hopping motion of entire C_{60} molecules in the progressive dynamical transition of solid C_{60} towards quasi-isotropic rotation above 260 K. Moreover and most importantly, PIMD, or less computationally-demanding PIGLET calculations, could provide a route towards the assessment of the role played by the NQEs in the carbon dynamics in this temperature range. However, the difficulty due to the lack of the full picture of nuclear dynamics in this important carbonaceous system, caused by the challenging computational requirements of the PIMD methodology, creates an opportunity for the geometrical analysis at the nanoscale to provide further insight, as will be demonstrated in the following section.

4. Nuclear kinetic energy and fractal structure of carbonaceous systems

The theoretical difficulties that are faced by the *ab initio* modelling of dynamical transitions in C_{60}, described in the section above, motivate even further to find correlations between the values of the kinetic energy of a given nucleus and the degree of complexity and ordering of the structural motif characterising the surrounding environment felt by this particular nucleus. The overarching aim here is to predict, based on purely geometrical signatures of a given system, the value of the nuclear kinetic energy and avoid the expensive and complicated *ab initio* simulations altogether. In this spirit, in what follows we will analyse both past and present NCS results on carbon systems. In doing so, we will order them according to the increased geometrical dimension. To this end, motivated by recent methodological advances in materials science at the nanoscale, we will attempt to categorise system properties under study via application of the concept of the Hausdorff fractal Dimension (HD) [49]. HD serves as a measure of the local size of a space, taking into account the distance between its points. For the dimension of fractals, HD is based solely on its properties of scaling and self-similarity, and gives a non-integer result. This concept, widely used in topological analysis in mathematics, is also widely applied in many areas of physical chemistry, with theoretical and experimental studies of diffusion, percolation, phase transitions, bonding to name but a few [50–52].

HD analysis has already been performed successfully in the case of carbon systems at the nanoscale, concretely in the description of conductance fluctuations in high mobility monolayer graphene [51]. This study is a perfect example of the fact that the HD related to the geometry of the system may, in general be different from the HD attributed to a physical process that takes place in this system. Concretely, for a typical graphene sample fabricated on a SiO_2 substrate the authors quote HD of 1.3. However, the Higushi algorithm, applied to compute the HD dimension of conductance fluctuations, yields a HD value of ca. 1.2 at 0 K to ca. 0.8 at 30K [51]. Moreover, unlike its geometrical counterpart, the HD of the conductance fluctuations exhibits a progressive increase for low temperatures. The authors attribute these increase to the fact that the conductance fluctuations are not purely random but have a statistical fractal structure associated with an increase in the complexity of the mechanisms that create them [51]. Above approximately 10 K, they observe a very simple, nearly binary, mechanism, whereas for lower temperatures the fluctuations become more complex approaching the HD of a pure Wiener process (HD = 1.5).

As noticed by Zhang *et. al.* [52], macroscopic geometrical methods can hardly describe events at the nanoscale and there is an upsurge of studies employing nano-geometry as an effective mathematical tool for exploring nanoscale physical and mechanical phenomena. To this end, the atomic lattice structure of perfect single-layer graphene can actually be regarded as a kind of hierarchical fractal structure [52]. The interplay between geometrical parameters and the properties of graphene has been shown in a number of recent studies. Substrate-induced sub-lattice symmetry breaking, giving rise to energy gaps in electronic structures [53], strain causing spatial modulations in the local conductance [54], local uniform magnetic field creation that exceeds 10 Tesla in magnitude [55], and curvature-induced changes in heat and electrical conductivities [56] are to be mentioned but a few. Contrary to the work of da Cunha *et. al.* [51], employing the Higushi algorithm, Zhang *et. al.* [52] found, quite surprisingly, that the outer boundary curve of graphene has a fractal dimension equal to one, quite differently from other normal curves whose fractal dimensions vary from one to two. Moreover, a multi-fractal structure was found comprising a series of strict self-similar hexagons comprised a rotating fractal set slewing at a constant counterclockwise angle of 19.1 degrees when observed from one level to the next higher level. The interior fractal structure was found to have HD of two.

For NQEs, as observed by the NCS, HD analysis may aid the analysis of correlations between the bonding topology and nuclear kinetic energy. To this end, the presented work draws from the pioneering work exploring the use of NCS in the investigation of H_2 adsorption in KC_{24}, a nanoporous graphite intercalation compound (GIC) [57]. This relatively simple exercise changed the way in which the binding scenarios are portrayed in adsorption studies. Namely, it helped identifying those modes which contribute the most to the binding, as measured relative to completely unbound case of two physically separate systems before the physisorption has taken place.

The link between NCS observables and the HD of graphene can be perhaps best emphasized by first resorting to the HD calculation algorithm. In their work on graphene Zhang et. al. [52] use the mass-radius method, which is based on the relationship of the mass distribution of an object with its radius. The nuclear-mass distribution can be a very good approximation of the graphite HD. Owing to the simple property of the Fourier transform that links position and momentum distribution, the link between NMDs and HD becomes straightforward. As the fractal is constructed by means of step-by-step paving of uniform hexagons, more and more contributions to nuclear kinetic energy of a given nucleus from the nearest and next-nearest neighbours is added. Thus, a natural hierarchical mapping is being established between the geometry of the binding motif and the nuclear kinetic energy involving the same nucleus. Moreover, the dissection of the total nuclear kinetic energy into translations, librations and vibrations adds more degrees of freedom into this mapping. Classically, this mapping is a mere consequence of the the moment of inertia corresponding to a given mass distribution. Quantum mechanically, however, the bonding strength and topology may complicate the picture and render

IOP Conf. Series: Journal of Physics: Conf. Series **1055** (2018) 012007 doi:10.1088/1742-6596/1055/1/012007

the self-scaling property of the interaction strength, typical for classical fractal structures, less apparent in practice.

Encouraged by the NCS work on GICs [57] and the success of the HD analysis in graphene, we will proceed to categorise NCS results by the HD dimension starting with a system of the lowest dimensionality in this respect, graphene and its compounds. Nemirovsky *et al.* [58] have measured the kinetic energy of He atoms adsorbed on activated carbon fibres at sub-monolayer coverage. Even though the authors do not quote any value of the carbon kinetic energy, digitising their data recorded for backscattering detector located at the scattering angle of 142 degrees, shown in Fig. 1 of Ref. [58], and fitting it with a purely Gaussian NMD model yields the carbon NMD width, σ_C of 14.0 $\pm$ 0.5 Å^{-1}. This value is obtained with relatively large standard deviation error due to the fact that care had to be taken to provide an upper conservative bound for data that was not obtained through the direct recording. Namely, as Nemirovsky *et al.* point out [58], for the active carbon fibre sample, ACF3000, the surface area was larger than the maximum theoretical limit of a graphite sheet assumed to consist of a single graphene layer adsorbing on its two sides. It was found to be due to the fact that the c-spacing was slightly larger than that of graphite due to lack of alignment of the graphitic layers. ACF pores could thus still be modelled as consisting of three-walled infinite graphitic sheets because the size of the crystallites of graphite are larger than 20 nm and the number of parallel sheets in each crystallite is far larger than that constituting the ACF. Also, it was shown through the analysis of NCS data obtained for He that a small (of the order of few percent) difference in distance between the graphitic layers between pure graphite and the ACF had no influence on the calculated He kinetic energies [58].

The next interesting quasi-2D carbon system investigated by means of the NCS method was the GIC [57]. Geometrically, GICs are characterised by a well-defined composition and regular structure [59,60]. Concretely, in the case of KC_{24}, investigated in the above mentioned work [57], below T=100 K, the system adopts a commensurate $\sqrt{7} \times \sqrt{7}$ M super-lattice bounded by a higher metal density in the inter-domain regions. These topological constraints have interesting consequences for the magnitude and symmetry of the carbon NMDs obtained from the NCS data analysis. For carbon atoms, the NMD, measured below 5K, exhibited cylindrical symmetry about the c-axis, yet the carbon NMD anisotropy was reversed compared to hydrogen. The value of the carbon NMD width of 13.74 Å^{-1} for in-plane bonding in the GIC under study was found to be very similar to the experimental value of $\sim$ 14.0 Å^{-1}, measured by Finkelstein *et al.* [61] for the base graphite material, registered under the trademark name as Papyex.

Interestingly in the context of the work of Nemirovsky *et al.* [58] on the ACF carbon, the average carbon out-of-plane width of 8.23 Å^{-1} in the GIC was significantly smaller than $\sim$11 Å^{-1} in pristine graphite [57]. Thus, unlike in the ACF the carbon NMD width was found to be quite sensitive to a decrease of the interaction between graphite planes, now separated by 5.6 Å in comparison to 3.35 Å in graphite. Such a reduction in momentum width was rationalised as a consequence of the softening of out-of-plane carbon vibrations by a factor of $(13.74/8.23)^2 \sim 2.8$ and was is in good agreement with Raman data on potassium-doped graphene [62].

The next carbon system to be mentioned in the order of dimensionality progression is C_{60}. This molecule, enjoying almost iconic status in chemistry, has a cage-like fused-ring structure (truncated icosahedron) that resembles a football, made of twenty hexagons and twelve pentagons, with a carbon atom at each vertex of each polygon and a bond along each polygon edge. The closest topological cousin of C_{60}, an icosahedron flake, or Sierpinski icosahedron, is formed by successive flakes of twelve regular icosahedrons. Each flake is formed by placing an icosahedron scaled by $\frac{1}{1+\phi}$ in each corner. The HD of C_{60} is equal to $\frac{\log 12}{\log(1+\phi)} = \approx 2.5819$.

On the whole, the quantum nature of the chemical binding renders the correlation between the carbon NMD widths and their HD not very clear. This lack of apparent correlation is additionally caused by discrepancies arising in the actual HD values for given systems. These discrepancies are the consequence of employing different algorithms to compute the HD. Also, sometimes simplifying assumptions are used in the calculations of HD for given physical structures. For example, in fact, C_{60} is a truncated Sierpinski icosahedron and a separate calculation of the HD must be carried out that does not simply use the exact icosahedron symmetry as a starting point. Moreover, different HD calculation algorithms rely on different pattern-recognition methods with different accuracy and image edge detection thresholds.

To avoid problems related to the HD calculation mentioned above, in this work we have calculated the HD using the same computational protocol for all systems under investigation, employing the box-counting method to compute the HD. Box-counting is a method of quantitative image analysis of perimeter roughness, routinely applied in the field of image analysis to characterise the degree of roughness of input images. The box-counting algorithm gives the aggregate perimeter roughness which implies parasitic emission sites for extremely rough perimeters [49]. On the HD scale, a dimension of one equates to a smooth line, while an HD of two implies fractal complexity like that of a Julia set [49]. In other words, in the box-counting method, the number N of boxes of size R needed to cover a fractal set follows a power-law, $N \sim R^{-DF}$, with $DF <= D$ (D is the dimension of the space, usually D=1, 2, 3). Because fullerenes are considered to be categorised as truncated fractals, the fractal dimension calculated for this family of compounds is bound by the above limits, *i.e.,* $1 <$ H.D. < 2.

The box-counting algorithm, implemented in Matlab [63], was used here. The algorithm started with an input image to which the Canny algorithm [64] was employed to find the edge within the image and superimposes a grid of N squares over the edge, while counting the number of occupied squares that the edge passes through, $N(s)$. This procedure was continued for an increasing number of squares and the HD was given by the gradient of the logarithm of the number of squares $\log N$, over the number of squares occupied by the edge $\log N(s)$, as indicated by the equation 14:

$$HD = \frac{\log N}{\log N(s)} \tag{14}$$

In what follows we will present the results of the application of a 3D-variant of the box-count algorithm, employing a protocol that can be summarised in three simple steps. First, the geometry-optimised structure, corresponding to the unit cell used as an input for the *ab initio* program solving the electronic structure problem and computing apVDOS within DFT, was saved in XYZ format with atomic coordinates expressed in units of Å. Then, a conversion was performed from XYZ format to 3D numeric Matlab variable format. Subsequently, the 3D matrix variable served as input for the 3D box counting variant of the HD calculation algorithm, `boxcount` [65]. The whole protocol was repeated by adding subsequent unit cells to the initial structure until a given asymptotic size of the super-cell was obtained, for which adding more unit cells to the super-cell did not result in any further change of the HD value.

Applying this algorithm to the individual image of C_{60} yielded, for the case of the C_{60}, HD= 2.807. The results of the application of the algorithm are shown in Figure 5. Repeating the above analysis for the case of the GIC system was performed using the original atomic coordinates from the geometry-optimised structure [57], shown in the inset of Figure. 6. The HD analysis yielded HD=1.585. Finally, applying the same protocol for the graphene structure in Fig. 7, we obtained HD=1.585, a value slightly above the one obtained using the Higuchi method 1.3 [51].

Our next task, following the HD analysis, was the finding right figures of merit that would support correlations between the HD dimension and NMD shape. Drawing from the experience of the NCS work on cesium hydrogen sulfide [38], the first candidate figure of merit is the Quantum Energy Excess (QE), defined as the ratio of spatially averaged nuclear kinetic energy from an NCS experiment, $\frac{3\hbar^2\sigma^2}{2M}$ and its classical counterpart, based on the Maxwell-Boltzmann

IOP Conf. Series: Journal of Physics: Conf. Series **1055** (2018) 012007 doi:10.1088/1742-6596/1055/1/012007

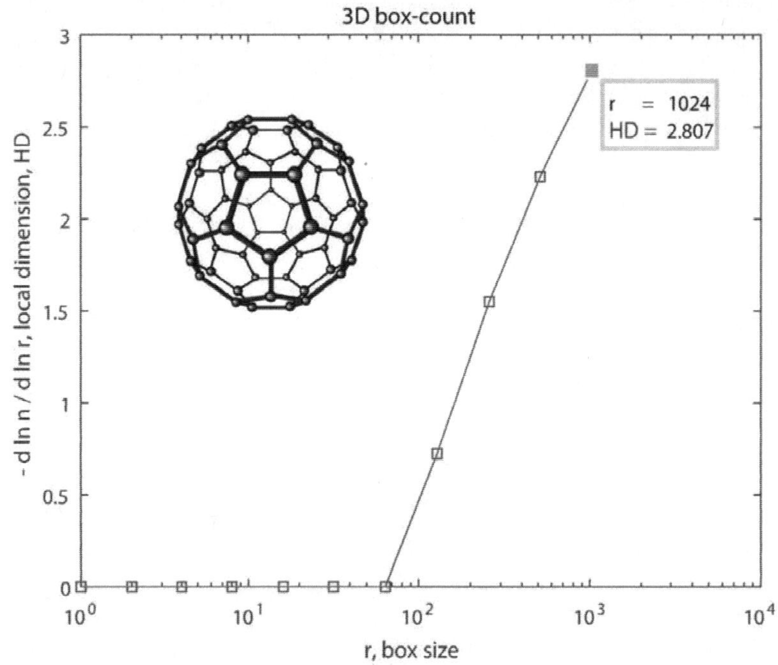

Figure 5. The results of the application of the box counting method to calculate the HD of C_{60}. See text for details.

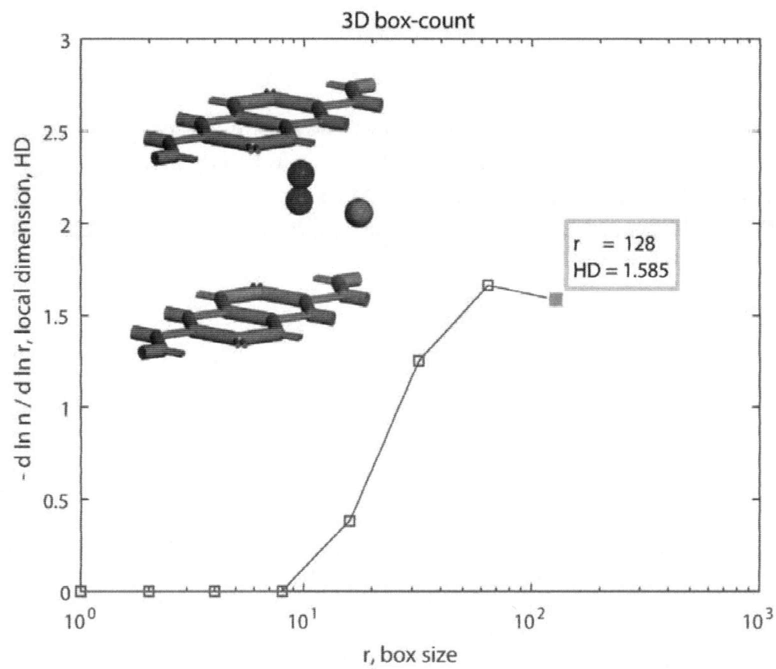

Figure 6. The results of the application of the box counting method to calculate the HD of the GIC system, KC_{24}. See text for details.

IOP Conf. Series: Journal of Physics: Conf. Series **1055** (2018) 012007 doi:10.1088/1742-6596/1055/1/012007

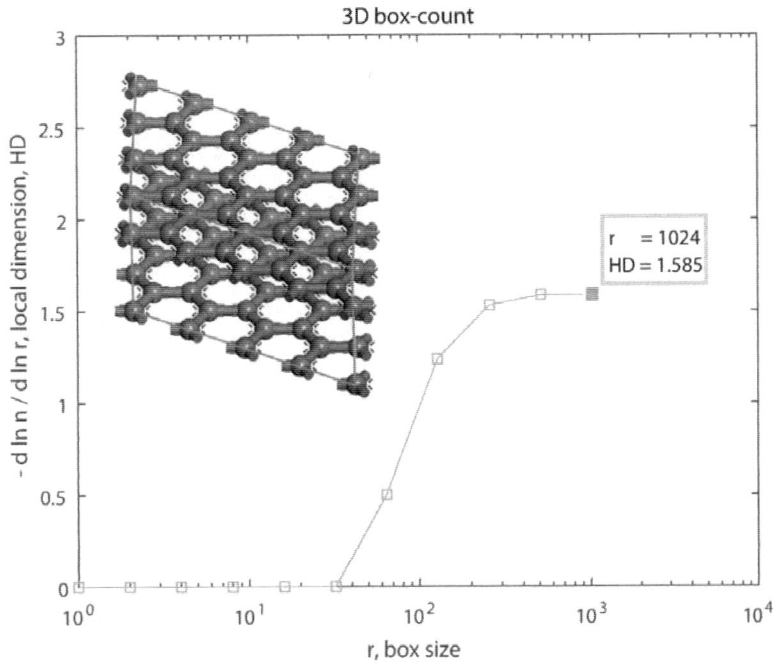

Figure 7. The results of the application of the box counting method to calculate the HD of the graphene. See text for details.

kinetic energy distribution, $\frac{3kT}{2}$

$$\mathrm{QE} = \frac{\hbar}{MkT}\sigma^2 = \frac{\sigma^2}{\sigma_{cl}^2}. \tag{15}$$

As discussed [38], the values of QE of unity provides the limit that corresponds to the situation of no chemical binding of an atom and no many-body correlations present in the system under consideration. Mathematically, this limit can be expressed using the classical limit value of the STD of an NMD, $\sigma_{cl} = \sqrt{MkT}/\hbar$: Consequently, any value of σ above σ_{cl} and any value of QE above the classical limit of one, obtained from an NCS experiment, signifies some degree of quantum chemical binding and associated atom confinement in given system under investigation [2, 3, 37].

Values of σ and σ_{cl} and QE, calculated for C_{60} at different temperatures, are shown in Table 2. Inspection of the QE values listed in Table 2 leads to the observation that a considerable scattering of QE values across various temperatures is observed. One possible reason behind such scattering of QE values is the fact that QE has been defined as the figure of merit characterising the degree of "quantumness" of a given condensed-matter system by normalising the quantum nuclear kinetic energy measured in an NCS experiment by the value of the corresponding classical counterpart based on the Maxwell-Boltzmann kinetic energy distribution which obeys the classical energy equipartition amongst the classical dimensions in space [38]. The consequence of such definition is that the QE magnitude is simultaneously affected by both the quantum nature of chemical binding and the structural motif characterising the surrounding environment felt by a particular nucleus.

Table 2. First-principles predictions for NMD observables of C in C_{60} at various temperatures. The columns correspond to the classical limit of σ and QE, as discussed in the text.

T [K]	σ [Å^{-1}]	σ_{cl} [Å^{-1}]	QE
5	12.63	1.112	129
80	12.65	4.449	8.08
100	12.66	4.974	6.48
250	12.85	7.865	2.67
270	12.90	8.174	2.49

Motivated by the fractal dimension analysis presented above, one can introduce a new figure of merit, QE_H, the quantum kinetic energy excess of a given nucleus in a system characterised by a given value of fractal dimension

$$QE_H = \frac{HD}{3}\frac{\sigma^2}{\sigma_{cl}^2} = \frac{HD}{3}\frac{\hbar^2}{MkT}\sigma^2 \tag{16}$$

In the definition, given by Eq. (16), the factor $HD/3$ provides the necessary normalisation by expressing the quantum energy excess per "effective degree of freedom". Such normalisation should lead to the smoothing of the correlation between the quantum energy excess and the fractal dimension when comparing a family of condensed matter systems within the same chemical binding motif (for instance, carbonaceous systems characterised by the sp^2 C hybridisation, whereby the C atom is attached to 3 groups and so is involved in 3 σ bonds).

As the consequence of the introduction of a new figure of merit, QE_H, one can also define a QE_H-derived observable, the STD of NMD normalised by the 'effective degree of freedom', σ_H:

$$\sigma_H = \sqrt{\frac{HD}{3}}\sigma_{cl} = \sqrt{\frac{HD}{3}}\frac{\sqrt{MkT}}{\hbar} \tag{17}$$

The values of the new figures of merit, QE_H, and σ_H, are listed in Table 3. One can see that the scaling of σ_{cl} and QE, by HD=2.807, leads to a slightly more pronounced quantum character of the carbon momentum distribution in C_{60}, especially below 100 K. The reason for this temperature effect is that at cryogenic temperature all motion ceases and the fractal dimension is expected to be dictated only by the system geometry, not the symmetry of the molecular motion. Moreover, the effective system dimensionality is reduced reduced by the HD reflecting the fact that solid C_{60} can be thought of a quasi-2D system of folded 2D sheets. Thus, the nuclear kinetic energy of carbon in solid C_{60} is closer to the value exhibited by a 2D-carbon system like graphene.

On the whole, the calculation of the fractal dimension, HD, for a family of quasi-2D carbonaceous systems, all characterised by the sp^2 C hybridisation, proves to be largely insensitive to the degree of complexity and ordering of the structural motif characterising the surrounding environment felt by the carbon. Chemically, this property may be dictated by the presence of strong C=C bonds which largely dominate both the energetics and anisotropy of the bonding. For instance, the fact that in GIC the graphene layers are intercalated with potassium and the hydrogen molecules are pinned, forming a preferential T-shaped geometry with respect to the graphene layers [57] does not seem to change the overall HD of the system markedly. Moreover, in order to corroborate this result, we have re-calculated the HD for graphene layers,

Table 3. First-principles predictions for NMD observables of C in C_{60}, σ, σ_H, and QE_H, calculated between 5 K and 270 K. σ_H and QE_H correspond to σ_{cl} and QE, scaled by the HD=2.807. See text for further details.

T [K]	σ [Å^{-1}]	σ_H [Å^{-1}]	QE_H
5	12.63	1.058	142
80	12.65	4.233	8.93
100	12.66	4.732	7.16
250	12.85	7.482	2.95
270	12.90	7.776	2.75

by systematically increasing the lattice spacing along the crystallographic c-axis from the value characteristic of pure graphene to the value which is larger by 20%. The only effect visible was a slight increase in the local maximum (a kink) on the penultimate point in the graph showing the result of the HD calculation, with the asymptotic HD value that remained unchanged. Calculating the QE_H figure of merit for the ACF graphite system measured by Nemirovsky *et al.* [58] and to those obtained by Finkelstein *et al.* [61] for Papyex, using HD=1.585, we obtain QE_H of 5.03. The average value of carbon NMD width for the GIC, taking into account in and out of the plane components, yields 12.14 Å^{-1}. However, the GIC measurement was performed at 5 K and thus σ_{cl} = 1.112 Å^{-1}. This yields the value of QE_H of 225 for the GIC.

The final carbon structure that remains to be examined is a pure bulk 3D system. As an example of such structure we will take the pyrolytic (oriented) graphite (PG). The PG structure and the HD analysis of it are shown in Figure 8. A value of 13.6 Å^{-1} was obtained by Fielding *et al.* for the PG graphite using NCS at 300 K [21]. Applying the HD renormalisation to carbon NCS data on bulk carbon systems we have to use the HD for the PG, being HD=2. Thus, for the PG we get the value of QE_H of 3.76.

Having collected figures of merit for a number of carbonaceous systems one has to consider which figure of merit is most sensitive to the fractal dimension. For this, the following remarks are in order. Firstly, the HD changes with temperature. These changes are due to symmetry changes induced by lattice expansion and as well as due to phase transitions.

Moreover and most importantly, however, the effective fractal dimension of a given system is the result of the interplay of the fractal nature of the system geometry and the motion that the system is undergoing at a given temperature. This last remark holds especially in case of random or diffusive motion like Brownian motion or random walk with no self-intersection, calculated to have HD=2 and 1.55, respectively [66]. Practically all types of diffusive motions have their fractal signatures [67]. Thus, the only way to disentangle the motional and geometrical contributions to the fractal dimension is to perform spectroscopic measurements at cryogenic temperature where all classical motion ceases and only ZPE effects are present. Thus, the figure of merit sought for must be temperature-normalised or calculated in the limit of no classical motion. For the same reason, the scaling cannot involve the classical prediction for the widths of NMDs, σ_{cl}, as it already contains the temperature scaling. Moreover, σ_{cl} is only an approximation based on the equipartition theorem. Taking all those constraints into account, let us see if the system fractal dimensionality, expressed via HD, correlates with the QE observable that is normalised by the classical temperature and in principle enables to compare kinetic energies of different systems at different temperatures.

The values, HD and QE, calculated for all carbonaceous system considered, are listed together in Table 4. It can be observed that, with decreasing system fractal dimensionality from nearly 3 towards nearly 1, the quantum nature of the system is getting more pronounced. Inspecting the

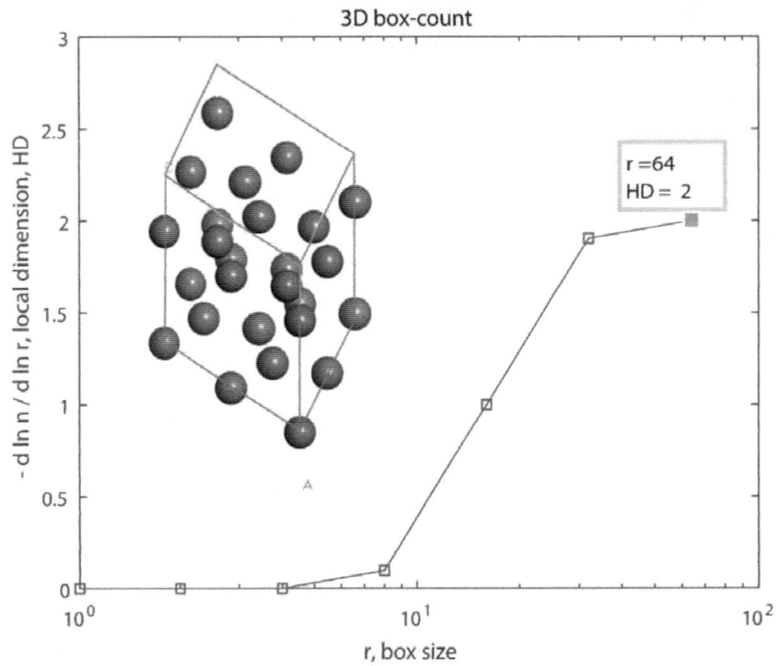

Figure 8. The results of the application of the box counting method to calculate the HD of the pyrolytic graphene. See text for details.

values listed in Table 4 one has in mind a picture in which, with increasing number of classical dimensions, subsequent bonds are added to sites, hence increasing the number of ways the kinetic energy of a quantum harmonic oscillator may be distributed over bonds. However, one has to seek an answer to the question how the quantum kinetic energy is redistributed over dimensions, not over bonds, modes or sites. To shed more light on this question, one may note a seminal work on the foundations of mesoscopic quantum physics by David Thouless and collaborators in the context of Anderson localisation [68, 69]. Thouless showed that any quantum system can be attributed a fundamental energy scale, referred as Thouless energy, E_T [70]. It is a measure of how sensitive the eigenstates in a quantum system are to a change in boundary conditions, thus also a criterion of the degree of quantum localisation of the wavefunctions in the whole system under investigation [68, 70]. Mathematically, the measure of the localisation, introduced by Thouless, is the participation ratio α which has its minimum value for a Bloch wave (the participation ratio is unity when all sites participate equally), and has its maximum value, unity, when the wavefunction is completely localized on one site [68]. Originally, Thouless applied his theory to wavefunctions describing electrons near the Fermi surface [68]. To this end, in an insulator, with localised eigenstates, the E_T is zero because if the wave functions do not extend to the boundaries their energies will be independent of boundary conditions [70]. Furthermore, Thouless showed in his seminal work [68] that for the 2D-square lattice the onset of localisation is reasonably sharp, contrary to the case of the diamond lattice where the onset of localization is less sharply defined, and localization occurs less easily than for the square lattice, but more easily than in Andersons theory.

Despite the fact that the Thouless theory was originally developed for electronic states, it can be applied to any eigenstates, including phonons, to characterise the degree of distinction

Table 4. The values of HD and QE calculated for all carbonaceous system considered at T=5 K. See text for further details.

system	HD	QE
graphene	1.3 –2	158.4
GIC	1.585	120.3
C_{60}	2.807	128.9
PG graphene	2	149.7

between systems with different dimensionality via corresponding apVDOS, $G(\omega)$. For phonons in D dimensions, $G(\omega) \propto \omega^{\frac{D}{\alpha}-1}$ and α=1, *as long as the dispersion relation remains linear and only one transverse mode or sub-band is present* [70]. However, in case of systems with non-integer fractal dimension, there is always some characteristic length scale which limits the extent of a given system either geometrically, or physically (*i.e.*, a characteristic length scale of spatial correlations along a given dimension). In consequence, this finite length scale introduces effectively a degree of localisation along a given dimension. Drawing from the intuition developed based on the study of the Thouless work on the electronic eigenstates [68] one can expect that this localisation will manifest itself more sharply in systems with HD between 2 and 3 than for bulky 3D systems. Thus, phonon localisation-induced, value of $\alpha < 1$, and the effective phonon density of states will scale faster than linear with the dimensionality of a system, as $\frac{D}{\alpha} > D$. In consequence, the first moment of the VDOS and hence the kinetic energy of a given nucleus in the lattice, will grow as $\omega^{\frac{D}{\alpha}}$, thus faster than the dimensionality. Applying a phonon picture one can anticipate that the shorter is the characteristic length scale the more pronounced is the bending of the phonon dispersion relations from a purely linear (acoustic phonon-like) towards more optic phonon-like. The consequence of this dispersion relation bending is the increase of the residual kinetic energy (mode energy in the limit of zero momentum transfer) of extra modes added to the density of states, and thus increase of the first moment of the apVDOS, the nuclear kinetic energy. The net effect of this may be the increased quantum kinetic energy excess in mesoscopic systems with intermediate (HD = 2–3) dimensionality. As mentioned above, the envisaged scaling law may be not universal, *i.e.*, it may be limited to the case of $2 < D < 3$. Moreover, trying to retrieve such relationship from a series of NCS and/or INS measurements may be a difficult task, as the overall trend may be subject to scattering due to system, not the dimensionality-specific diversity. To this end, applying the fractal dimension analysis to the INS and/or NCS results on a series of systems may help in the elucidation of an otherwise barely visible trend, as the fractal dimension-based scale is smoother than an integer-based scale.

5. Summary and Outlook

In this work we have presented a much-needed topological analysis of the nuclear kinetic energy of carbonaceous systems. NMD-inspired figures of merit, computed for a series of historic NCS results on graphene and graphite as well as new results on solid C_{60} system, correlate with the topological measure of surface roughness and bending, described quantitatively by the Hausdorff fractal dimension. The main result obtained in this study is that, with decreasing fractal dimensional from nearly 3 towards nearly 1, the quantum nature of the system is being more pronounced. An explanation of this trend is based on the analysis of the localisation of the phonon eigenstates and the consequences of this localisation for the magnitude of the first moment of the atom-projected vibrational density of states, the nuclear kinetic energy. Drawing from the seminal work on eigenstate localisation by Thouless, we conclude that this increased localisation should be manifested more as increased quantum kinetic energy excess in mesoscopic systems with intermediate (HD = 2–3) dimensionality. This conclusion has very important ramifications for the analysis of the nuclear quantum kinetic energy in a variety of condensed matter systems, including but not limited to, crystals, glasses, alloys, liquids, and, last but not least, biological systems.

Further work is necessary to understand the ramifications of the observed correlations. Topological analysis, aided by fractal-geometry analysis tools, may aid the mapping between the chemical bonding motifs and force matrix symmetry and sparsity, characterising the phonon density of states and hence the nuclear kinetic energy. Such topology-augmented analysis may help to find the smallest geometrical representation of a given system with the smallest number of force constants that can reproduce its vibrational properties within a given precision. In extreme cases of very sparse force constant matrices, extreme speed-up of the LD calculation can be achieved, rendering the *ab initio* calculation utterly unnecessary, as demonstrated by the success of the recursive method of phonon dispersion calculation in solid C_{60} system.

Further work on other families of condensed matter systems, chosen to exhibit the same chemical binding motif but different fractal dimensions, may shed more light on the geometrical scaling properties of vibrational densities of states and nuclear kinetic energy and may reveal hitherto unexplored links of these scaling laws to materials properties.

6. Acknowledgments

The authors gratefully acknowledge the UK Science and Technology Facilities Council for financial support and the ISIS Facility for access to beamtime.

[1] Fernandez-Alonso F and Price D L (eds) 2013 *Neutron Scattering – Fundamentals* (New York: Academic Press)
[2] Andreani C, Colognesi D, Mayers J, Reiter G F and Senesi R 2005 *Adv. Phys.* **55** 377
[3] Andreani C, Krzystyniak M, Romanelli G, Senesi R and Fernandez-Alonso F 2017 *Adv. Phys.* **66** 1–73
[4] Cazorla C and Boronat J 2017 *Rev. Mod. Phys.* **89**(3) 035003
[5] David W I F, Ibberson R M, Dennis T J S, Hare J P and Prassides K 1992 *Europhysics Letters* **18** 735
[6] Parker S F, Bennington S M, Taylor J W, Herman H, Silverwood I, Albers P and Refson K 2011 *Phys. Chem. Chem. Phys.* **13**(17) 7789–7804
[7] Rapcewicz K and Przystawa J 1994 *Phys. Rev. B* **49**(18) 13193–13196
[8] Jiang Q, Xia H, Zhang Z and Tian D 1992 *Chem. Phys. Lett.* **192** 93
[9] Krzystyniak M and Fernandez-Alonso F 2011 *Phys. Rev. B* **83**
[10] Colognesi D, Degiorgi E and Pace E 2001 *Physica B* **293** 317
[11] Andreani C, Degiorgi E, Senesi R, Cilloco F, Colognesi D, Mayers J, Nardone M and Pace E 2001 *The Journal of Chemical Physics* **114** 387–398
[12] Colognesi D, Degiorgi E and Pace E 2001 *Physica B* **293** 317
[13] Sears V F 1984 *Phys. Rev. B* **30** 44
[14] Watson G I 1996 *J. Phys.: Condens. Matter* **8** 5955
[15] Mayers J, Andreani C and Baciocco G 1989 *Phys. Rev. B* **39** 2022
[16] Mayers J 1990 *Phys. Rev. B* **41** 41
[17] Mayers J 1993 *Phys. Rev. Lett.* **71** 1553
[18] Andreani C, Colognesi D and Pace E 1999 *Phys. Rev. B* **60** 10008
[19] Mayers J and Abdul-Redah T 2004 *J. Phys.: Cond. Matter* **16** 4811
[20] Mayers J and Reiter G F 2012 *Meas. Sci. Technol.* **23** 045902
[21] Fielding A L, Timms D N and Mayers J 1998 *Europhysics Letters* **44** 255
[22] Krzystyniak M 2010 *J. Chem. Phys.* **133** 144505
[23] Krzystyniak M and Fernandez-Alonso F 2011 *Phys. Rev. B* **83** 134305
[24] Krzystyniak M, Druzbicki K, Romanelli G, Gutmann M J, Rudić S, Imberti S and Fernandez-Alonso F 2017 *Phys. Chem. Chem. Phys.* **19** 9064–9074
[25] Pinna R S, Rudić S, Capstick M J, McPhail D, Pooley D E, Howells G D, Gorini G and Fernandez-Alonso F 2017 *Nucl. Instr. Meth. A* **870** 79
[26] Pinna R S, Rudić S, Parker S F, Gorini G and Fernandez-Alonso F 2015 *EPJ Web of Conferences* **83** 03013
[27] Parker S F, Fernandez-Alonso F, Ramirez-Cuesta A J, Tomkinson J, Rudić S, Pinna R S, Gorini G and non J F F C 2014 *J. Phys.: Conf. Ser.* **554** 012003

[28] Oram E 2016 Simulation support and further indirect inelastic specific improvements within mantid Tech. Rep. RAL-TR-2016-019 Rutherford Appleton Laboratory
[29] `http://www.isis.stfc.ac.uk/instruments/vesuvio/`
[30] Pietropaolo A and Senesi R 2011 *Physics Reports* **508** 45
[31] Monserrat D, Vispa A, Pardo L C, Tolchenov R, Mukhopadhyay S and Fernandez-Alonso F 2015 *J. Phys.: Conf. Ser.* **663** 012009
[32] Romanelli G and Krzystyniak M 2016 *Nucl. Instrum. Meth. A* **819** 84 – 88
[33] Schomaker V and Trueblood K N 1968 *Acta Crystallographica Section B* **24** 63–76
[34] Willis B T M and Howard J A K 1975 *Acta Crystallographica Section A* **31** 514–520
[35] Whitten A E and Spackman M A 2006 *Acta Crystallographica Section B* **62** 875–888
[36] Reilly A M, Wann D A, Gutmann M J, Jura M, Morrison C A and Rankin D W H 2013 *Journal of Applied Crystallography* **46** 656–662
[37] Andreani C, Senesi R, Krzystyniak M, Romanelli G and Fernandez-Alonso F 2017 *Neutron Scattering - Applications in Biology, Chemistry, and Materials Science* (*Experimental Methods in the Physical Sciences* vol 49) ed Fernandez-Alonso F and Price D L (Academic Press) pp 403 – 457
[38] Krzystyniak M, Druzbicki K and Fernandez-Alonso F 2015 *Phys. Chem. Chem. Phys.* **17**
[39] Ceriotti M, Fang W, Kusalik P G, McKenzie R H, Michaelides A, Morales M A and Markland T E 2016 *Chem. Revs.* **116** 7529–7550
[40] Morrone J A, Srinivasan V, Sebastiani D and Car R 2007 *The Journal of Chemical Physics* **126** 234504
[41] Morrone J A, Lin L and Car R 2009 *The Journal of Chemical Physics* **130** 204511
[42] Ceriotti M, Bussi G and Parrinello M 2010 *Journal of Chemical Theory and Computation* **6** 1170–1180
[43] Baameur F, Krallafa A, Aourag H and Khelifa B 2000 *Physica Status Solidi (B)* **220** 821–828
[44] Cheng A and Klein M L 1992 *Phys. Rev. B* **46** 4958–4962
[45] Ehrenreich H and Spaepen F 1994 *Fullerenes* Solid State Physics (Elsevier Science) ISBN 9780080865126
[46] Lu J P, Li X P and Martin R M 1992 *Phys. Rev. Lett.* **68**(10) 1551–1554
[47] Lu J P, Li X P and Martin R M 1992 *Phys. Rev. B* **46**(7) 4301–4303
[48] Sprik M, Cheng A and Klein M L 1992 *The Journal of Physical Chemistry* **96** 2027–2029
[49] Dodson M M and Kristensen S 2003 *ArXiv Mathematics e-prints* (*Preprint* `math/0305399`)
[50] Ding R G, Finnerty J J, Zhu Z H, Yan Z F and Lu G Q 2004 *Encyclopedia of Nanoscience and Nanotechnology* vol 4 (Valencia: American Scientific Publishers) pp 13–33
[51] da Cunha C R, Mineharu M, Matsunaga M, Matsumoto N, Chuang C, Ochiai Y, Kim G H, Watanabe K, Taniguchi T, Ferry D K and Aoki N 2016 *Nature, Scientific Reports* **6** 33118
[52] Zhang T and Ding K 2013 *Front. Mech. Eng.* **8** 371
[53] Guinea F, Katsnelson M I and Geim A K 2009 *Nature Physics* **6** 30
[54] Levy N, Burke S A, Meaker K L, Panlasigui M, Zettl A, Guinea F, Castro N A H and Crommie M F 2010 *Science* **329** 544
[55] Bao W, Miao F, Chen Z, Zhang H, Jangand W, Dames C and Ning L C 2009 *Nanotechnology* **4** 562
[56] Pereira V M, Castro N A H, Liang H Y and Mahadevan L 2010 *Phys. Rev.Lett.* **105** 156603
[57] Krzystyniak M, Adams M A, Lovell A, Skipper N T, Bennington S M, Mayers J and Fernandez-Alonso F 2011 *Farady Discussions* **151**
[58] Nemirovski D, Moreh R, Andersen K H and Mayers J 1999 *J. Phys.: Condens. Matter* **11** 6653
[59] Dresselhaus M S and Dresselhaus G 2002 *Adv. Phys.* **52** 1
[60] Solin S A and Zabel H 1988 *Adv. Phys.* **37** 87
[61] Finkelstein Y, Nemirovski D, Moreh R and Kimmel G 2000 *Physica B* **291** 213–218
[62] Howard C A, Dean M P M and Withers F 2011 *Phys. Rev. B* **84**(24) 241404
[63] `http://www.ee.ucl.ac.uk/ pfrench`
[64] Canny J A 1986 *IEEE Trans. Pattern Analysis and Machine Intelligence,* **8** 679
[65] Revision no. 2.10, authored by F. Moisy, created on 2008/07/09, and downloaded from the Matlab File Exchange website on 2018/02/01
[66] Saberi A A 2011 *Phys. Rev. E* **84**(2) 021113
[67] Meerschaert M M, Nane E and Xiaoa Y 2013 *Statistics and Probability Letters* **83** 1083 – 1093
[68] Edwards J and Thouless D J 1972 *Journal of Physics C: Solid State Physics* **5** 807
[69] Last B J and Thouless D J 1974 *Journal of Physics C: Solid State Physics* **7** 699
[70] Geller M R (ed) 2018 *Quantum Phenomena in Low-Dimensional Systems* (New York: UNESCO)

IOP Conf. Series: Journal of Physics: Conf. Series **1055** (2018) 012008 doi:10.1088/1742-6596/1055/1/012008

Enhancement of counting statistics and noise reduction in the forward-scattering detectors on the VESUVIO spectrometer

P Ulpiani[a,b], G Romanelli[c,b], L Arcidiacono[d,b,e], D Onorati[f,b], G Festa[d,b], M Krzystyniak[c], E Schooneveld[c], F Fernandez-Alonso[c,g], C Andreani[f,b,d,h] and R Senesi[f,b,d,h]

[a] Università degli studi di Roma Tor Vergata, Dipartimento di Scienze e Tecnologie Chimiche, Via della Ricerca Scientifica 1, Rome, 00133 Italy
[b] Università degli studi di Roma Tor Vergata, Centro NAST, Via della Ricerca Scientifica 1, Rome, 00133 Italy
[c] ISIS Facility, Rutherford Appleton Laboratory, Chilton, Didcot, Oxfordshire OX11 OQX, UK
[d] Centro Fermi - Museo Storico della Fisica e Centro Studi e Ricerche "Enrico Fermi", Piazza del Viminale 1, Rome, 00184 Italy
[e] UCL-Institute of Archaeology, University College of London 31-34 Gordon Square London WC1H 0PY United Kingdom
[f] Università degli studi di Roma Tor Vergata, Dipartimento di Fisica, Via della Ricerca Scientifica 1, Rome, 00133 Italy
[g] Department of Physics and Astronomy, University College London, Gower Street, London WC1E 6BT, UK
[h] CNR-IPCF Sezione di Messina, Viale Ferdinando Stagno d'Alcontres 37, Messina, 98158 Italy

E-mail: pierfrancesco.ulpiani@uniroma2.it

Abstract. New evidence of the increased count rate in deep inelastic neutron scattering spectra is provided. Experiments were performed using photon-sensitive yttrium-aluminum-perovskite detectors, installed on the VESUVIO spectrometer at the ISIS pulsed neutron and muon source. At present, these detectors work with a low-level discrimination threshold measuring photons with energy greater than 600 keV in order to avoid background photons from the boron in the beam stop, and other environmental sources. We discuss the advantage in reducing the level of the threshold so as to detect some high-intensity low-energy prompt-gamma rays emitted after the radiative capture of 4.9 eV neutrons by gold, used as energy analyser on the VESUVIO spectrometer. This work shows an improvement of the statistical poissonian error bars and noise on the difference of spectra recorded with and without the energy analyser. The application of such new acquisition strategy discussed here will improve the detection limit of hydrogen atoms in samples, as well as allow a more precise line-shape analysis of nuclear momentum distributions, mentioning just few applications of deep inelastic neutron scattering experiments on VESUVIO.

1. Introduction

The quantitative and non-destructive assessment of hydrogen content in bulk samples is a topic of increasing interest in materials science [1, 2, 3]. Neutrons are the probe of choice for the investigation of the hydrogen dynamics in materials, owing to a higher scattering cross section compared to X-rays. Amongst the many neutron-based techniques, electron-Volt (eV)

IOP Conf. Series: Journal of Physics: Conf. Series **1055** (2018) 012008 doi:10.1088/1742-6596/1055/1/012008

spectroscopy allows for a direct counting of hydrogen atoms in a sample. For high values of energy and momentum transfer, the spectral response of a system is composed of a mass-resolved collection of peaks which are related to the nuclear momentum distributions of every element in the system [4, 5, 6], with the hydrogen peak being the most intense and the best resolved signal. The described regime is referred to as impulse approximation, and the technique is named Deep Inelastic Neutron Scattering (DINS). The unique information available using epithermal neutrons comes at two costs: first, neutron facilities tend to moderate the probe's energy to maximise the flux in the cold and thermal regions; second, the efficiency of traditional neutron detectors is inversely proportional to the neutron velocity, thus poor for epithermal neutrons. Therefore, bespoken detection strategies need to be identified, and many efforts have been devoted to develop novel detection technologies [7, 8, 9]. The development of pulsed accelerator-driven neutron facilities provided an optimal framework for the development of eV neutron spectroscopy, owing to a substantial epithermal flux as opposed to research reactors. In such cases, inverted- (direct-) geometry instruments select the final (initial) energy of neutrons using an energy analyser, while the initial (final) energy and the energy transfer are calculated using the Time-Of-Flight (TOF) technique. The detection difficulties are overcome by using foils with narrow and intense nuclear resonances by which a neutron with a certain energy is captured and a prompt-gamma ray cascade is emitted and efficiently recorded by gamma-sensitive detectors. In order to be a suitable resonance filter, the material should be easy to handle and characterised by a narrow nuclear resonance at the eV energy isolated from other resonances. Several materials have been discussed in pioneering experiments at the Los Alamos National Laboratory, such as plutonium, rhenium, uranium and gold [10].

VESUVIO is an inverted-geometry spectrometer at the ISIS pulsed neutron and muon source [11, 12], using the gold resonance at ca. 4.9 eV to fix the final energy of scattered neutrons. The former incarnation of the instrument, the eV Spectrometer (EVS), was equipped with foils of ^{238}U, with intense prompt-gamma emission peaks at high energies (4060 keV and 6395 keV [13]). In order to discard the high photon background from boron present in the beam dump at ca. 500 keV, a Low-Level Discrimination Threshold (LLDT) was set to 600 keV.

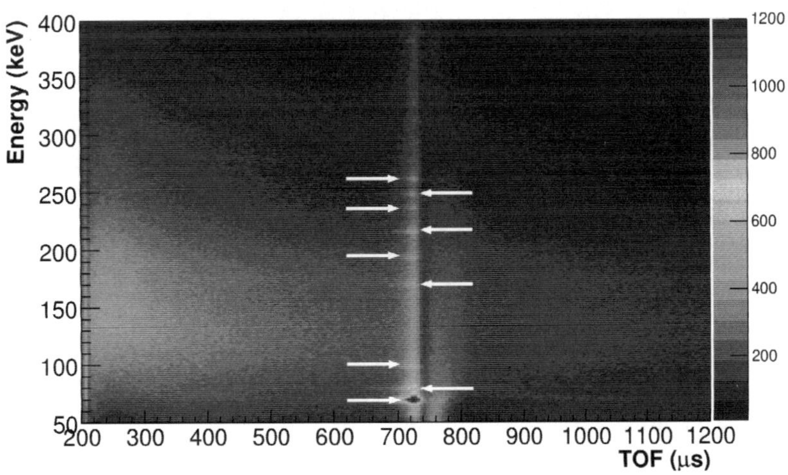

Figure 1: Intensity contour map of γ emission for radiative neutron capture of the gold foil in a bi-parametric (neutron TOF - photon energy) map, reproduced from reference [14]. It is evident the predominance of signal peaks at energies lower than the actual energy threshold of 600 keV at ca. 700 μs which corresponds to the resonant gold capture at 4.9 eV.

Recent time-resolved prompt-gamma activation analysis (T-PGAA) results [14], reported in Figure 1, show how the most intense γ peaks from radiative neutron capture of gold, used at

IOP Conf. Series: Journal of Physics: Conf. Series **1055** (2018) 012008 doi:10.1088/1742-6596/1055/1/012008

present as energy analyser on VESUVIO, are below the LLDT used, therefore suggesting that a better choice of the threshold could be identified. Our results show that decreasing the threshold there is an improvement in the spectra acquired, therefore demonstrating that the actual energy threshold could be improved for this detection configuration. We present preliminary results about such investigation.

2. Methods

We acquired DINS spectra [4, 15] on VESUVIO from a standard polyethylene sample with a thickness of 0.25 mm. Since 2007, VESUVIO is equipped with photon-sensitive Yttrium-Aluminum-Perovskite (YAP) detectors [16] in forward scattering. Two gold resonant filters are used: one of them is positioned on the detector surface in order to convert neutrons into detectable photons [17], while the second foil can be cycled along (foil in) or away from (foil out) the trajectory between the sample and the detector. The difference of the spectra recorded in the two configurations allows for the subtraction of the environmental gamma-background, in a procedure referred to as Foil Cycling Technique (FCT) [16, 18]. The expression of the count rate, $C(t)$, as a function of TOF t within the impulse approximation and in the assumption of the same background in the foil-in and -out configurations, and after the foil difference is

$$C(t) = 2\left(\frac{2}{m}\right)^{1/2} \frac{E_0^{3/2}}{L_0} I(E_0) A^2(E_1) \eta(LLDT) N_M \sum_M b_M^2 \sqrt{\frac{E_1}{E_0}} \frac{M}{\hbar Q} J_M(t) \Delta\Omega, \quad (1)$$

where m is the neutron mass, $I(E_0)dE_0$ is the number of incident neutrons with energy between E_0 and $E_0 + dE_0$ per unit time, $\eta(LLDT)$ is the probability that the γ-ray cascade is detected for a fixed LLDT, A(E_1) is the n-γ conversion efficiency of neutron with energy E_1, N_M is the number of scattering centres with mass M in the sample, $J_M(t)$ is the Neutron Compton profile, b_M the scattering length and $\Delta\Omega$ the solid angle covered by a given detector.

The threshold of five YAP detectors was decreased from the original value $V = 600$ keV operating on the individual detector discriminator card. As the lowest value of the settable threshold $V_0 > 0$ for the discriminator was unknown, the threshold for each detector was set dividing the available range $\Delta V = V - V_0$ in five equal intervals. Table 1 shows the thresholds chosen for used detectors. All measurements were performed both with changed set-up and standard configuration.

Table 1: Threshold chosen for the five used detectors. The available range ΔV, from the lowest energy value V_0 to the actual threshold $V = 600$ keV, was divided in five parts, one for each detector. The detector position is also reported; r is the distance between the sample and detector, θ the polar angle while ϕ is the azimuthal angle.

Detector	Estimated threshold	r [m]	θ	ϕ
a	V_0	0.51	52	131
b	$V_0 + 20\%\ \Delta V$	0.50	47	140
c	$V_0 + 40\%\ \Delta V$	0.50	44	151
d	$V_0 + 60\%\ \Delta V$	0.51	41	163
e	$V_0 + 80\%\ \Delta V$	0.52	52	-132

3. Results and Discussion

The spectra acquired with different thresholds were compared in order to quantify the improvement obtained. Figure 2 shows the largest difference in DINS spectra recorded after (red line) and before (black line) the change of the LLDT with a total proton charge of ca. 1530 µAh . The spectra reported for detector a are normalised by the proton change accumulated in the synchrotron. We can see the polyethylene recoil peaks of hydrogen (lower TOF) and carbon (higher TOF). It is evident that lowering the threshold there is a count rate increase, in particular there is about a factor 3 at the hydrogen peak position. This increase is very significant in detector a and it decreases gradually for the other detectors where the modified threshold approaches the original value.

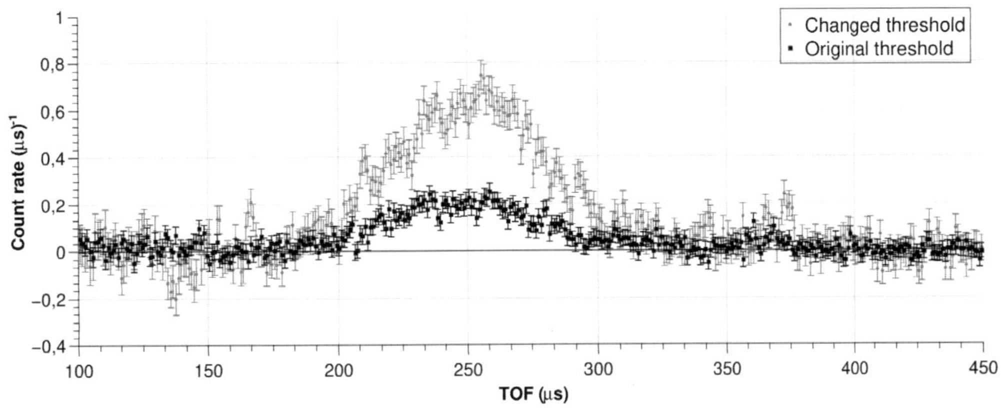

Figure 2: Comparison of polyethylene spectra acquired with original (black line) and changed (red line) threshold from detector a normalized to the proton charge accumulated in the synchrotron. The FCT for signal isolation is used. The peak at lower TOF is associated to hydrogen and the one at higher TOF to carbon. An enhancement of the count rate is evident for the changed configuration compared to the original one.

In order to show the decrease of the relative error bars, it is useful to calculate the ratio

$$\delta(t) = \frac{\text{Relative error bars for changed threshold}}{\text{Relative error bars for original threshold}} = \frac{\Delta_c(t)}{\Delta_o(t)}. \tag{2}$$

As shown in Figure 3 the ratio between relative error bars is qualitatively lower than one for all

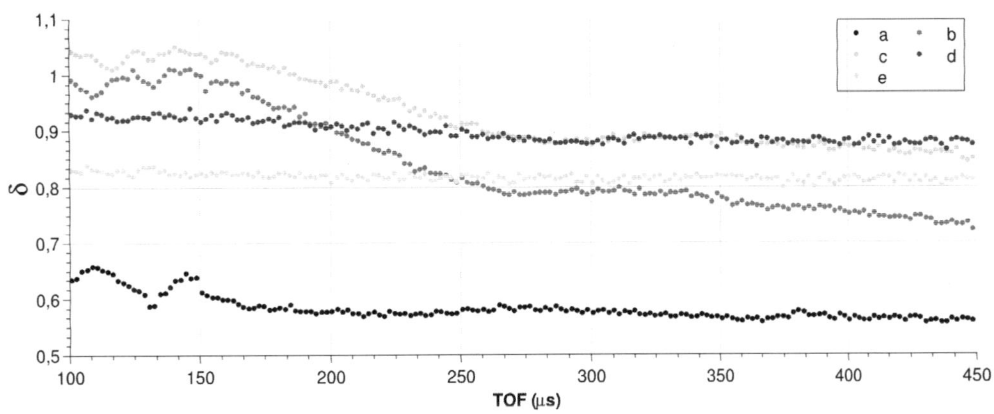

Figure 3: Comparison between the ratio δ for the five different detectors associated to the polyethylene spectra. The ratio represented implies an error bar reduction from 10% (detector c) to more than 40% (detector a).

the five detectors considered. In particular, detector a, d and e show a relatively flat ratio over the TOF range as opposed to detector b and c where the ratio approaches one for low TOF, then decreases and becomes almost constant for TOF larger than 250 μs. The reduction of the relative error bars for TOF higher than 250 μs ranges between 10% for detector d and more than 40% for detector a. The features that one can see between 100 and 150 μs in Figure 3 can be related to the structures of the raw data shown in Figure 4 where the spectra acquired in foil in and out configuration are represented for original and changed threshold of detector b. As we can see even if the background is satisfactorily subtracted through the difference method, the rise of count rate at low TOF for the changed threshold brings to a not-constant δ factor. The decrease of the delta factor is a consequence of the increased number of photons detected as the threshold is lowered. Of all the photons detected, those coming from the neutrons scattered by the sample and captured by the gold foils would contribute to a constant lowering of δ over the entire TOF range, as they would correspond to a constant increase of the signal from neutron Compton profiles. The features of the δ factor at low TOF values therefore suggests that additional environmental background is included in the count rate when the LLDT is lowered. Looking at the TOF where the new background peaks are, we can associate them to the resonant gamma emission from cadmium which covers the forward scattering detectors.

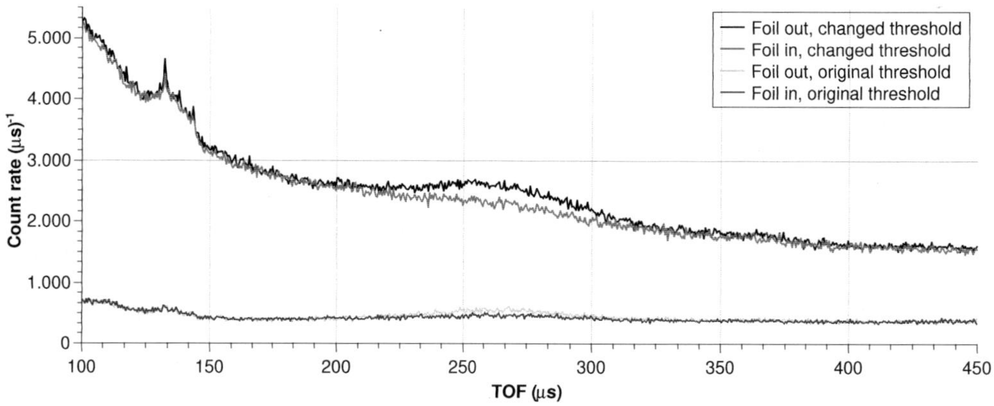

Figure 4: Comparison between the polyethylene spectra acquired in foil in and out configuration with detector b using changed and original threshold. It is evident for low TOF and changed threshold an increase in background counts which is not present for original threshold. It is perfectly subtracted using the difference method.

In addition to the showed decrease in the relative error bars, the change of the LLDT provides a reduction of the noise in DINS spectra. In order to show this, we focus our attention on a certain region of the TOF where there are no neutron Compton profile peaks and the count rate is expected to be vanishing. In this region we calculate the standard deviation σ, around the average value ρ, of the dispersion of count rate $C(t_i)$ for the original and changed setting. The ranges chosen are

$$range\ 1: \quad (100 < t < 160)\ \mu s$$
$$range\ 2: \quad (390 < t < 450)\ \mu s$$

respectively to the left of the proton peak (range 1) and to the right of the carbon peak (range 2). In the assumption of constant binning, the standard deviation, is calculated as

$$\sigma = \sqrt{\frac{\sum_i (C(t_i) - \rho)^2}{N-1}}, \tag{3}$$

Where i is the bin index and N is the total number of bins in the TOF range analysed. The result for the two ranges is shown in Figure 5 where is represented the ratio $\chi = \frac{\sigma_{changed}}{\sigma_{original}}$

IOP Conf. Series: Journal of Physics: Conf. Series **1055** (2018) 012008 doi:10.1088/1742-6596/1055/1/012008

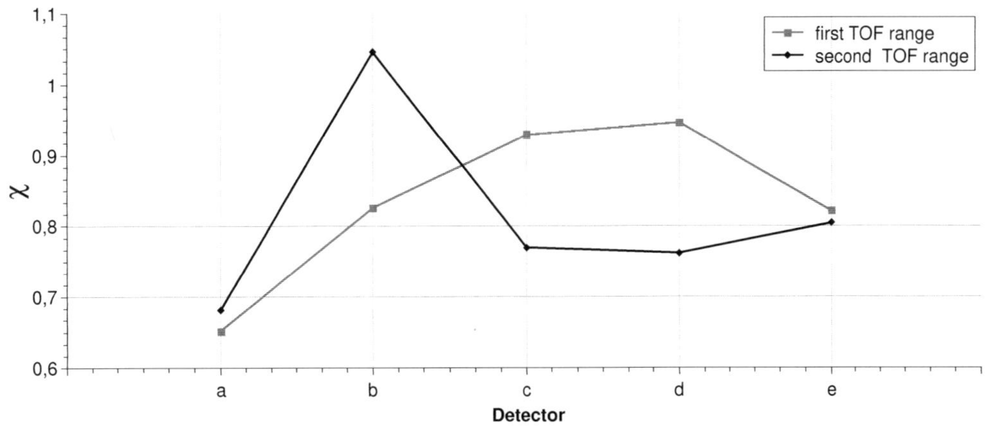

Figure 5: Ratio of the statistical noise $\chi = \frac{\sigma_{changed}}{\sigma_{original}}$ in the first (left of hydrogen peak) and second (right of carbon peak) TOF range. The ratio is smaller than one in the most cases.

for each detector. One can notice how the noise is reduced in the majority of detectors, a particularly relevant result for the analysis of the line-shape of neutron Compton profiles in DINS experiments. The change of the LLDT has brought an increase of the total count rate that is due not only to signal but also to background counts. What is interesting is that the signal to background ratio remains in the same order of magnitude even if the there is an increasing effect in the signal counts. The optimizations showed are reflected also on the physical parameters obtained with DINS experiments. In particular, fitting the hydrogen experimental Neutron Compton profile of polyethylene sample in both the changed and original configurations, we calculated the values of the hydrogen mean kinetic energy $< E_k >$ reported in Table 2. The results are obtained summing all the detectors from a to e to increase the statistics. As we can see the error is decreased by ca. 50% for the changed configuration.

Table 2: Values of hydrogen mean kinetic energy $< E_k >$ in polyethylene obtained fitting the experimental Neutron Compton profile for both changed and original threshold.

	Original configuration (meV)	Changed configuration (meV)
$< E_k >$	148 ± 11	145.2 ± 5.9

4. Outlook and Conclusions

In this work we have shown how the change of the low-level discrimination threshold provides a reduction of the relative statistical error bars and of the noise in deep inelastic neutron scattering experiments which make use of YAP detectors. In particular, the reduction on statistical error bars was found in range between 10 % for small changes of the threshold and 40 % for a nominal removal of the threshold. Further studies will be devoted to evaluate, as acquisition method, an upper level discrimination threshold and its effect on the quality of the spectra acquired detecting only photons with energy lower than 300 keV. Alternatively the implementation of detectors sensitive to low-energy photons such as gas-electron multiplier [19, 20] will be considered. In future investigations, the linearity relation between the LLDT and the photon energy will be tested, and the value of V_0 will be assessed, moreover, the response of a given detector to a change of LLDT due to its electronics will be analysed. Our results show a promising potential for

VESUVIO science programme, as they can improve the line-shape analysis of neutron Compton profiles [21].

Acknowledgements
This work was partially supported within the CNR-STFC Agreement (2014-2020) concerning collaboration in scientific research at the ISIS pulsed neutron and muon source.
The authors gratefully acknowledge the UK Science & Technology Facilities Council for financial support and access to beam time at the ISIS Facility.

References
[1] Silverwood I P, Parker S F and Catlow C R A 2016 *Physical Chemistry Chemical Physics* **18** 17140–17140
[2] Kaplan D, Sol N, Velasco G and Thomas P 1978 *Applied Physics Letters* **33** 440–442
[3] Philipps V, Wienhold P, Kirschner A and Rubel M 2002 *Vacuum* **67** 399–408
[4] Andreani C, Krzystyniak M, Romanelli G, Senesi R and Fernandez-Alonso F 2017 *Advances in Physics* **66** 1–73
[5] Pietropaolo A and Senesi R 2011 *Physics Reports* **508** 45–90
[6] Tardocchi M, Pietropaolo A, Andreani C, Bracco A, D'Angelo A, Gorini G, Imberti S, Senesi R, Rhodes N and Schooneveld E 2004 *Nuclear Instruments and Methods in Physics Research Section A: Accelerators, Spectrometers, Detectors and Associated Equipment* **526** 477–492
[7] Belgya T, Kis Z, Szentmiklósi L, Kasztovszky Z, Festa G, Andreanelli L, De Pascale M, Pietropaolo A, Kudejova P, Schulze R *et al.* 2008 *Journal of Radioanalytical and Nuclear Chemistry* **278** 713–718
[8] Pietropaolo A, Andreani C, Rebai M, Giacomelli L, Gorini G, Cippo E P, Tardocchi M, Fazzi A, Rinati G V, Verona C *et al.* 2011 *EPL (Europhysics Letters)* **92** 68003
[9] Tardocchi M, Gorini G, Pietropaolo A, Andreani C, Senesi R, Rhodes N and Schooneveld E 2004 *Review of scientific instruments* **75** 4880–4890
[10] Brugger R, Taylor A, Olsen C, Goldstone J and Soper A 1984 *Nuclear Instruments and Methods in Physics Research* **221** 393–407
[11] Isis facility website `www.isis.stfc.ac.uk`
[12] Senesi R, Andreani C, Bowden Z, Colognesi D, Degiorgi E, Fielding A, Mayers J, Nardone M, Norris J, Praitano M *et al.* 2000 *Physica B: Condensed Matter* **276** 200–201
[13] Molnar G 2004 *Handbook of Prompt Gamma Activation Analysis*
[14] Onorati D Master's thesis University of Rome Tor Vergata URL `http://purl.org/net/epubs/work/33481015`
[15] Andreani C, Colognesi D, Mayers J, Reiter G and Senesi R 2005 *Advances in Physics* **54** 377–469
[16] Schooneveld E, Mayers J, Rhodes N, Pietropaolo A, Andreani C, Senesi R, Gorini G, Perelli-Cippo E and Tardocchi M 2006 *Review of scientific instruments* **77** 095103
[17] Andreani C, Pietropaolo A, Senesi R, Gorini G, Tardocchi M, Bracco A, Rhodes N and Schooneveld E 2002 *Nuclear Instruments and Methods in Physics Research Section A: Accelerators, Spectrometers, Detectors and Associated Equipment* **481** 509–520
[18] Mayers J, Tomkinson J, Abdul-Redah T, Stirling W, Andreani C, Senesi R, Nardone M, Colognesi D and Degiorgi E 2004 *Physica B: Condensed Matter* **350** E659–E662
[19] Sauli F 2016 *Nuclear Instruments and Methods in Physics Research Section A: Accelerators, Spectrometers, Detectors and Associated Equipment* **805** 2–24
[20] Kaneko K and Kohmura T 2012 *Journal of Instrumentation* **7** C07001
[21] Ulpiani P Master's thesis University of Rome Tor Vergata URL `https://epubs.stfc.ac.uk/work/35035571`

IOP Conf. Series: Journal of Physics: Conf. Series **1055** (2018) 012009 doi:10.1088/1742-6596/1055/1/012009

Gamma background characterization on VESUVIO: before and after the moderator upgrade

D Onorati[1,3], C Andreani[1,2,3,4], L Arcidiacono[2,3,5], F Fernandez-Alonso[6,7], G Festa[2,3], M Krzystyniak[6], G Romanelli[6], P Ulpiani[8,3] and R Senesi[1,2,3,4]

[1] Università degli studi di Roma "Tor Vergata", Department of Physics, Via della Ricerca Scientifica 1, Rome, 00133 Italy

[2] Centro Fermi - Museo Storico della Fisica e Centro Studi e Ricerche "Enrico Fermi", Piazza del Viminale 1, Rome, 00184 Italy

[3] Università degli studi di Roma "Tor Vergata", NAST Center, Via della Ricerca Scientifica 1, Rome, 00133 Italy

[4] CNR-IPCF Sezione di Messina, Viale Ferdinando Stagno d'Alcontres 37, Messina, 98158 Italy

[5] UCL - Institute of Archaeology, University College of London 31-34 Gordon Square London WC1H 0PY UK

[6] ISIS Facility, Rutherford Appleton Laboratory, Chilton, Didcot, Oxfordshire OX11 OQX, UK

[7] Department of Physics and Astronomy, University College London, Gower Street, London WC1E 6BT, UK

[8] Università degli studi di Roma "Tor Vergata", Department of Chemical Sciences and Technologies, Via della Ricerca Scientifica 1, Rome, 00133 Italy

E-mail: giulia.festa@centrofermi.it

Abstract. The VESUVIO spectrometer at the ISIS pulsed neutron and muon source is a unique instrument which makes use of eV neutrons and inverted geometry, allowing deep inelastic neutron scattering experiments with high values of energy and wavevector transfers. The neutron detection techniques on the VESUVIO forward-scattering detector banks is based on (n,γ) conversion, therefore neutrons are indirectly detected and the signals produced by scattered neutrons, accordingly the photons, is recorded using gamma scintillators. The use of γ-sensitive detectors make γ-background one of the main limiting factors affecting the data quality and instrument sensitivity on VESUVIO. This work aims to assess how the sample-independent gamma background has changed after the recent upgrades to the water moderator viewed by the instrument, which resulted in a twofold increase of the thermal neutron flux. Here we show that the gamma background is mainly influenced by the thermal neutron flux and that the recent upgrade results in a fivefold increase in the gamma background in the photon energy range 300 keV-3 MeV. We point out the possibility of providing a thermal-neutron filter along the incident beam in order to suppress this background source.

1. Introduction

Currently, the ISIS pulsed Neutron and Muon Source [1] has over 30 neutron and muon instruments distributed in two main buildings, Target Station 1 (TS1) and Target Station 2 (TS2). Amongst the many techniques available, beamlines at ISIS routinely allow neutron imaging techniques for applications in Engineering or Material Science and Cultural Heritage, neutron diffraction techniques to study crystalline solids or liquids, and tests microchip's response to neutron radiation. Other beamlines at ISIS allow the investigation of atomic and magnetic motions of atoms. The VESUVIO spectrometer installed at TS1, makes use of Deep Inelastic Neutron Scattering (DINS)[2], to measure atomic momentum distributions and nuclear quantum effects in condensed matter systems [3]. A water moderator, decoupled and poisoned, at 295 K serve several beamlines on TS1, providing neutrons with the energy useful for the various fields of application. The "poisoning"of the moderator with gadolinium foils is used to lower the Maxwellian component of the neutron velocity distribution and reduce the intensity of the over-thermalized neutrons in the reflectors system [4]. On February 2016, the TS1 water moderator has undergone an upgrade [5, 6] which consisted in the removal of one of the two gadolinium poisoning foils. As a result, several beamlines on TS1 like MAPS, SXD and, in particular VESUVIO, benefit from an increase factor of up to two in thermal neutron flux. Figure 1 shows the predicted gain in the neutron flux as obtained from MCNPX [7] simulations [5, 6]. The experimental results has shown an even large increase of 10 - 20% , as reported in ref. [6], compared to the simulated results, possible due to simplified moderator model which cannot include all eventual sources of performance deterioration over the years. The benefits of the extra flux for these instruments include the capability to measure smaller samples (SXD), improve the speed and statistical quality of low energy excitations measurements (MAPS) and improve diffraction and transmission data collection done in parallel with spectroscopy experiments (VESUVIO) [8].

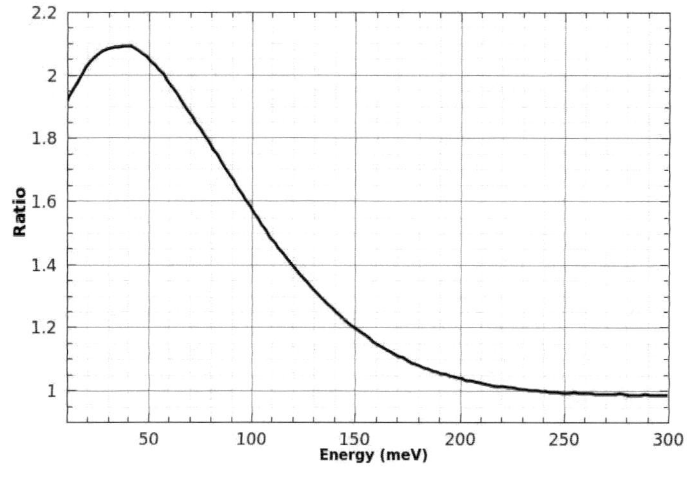

Figure 1. Simulated flux ratio before and after water moderator upgrade as a function of neutron energy [5].

Forward-scattered neutrons on VESUVIO are converted into prompt-gamma radiations through an (n,γ) process in the resonance condition, and the photon cascade, rather than the scattered neutron, is therefore detected. Gold resonant filters select the neutrons final energy at ca. 4.9 eV and the γ-ray emitted following radiative capture process are detected using YAP (Yttrium-Aluminium-Perovskite) scintillators. The use of gamma-sensitive detectors means that the detected signal contains all photons triggered by the neutron scattered from the sample as well as generated in (n,γ) processes by the sample environment and beamline components elements, such as beam stop, moderator, or collimators. This background signal affect DINS data and results as the main component of the raw spectra acquired on VESUVIO. DINS spectra are indeed obtained by the foil cycling technique [9], that envisages the use of two foils of the same neutron absorbing material, one is the analyzer foil itself, fixed in front of the photon detector, the second one is placed between the sample and the analyzer. During the scattering experiments, the latter alternates in and out of the scattered neutron beam. The difference between filter-in and filter-out measurements allows to obtain a subtraction of the background. However, gamma background remains one of the principal factors affecting both the data quality and instrument sensitivity on VESUVIO. Therefore, a characterization of the gamma background is relevant for beamlines that employ gamma-sensitive detectors to identify optimization strategies of signal-to-background ratio.

2. Experimental set-up and data collection

Sample-independent background spectra were collected in February 2017 with an High-Purity Germanium detector (HPGe) designed to be resistant to damage from fast neutrons, whose characteristics are shown in table 1. The detector resolution indicated by the producer is 2.5 keV (FWHM) measured on the ^{60}Co peak at 1.33 MeV and 760 eV on the ^{56}Fe peak at 5.9 keV. This might be dependent on the experimental conditions; Its experimental evaluation was 3.5 keV on the ^{115}In at 1.29 MeV. This variation may be due to many factors including an incomplete polarization of the diode. The high resolution of this kind of detector allows to identify and to assign the gamma peaks to the corresponding chemical elements and isotopes. According to the VESUVIO layout, the neutron beam travels in vacuum from the moderator to the sample located inside an aluminium sample tank, passing through a set of collimators. The beam dump situated ca. 5 m from the sample position is mainly composed of elements such as hydrogen, boron, and iron. The HPGe detector was placed at an angle $\theta \simeq 45°$ with respect to the centre of neutron beam and at a distance of approximately 0.7 m from the sample position. Due to geometrical constrains, the detector was placed between the sample tank and the beam dump as shown in figure 2. The HPGe was powered in order to provide high voltage to diode and low voltage to the preamplifier. The pulse-height spectra were acquired using the ORTEC MAESTRO acquisition software[10]. The internal lower level discriminator was set to 100 keV in order to limit the deadtime. The spectra recoded, for an IPC (Integrated Proton Current) of 75 μA, correspond to the prompt-gamma activation spectra from the elements presents in the beamline.

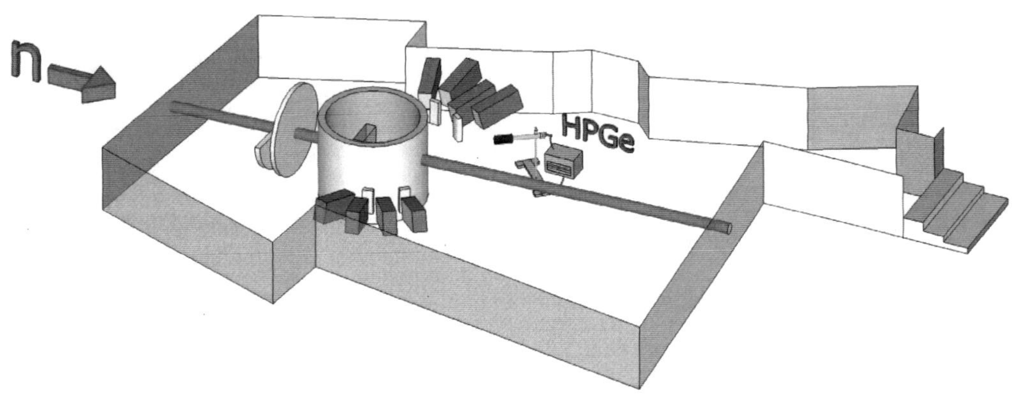

Figure 2. Sketch of the HPGe detector position in the layout of VESUVIO blockhouse used both in 2013 and 2017 measurements. The blue square inside the tank represent the nominal sample position during routinely DINS experiments.

Table 1. General characteristics of the HPGe detector used.

	Detector details
Material	HPGe, n-type
Crystal configuration	Coaxial
Crystal diameter	83 mm
Cup length	170.5 mm
Crystal length	168 mm
High Voltage bias	-3800 V
Relative efficiency	44.5 %
Endcap diameter	93 mm
Manufacturer and year	ORTEC 2013

3. Results and discussion

The energy calibration, to convert electronic channels into energy bins, was performed using well-known peaks in the background spectra, arising from radiative capture in elements present in the beamline. For example, by exploiting the gamma-ray line at 478 keV emitted by the decay of the excited ^{7}Li to the ground state in the reaction n + ^{10}B $\rightarrow$ ^{7}Li* + α , clearly recognizable by its shape, the neighbouring gamma ray corresponding to 511 keV, due to electron-positron annihilation following pair production, and the 1292 keV line of ^{115}In present in the detector contact.

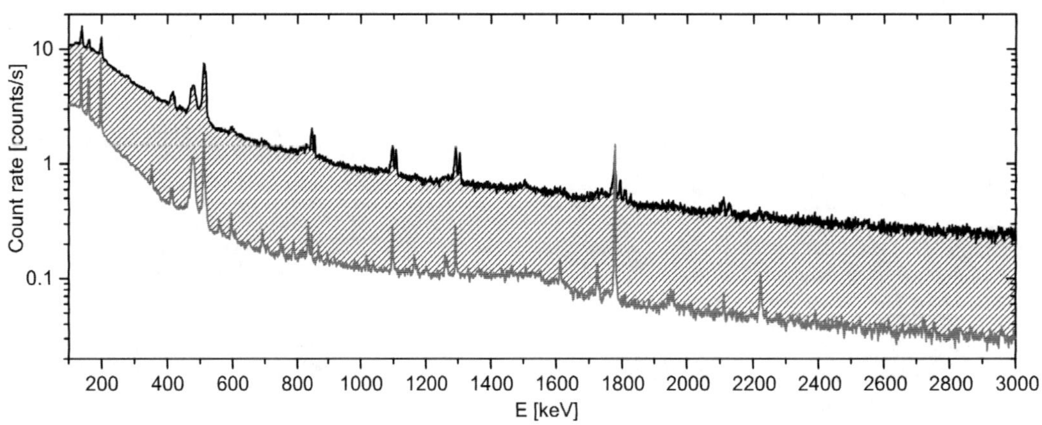

Figure 3. Comparison between normalized γ-ray spectra background acquired on February 2017 (black line)and on May 2013 (red line). The dashed area represents the increase in the count rate.

Figure 3 shows the comparison between the background spectra acquired on February 2017 and on May 2013 in the 100 keV-3 MeV energy region. Both spectra are normalized to their own acquisition time (live time), and the differences in the peak-to-background ratio are attributable to the logarithmic scale used to emphasize the increment in the baseline on the entire energy range, showing that there is a rigid shift of the background, as expected. The ratio of the count rate of the two spectra in the entire energy range is about five, demonstrating that an increase in the flux of thermal neutrons results in an increase in the gamma background, as shown in figure 3. Despite this, the off-peak continuum has the same overall shape in the two spectra. Both background spectra have high count-rate, especially in the low energy region, due to Compton scattering processes between photons and electrons in the walls of the block house or between electrons inside the detector. The main effect of the Compton continuum is the loss of correlation between the intensity in a particular energy channel and the parent prompt gamma emission line. This limits the signal to background ratio hindering an optimised quantification of elements content in the prompt-gamma activation analysis [11]. Figure 4 shows background spectra from 2013 and 2017 (before and after moderator upgrade), normalized to their respective total area for a better comparison between the peaks in the energy region 100 keV-825 keV. Blue dashed lines are a guide to identify significant gamma peaks and each one was labelled for the corresponding element identification [12].

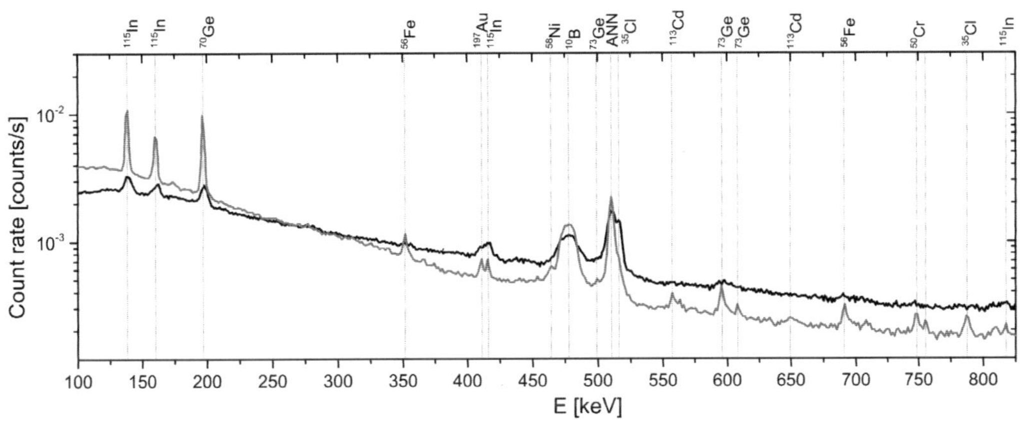

Figure 4. Background spectra obtained during 2013 (red line) and 2017 (black line), that is before and after moderator upgrade, in the γ-energy range 100 keV-825 keV. Blue dashed lines identify the most significant gamma peaks and each one was labelled for the corresponding element.

The latter made use of databases in refs.[13, 12] published by the Chemical Research Center of Budapest and by the International Atomic Energy Agency (IAEA), respectively. Figure 4 shows that the elemental composition assigned to the two spectra does not show differences. A similar analysis over the entire energy range confirms the above findings. The spectra show peaks from indium (that is one of the materials surrounding the diode [14]), germanium itself from the detector active material, peaks from iron contained in flanges and collimators, boron present in large quantities in the beam dump and in the walls of the block house, and gold peaks from analyser foils. Also, we are expecting that not all elements respond in the same way to the increase in thermal neutrons, due to the dependence of the cross-section (n,γ) from the energy of the neutrons . This is reflected, as shown in Figure 4, in the change of the relative intensities of the peaks attributed to the same element in the two compared spectra. In fact, there is an increase in the intensity of peaks due to elements with intense cross section (n,γ) in the thermal energy range at the expense of elements that have a large cross section in the epithermal range, the latter remaining unchanged after the upgrade.

4. Outlook and Conclusion

In this work we present gamma-background spectra acquired on the VESUVIO spectrometer in the years 2013 and 2017 in order to assess the effect of the moderator upgrade on the background measured by the YAP scintillators. The results demonstrate that the elemental composition assigned to the two spectra does not show differences, yet an increase in the background count rate due to the increase of thermal-neutron flux on the sample is evident. Figure 5 shows a direct comparison of the background measured by the YAP detectors before and after the moderator upgrade. This figure was obtained by calculating the ratio of the time of flight

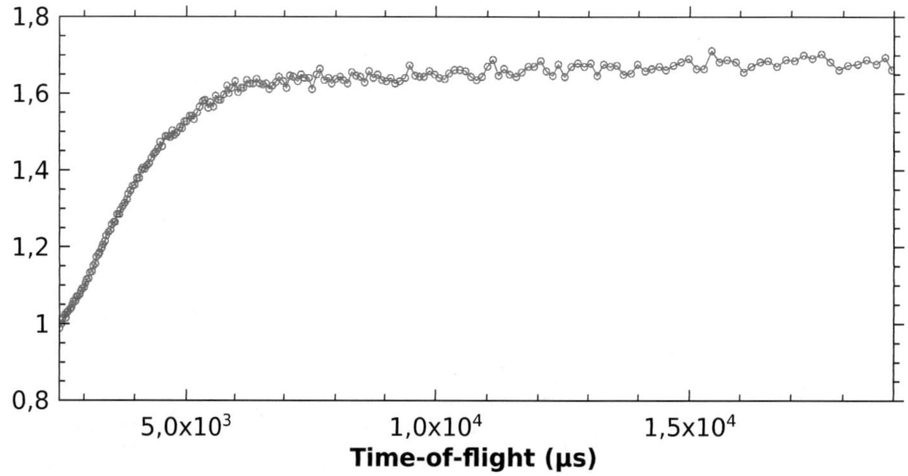

Figure 5. Ratio between the time of flight background spectra measured by the YAP on March 2015 and on May 2017, respectively before and after the moderator upgrade.

spectra, summed over the 64 forward scattering detectors (YAP), between two measurements with the same experimental set-up carried out before and after the upgrade. The two spectra were normalised to IPC. As can be seen in the ToF region 2500-19000 μs, corresponding to neutron energy less than ca. 100 meV, there is a marked increase in the measured background which saturates to a value of 1.6 $\pm$ 0.1. In order to obtained other information about the Time of Flight (ToF) structure of the observed background, we plan to set up a biparamentric acquisition [15] ToF vs Energy to understand if this increase may be significant in the Time of Flight range 50 μs -600 μs of interest for DINS measurements. In addiction, despite the background has increased by a constant factor in the whole range considered, the efficiency of YAP is not constant but decreases with increasing energy. Therefore in the DINS spectra this increase in the background is mitigated by the efficiency of the YAP. Background spectrum is strongly sensitive to the detector's position within the VESUVIO block house, therefore calling for additional measurements could be made by moving the germanium detector in different position in order to obtain a map of the background. We finally suggest the possibility to include a thermal-neutron filter along the incident beam, so as to suppress the component of the background induced by the absorption of thermal neutrons in the blockhouse and beam dump.

Acknowledgements
This work was partially supported within the CNR-STFC Agreement (2014-2020) concerning collaboration in scientific research at the ISIS pulsed neutron and muon source. The authors gratefully acknowledge the UK Science & Technology Facilities Council for financial support and access to beam time at the ISIS Facility. Finally, we thank Dr G. Škoro for useful and constructive discussions.

References

[1] https://www.isis.stfc.ac.uk/
[2] Andreani C, Krzystyniak M, Romanelli G, Senesi R and Fernandez-Alonso F 2017 *Advances in Physics* **66** 1–73
[3] Krzystyniak M, Druzbicki K, Romanelli G, Gutmann M J, Rudic S, Imberti S and Fernandez-Alonso F 2017 *Phys. Chem. Chem. Phys.* **19**(13) 9064–9074
[4] Pietropaolo A, Cippo E P, Gorini G, Tardocchi M, Schooneveld E, Andreani C and Senesi R 2009 *Nuclear Instruments and Methods in Physics Research Section A: Accelerators, Spectrometers, Detectors and Associated Equipment* **608** 121–124
[5] https://www.isis.stfc.ac.uk/Pages/Moderator-upgrade-successfully-increases-flux-to-TS1
[6] Škoro G, Bewley R, Lilley S, Ewings R, Romanelli G, Gutmann M, Smith R, Rudić S and Ansell S *J. Phys.: Conf. Series (in press)*
[7] MCNPX 2.7.0 - Monte Carlo N-Particle Transport Code System for Multiparticle and High Energy Applications, https://mcnpx.lanl.gov/
[8] Romanelli G, Krzystyniak M, Senesi R, Raspino D, Boxall J, Pooley D, Moorby S, Schooneveld E, Rhodes N, Andreani C *et al.* 2017 *Measurement Science and Technology* **28** 095501
[9] Pietropaolo A, Tardocchi M, Schooneveld E and Senesi R 2006 *Nuclear Instruments and Methods in Physics Research Section A: Accelerators, Spectrometers, Detectors and Associated Equipment* **568** 826–838
[10] http://www.ortec-online.com/products/application-software/maestro-mca
[11] Tian K V, Festa G, Szentmiklósi L, Maróti B, Arcidiacono L, Laganà G, Andreani C, Licoccia S, Senesi R and Cozza P 2017 *Journal of Analytical Atomic Spectrometry* **32** 1420–1427
[12] Molnar G 2010 *Handbook of Prompt Gamma Activation Analysis: with Neutron Beams* (Springer US) ISBN 9780387233598
[13] https://www-nds.iaea.org/pgaa/databases.htm
[14] Roy J C, Côté J E, Durham R W and Joshi S R 1989 *Journal of Radioanalytical and Nuclear Chemistry* **130** 221–230 ISSN 1588-2780 URL `https://doi.org/10.1007/BF02037716`
[15] Festa G, Arcidiacono L, Pappalardo A, Minniti T, Cazzaniga C, Scherillo A, Andreani C and Senesi R 2016 *Journal of Instrumentation* **11** C03060

IOP Conf. Series: Journal of Physics: Conf. Series **1055** (2018) 012010 doi:10.1088/1742-6596/1055/1/012010

Absolute efficiency calibration of a coaxial HPGe detector for quantitative PGAA and T-PGAA

A Parmentier[1], L Arcidiacono[1,2,4], R Senesi[1,3,4,*], G Romanelli[5], C Andreani[1,3,4], J Moir[5], and G Festa[4]

[1] NAST Center and Dipartimento di Fisica, Università degli Studi di Roma "Tor Vergata", via della Ricerca Scientifica 1, 00133 Rome, Italy

[2] UCL-Institute of Archaelogy, University College London, 31-34 Gordon Square, London WC1H 0PY, UK

[3] CNR-IPCF Sezione di Messina, viale Ferdinando Stagno d'Alcontres 37, 98158 Messina, Italy

[4] Centro Fermi-Museo Storico della Fisica and Centro Studi e Ricerche "Enrico Fermi", piazza del Viminale 1, 00184 Rome, Italy

[5] ISIS Facility, Rutherford Appleton Laboratory, Chilton, Didcot, Oxfordshire, OX11 0QX, UK

E-mail: `roberto.senesi@uniroma2.it`

Abstract. Following international guidelines by the IEEE and a well-established calibration procedure developed at the Hungarian Institute of Isotopes in Budapest, the experimental absolute efficiency calibration of an Ortec n-type coaxial high-purity germanium (HPGe) detector (GMX Profile series) has been performed, in view of direct application to quantitative PGAA and T-PGAA investigations of Cultural Heritage multicomponent manufacts.

1. Introduction

Among the several characteristics of a γ-ray detector, the absolute full-energy-peak (FEP) efficiency (*i.e.*, the ratio of the number of counts under the photopeak to the total number of γ-rays emitted by the source) plays a vital role for quantitative spectroscopy, especially when complex γ-ray spectra involving many lines are under scrutiny. In particular, the development of Prompt-Gamma Activation Analysis (PGAA) and Time-resolved Prompt-Gamma Activation Analysis (T-PGAA) capabilities at pulsed neutron sources has allowed to expand the potential of PGAA at continuous sources for the study of functional materials and Cultural-Heritage samples [1–3]. These investigations require the use of high-resolution γ-ray detectors to resolve the complex spectra of multicomponent samples.

In this field, semiconductor detectors are much appreciated in comparison to inorganic scintillators such as NaI(Tl), due to their superior energy resolution when operated at low temperatures. However, this improved energy resolution comes along with a major drawback, since the generally smaller sizes of common germanium detectors, together with the lower atomic number of Ge with respect to iodine, lowers the FEP efficiency (ϵ_{fep}) by an order of magnitude in typical cases [4].

For this reason, the assessment of the absolute $\epsilon_{fep}(E)$ curve (*i.e.*, ϵ_{fep} as a function of photon energy) of a HPGe detector becomes a crucial analytic passage, also taking into account the unknown exact functional dependence of efficiency on energy (only partially fixed by the

Table 1: Source features, from source datasheet and LNE-LNHB/CEA nuclear-data tables [9]. The most important acquisition parameters are reported as well: A_0 = reference activity (3% uncertainty with 95% confidence), $t_{\frac{1}{2}}$ = half time, t_{live} = livetime, *i.e.*, total time of actual count acquisition with no dead time included, and d = source-detector distance.

			Set 1		Set 2	
	A_0	$t_{\frac{1}{2}}$	t_{live}	d	t_{live}	d
^{60}Co	(42.10 ± 1.26) kBq	5.2711 ± 0.0008 y	7200 s	11.5 cm	300 s	11.5 cm
^{133}Ba	(36.60 ± 1.10) kBq	10.539 ± 0.006 y	7200 s	11.5 cm	300 s	11.5 cm
^{137}Cs	(41.10 ± 1.23) kBq	30.05 ± 0.08 y	7200 s	11.5 cm - 1 cm	300 s	11.5 cm 3.7 cm 0.5 cm
^{241}Am	(39.60 ± 1.19) kBq	432.6 ± 0.6	7200 s	11.5 cm	300 s	11.5 cm

application of empirical or semi-empirical models [5]), as well as limits to the overall accuracy due to unavoidable sources of uncertainty hidden in the geometry of the experimental setup.

The 1996 IEEE protocol, which establishes terminology and standard test procedures for germanium γ-ray detectors [6], contains popular guidelines to perform various types of calibrations. Its integration to the novel efficiency-calibration procedure developed at the Hungarian Institute of Isotopes in Budapest [5,7] allows for the improvement of some practices, such as the assessment of the system non-linearity, which is mainly due to the analog-to-digital converter (ADC), and is cause to major FEP distorsions that cannot be neglected when (T-)PGAA is performed because of the very low amplifier gains employed in this technique [8]. Starting from a careful mix of such protocols, the experimenter is allowed to perform the ϵ_{fep} calibration of a HPGe detector over a wide energy range ($\sim$ 0.05-11.00 MeV), typically making use of a large set of calibrated and noncalibrated samples of radionuclides over the low-energy region (up to about 2 MeV), while exploiting cascades from thermal-neutron capture on non-calibrated, non-commercial light-nucleus targets (such as ^{14}N or ^{35}Cl) over the high-energy counterpart (above 2 MeV) [5].

In the present work, a practical application of this procedure is presented for the assessment, over the low-energy region, of the absolute $\epsilon_{fep}(E)$ curve of an Ortec n-type coaxial HPGe detector (GMX Profile series).

2. Experimental setup

The thermostated HPGe detector shown in Fig. 1 was employed to collect a double set of γ spectra of an ensemble of 4 calibrated radionuclides (reference time: t_0 = 01/11/2005) in a shielded area located in the Target Station 2 (TS2) building at the ISIS pulsed neutron and muon source [10], UK. Measurements included "beam-off" background (both measurement sets were collected in the course of a mainteinance shutdown of all ISIS beamlines, in order to minimize unwanted signals in the spectra).

A summary of source features and major acquisition parameters is reported in Table 1. In step with the IEEE protocol [6], discoidal standards were used with an active nucleus of ≤ 2 mm in diameter, and the livetime of the background collection was consistent with the one of any source spectrum. Distances from the crystal - which do not include the gap (2.6 cm) between the endcap and crystal - were shortened in comparison to the customary 25-cm value, due to the very low activities of the sources. These distances were chosen as a reasonable trade-off

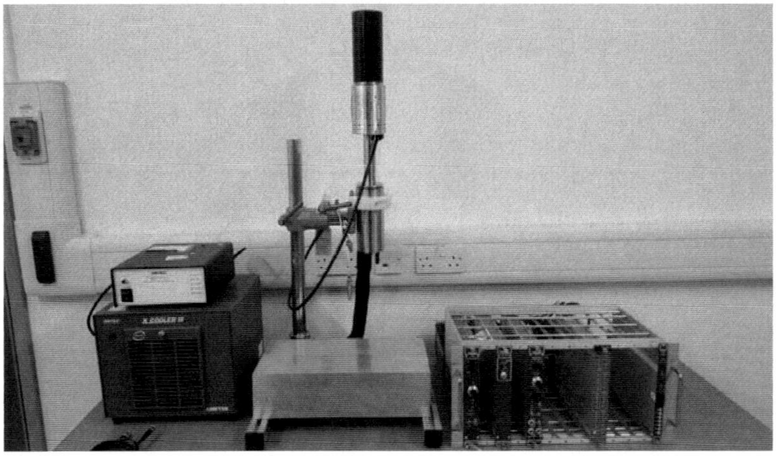

Figure 1: The Ortec coaxial HPGe detector used for measurements.

between maximization of signal-to-noise (S/N) ratio and minimization of coincidence summing (which induces FEP loss). Additional distances for ^{137}Cs were taken for future employment in simulations.

Following the same route, a second batch of measurements (Set 2) was taken in order to correct source placement (according to a coaxial geometry, Fig. 2) and fix the non-optimal polarization of the semiconductor junction by switching to -3800 V in the high-voltage (HV), compared to the initial -1800 V value.

No collimation and/or Compton-suppression arrangement were used.

3. Data reduction and analysis

After a channel-by-channel background subtraction from any source spectrum, corresponding spectra in Sets 1 and 2 were compared to one another. Though the former show larger S/N ratios (a neat example is shown in Fig. 3 for barium), the latter were chosen for subsequent analysis due to the significant reduction in the left skewing of peaks following HV correction (an example is presented in Fig. 4 for cobalt), with consequent enhancement of spectral resolution, lowering of the low-energy threshold down to about 20 keV, and removal of any possible multi-peak fitting of single lines.

The first part of the data analysis was performed by Hypermet-PC software [8, 11], which is currently considered the most reliable - though closed-source - package for the automated analysis of complex γ-ray spectra [12]. The procedure outlined below follows the indications reported in [5]. After direct editing and update of the Hypermet-PC database of nuclear data (originally dating back to the early 2000s), the following relationship

$$E = a + m \cdot Ch, \tag{1}$$

where a (in keV, like E) and m represent the intercept and slope, respectively, of the calibration line, while Ch is the position of the peak centroid (in channels), was used to perform an initial energy pair-calibration (*i.e.*, the conversion of channels into energies) for each spectrum with no consideration - at a first approximation - of non-linearity effects introduced by the ADC.

This calibration was used to perform a first fit of each spectrum ("peaklist 0"), taking into account that the ith peak profile in the jth channel is formalized as:

Figure 2: Source placement for Set 2.

$$p(j) = \Gamma_i e^{-\left(\frac{j-j_{i,0}}{\delta}\right)^2} + \frac{1}{2} A_i e^{\frac{j-j_{i,0}}{\beta}} erfc\left(\frac{\delta}{2\beta} + \frac{j-j_{i,0}}{\delta}\right), \quad (2)$$

where the first term in the right part of the equation is a Gaussian of amplitude Γ_i and variance $\frac{\delta}{\sqrt{2}}$, and the second one is an exponentially-modified Gaussian (EMG) - of amplitude A_i, variance $\frac{\delta}{\sqrt{2}}$, and exponential-decay time constant β - representing the left skewing.

An additional baseline correction of the form:

$$S_i\, erfc\left(\frac{j-j_{i,0}}{\delta}\right) + \frac{1}{2} W_i e^{\frac{j-j_{i,0}}{\tau}} erfc\left(\frac{\delta}{2\tau} + \frac{j-j_{i,0}}{\delta}\right) + c_0 + c_1 j + c_2 j^2, \quad (3)$$

was applied, including in progressive order:

- an error function of amplitude S_i, which accounts for photons scattered at small angles;
- a "tail"contribution (EMG with exponential-decay time constant $\tau \gg \beta$), which accounts for detector surface-effects, especially at very low energies;
- a parabolic contribution with polynomial coefficients c_0, c_1, and c_2.

A higher-order refinement of the above fit ("peaklist 1") was, then, performed by a non-linearity assessment accomplished by construction of a non-linearity curve via orthonormal-polynomial least-square (OPLS) fit of source peak-positions in peaklist 0. This procedure - which consists of measuring the positions of the FEP centroids in the spectrum of a multi-emitter, and then evaluating their displacement from a calibration linear curve [5,8] - is more accurate than the one suggested by the IEEE, which implies the use of a high-precision pulse generator at the test-input terminal of the pre-amplifier to "simulate" the electronic-noise contribution to the FEP FWHM [6].

Being only one multi-emitter (^{133}Ba) available in the calibration set, this initially limited result accuracy over the energy range $\geq$ 400 keV. The problem was bypassed via an *ad hoc* correction of centroid errors for ^{137}Cs and ^{60}Co by picking appropriate values reported in LNE-LNHB/CEA tables.

The ensemble of all peaklists 1 was finally used to perform the assessment of the intrinsic part of $\epsilon_{fep}(E)$, namely $\epsilon_{int}(E)$, that can be defined as [5]:

$$\epsilon_{int} = \frac{C}{P_\gamma N \frac{t_{live}}{t_{acq}}} \qquad \textit{(abs calibrated source)} \quad (4)$$

IOP Conf. Series: Journal of Physics: Conf. Series **1055** (2018) 012010 doi:10.1088/1742-6596/1055/1/012010

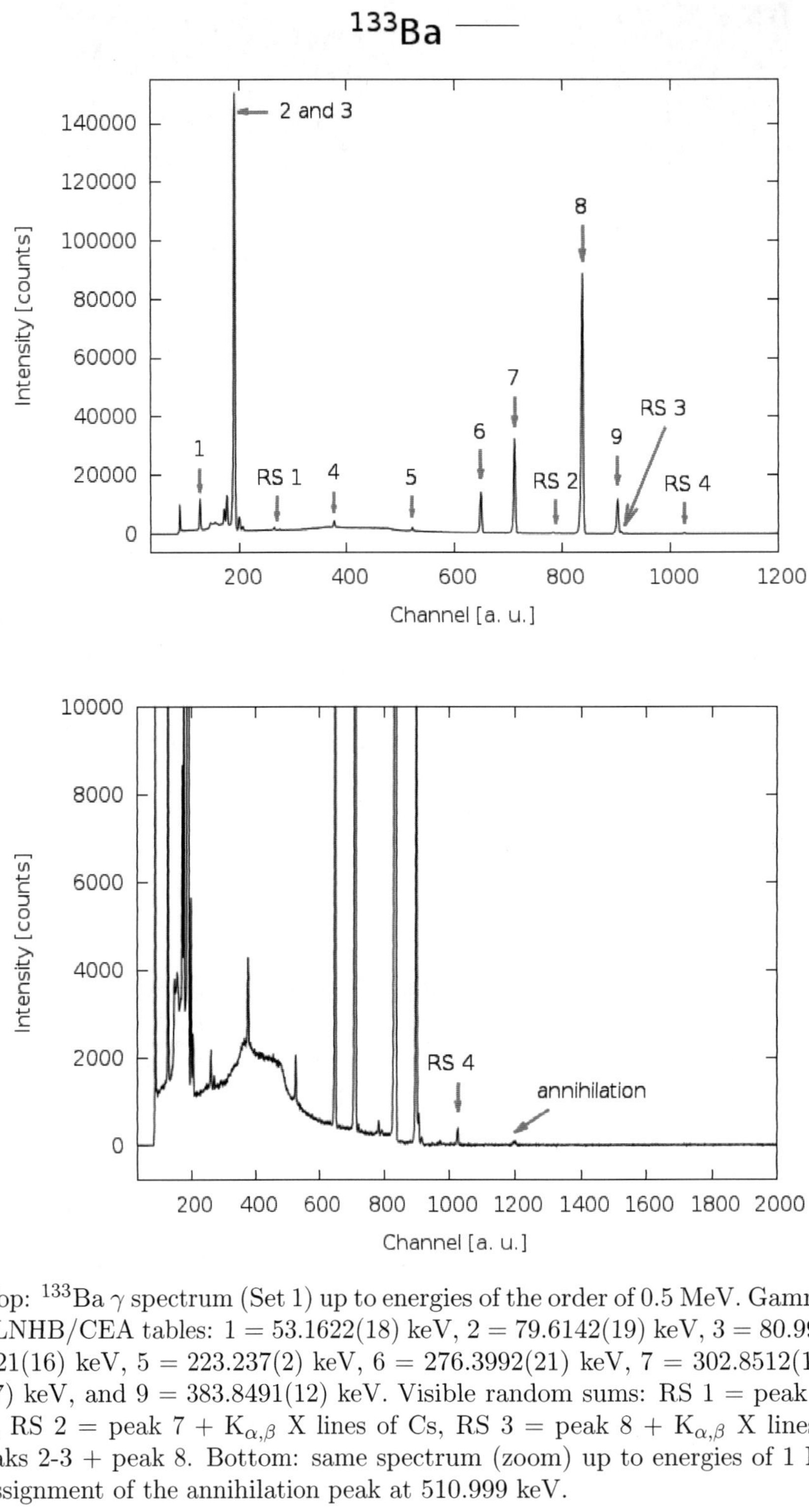

Figure 3: Top: ^{133}Ba γ spectrum (Set 1) up to energies of the order of 0.5 MeV. Gamma emissions from LNE-LNHB/CEA tables: 1 = 53.1622(18) keV, 2 = 79.6142(19) keV, 3 = 80.9979(11) keV, 4 = 160.6121(16) keV, 5 = 223.237(2) keV, 6 = 276.3992(21) keV, 7 = 302.8512(16) keV, 8 = 356.0134(17) keV, and 9 = 383.8491(12) keV. Visible random sums: RS 1 = peak 3 + $K_{\alpha,\beta}$ X lines of Cs, RS 2 = peak 7 + $K_{\alpha,\beta}$ X lines of Cs, RS 3 = peak 8 + $K_{\alpha,\beta}$ X lines of Cs, and RS 4 = peaks 2-3 + peak 8. Bottom: same spectrum (zoom) up to energies of 1 MeV, with a tentative assignment of the annihilation peak at 510.999 keV.

IOP Conf. Series: Journal of Physics: Conf. Series **1055** (2018) 012010 doi:10.1088/1742-6596/1055/1/012010

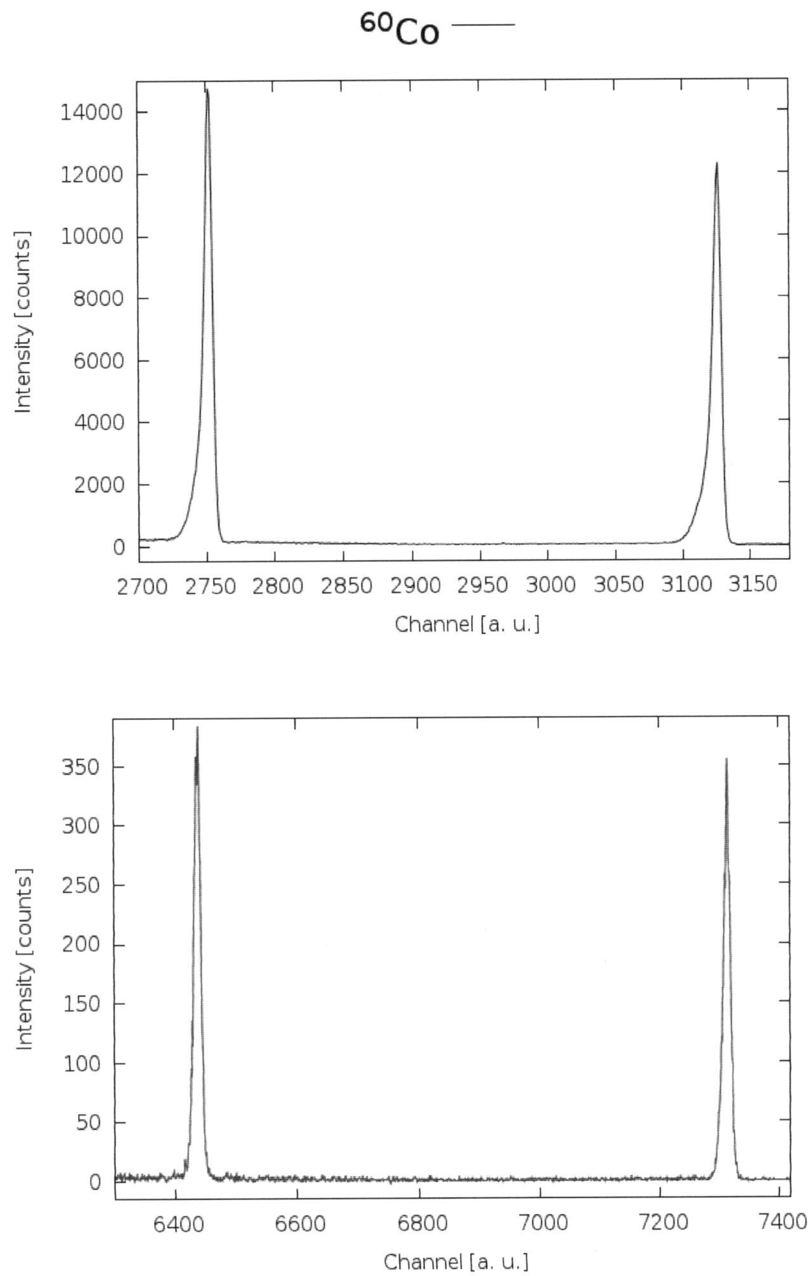

Figure 4: Top: 1.173- e 1.332-MeV lines of ^{60}Co in the Set-1 spectrum. Bottom: same lines in the correspondent Set-2 spectrum.

Table 2: Coefficients recovered by OPLS fitting (ROOT software) of discrete efficiency points associated with radionuclide peaks.

	a_0	a_1	a_2	a_3
grade 2	-1.979 ± 0.216	4.128 ± 0.082	-0.378 ± 0.007	-
grade 3	-29.776 ± 0.046	19.347 ± 0.014	-3.074 ± 0.002	0.1556 ± 0.0002

$$\epsilon_{int} = \frac{C}{P_\gamma} \qquad \textit{(non calibrated source)}, \tag{5}$$

where

- C = number of counts for each peak
- P_γ = γ-emission probability
- $N(t) = \frac{A_0}{\lambda} e^{-\lambda(t-t_0)}(1 - e^{-\lambda t_{acq}})$ = number of decays from the start of measurement
- t_{live} = livetime
- t_{acq} = acquisition time.

In this way, discrete efficiencies were calculated for each line of each radionuclide, selecting one source (^{133}Ba) as absolutely calibrated, and using it as a normalizing factor for the remaining standards (which practically corresponds, in the log-log scale, to an efficiency shift of ^{60}Co, ^{137}Cs, and ^{241}Am along the vertical axis to match ^{133}Ba values).

Then, an OPLS fit of the above discrete efficiency points, based on the commonly-used function [13, 14]

$$ln\,[\epsilon_{int}(E)] = \sum_{i=0}^{n} a_i [ln(E)]^i, \tag{6}$$

(linearized χ^2 method) was performed in order to get the desired $\epsilon_{int}(E)$ curve. Due to Hypermet-PC being proprietary software, the obtained curve could not be exported to a regular ASCII file. An external reconstruction of $\epsilon_{int}(E)$ was attempted, starting from fit coefficients a_i calculated by Hypermet-PC. Yet, even taking into account the remapping of energies onto the $(-1, 1)$ range operated by the software in order to improve the stability of the fitting algorithm [5, 15], no adequate correspondence was found, probably due to an unknown compression/expansion factor applied by Hypermet-PC to the curve. Consequently, a parallel fit (again by Eq. 6) of the same discrete efficiency points was performed via ROOT software [16], recovering the results summarized in Table 2 and Fig. 5. Low-grade polynomial functions were used in order to avoid any possible artificial oscillation of the resulting efficiency response curve in those regions covered by sporadic experimental points.

Finally, in Fig. 6 the grade-3 fit for $\epsilon_{int}(E)$ is shown together with its uncertainty region, after normalization to the mean value of the discrete intrinsic efficiency of the barium line at 160.6121 keV.

In order to get the global absolute FEP efficiency $\epsilon_{fep}(E) = \epsilon_{geom}\epsilon_{int}(E)$ for a point-source geometry, a multiplication must be performed by a geometric factor $\epsilon_{geom} = \frac{\Omega}{4\pi}$, with $\Omega = 2\pi\left(1 - \frac{d}{\sqrt{d^2+r^2}}\right)$ = solid angle "seen" by the detector, d = source distance from the crystal, and r = crystal radius. In our case, at $d = (11.5 + 2.6)$ cm and $r = 2.98$ cm, it is $\epsilon_{geom} = 0.011$, which explains the initial need for reducing the source-detector distance from 25 to 14.1 cm.

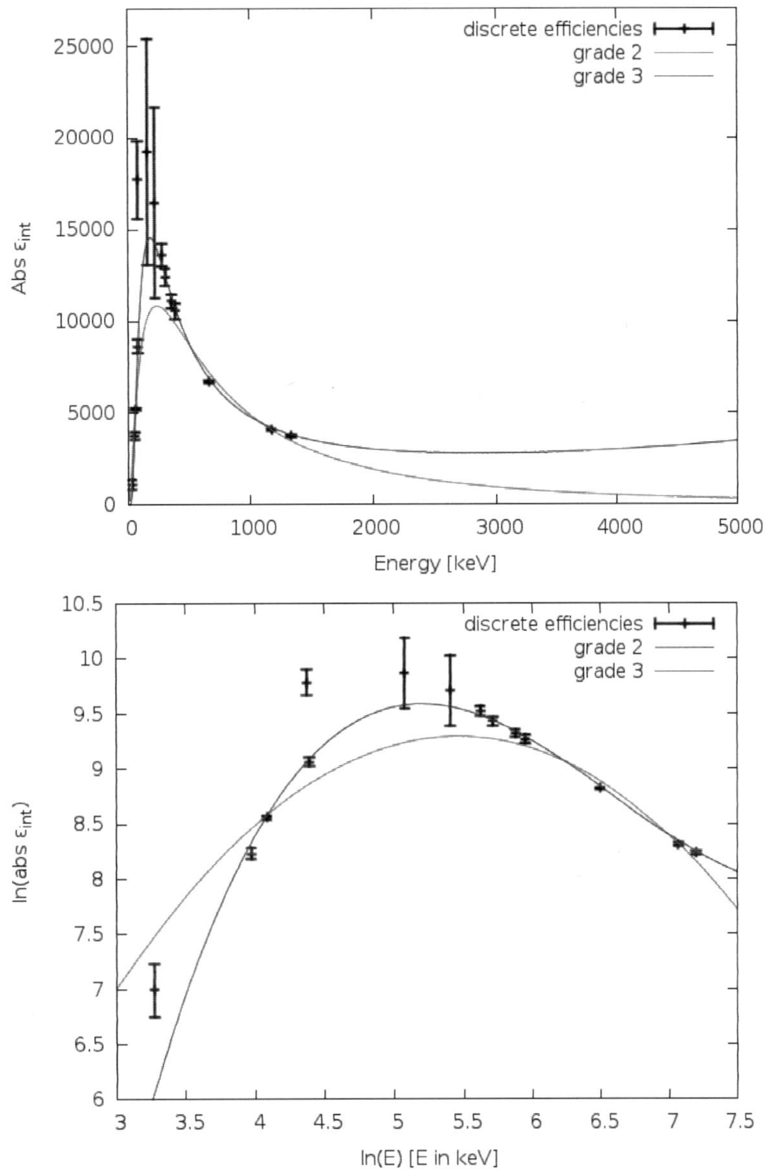

Figure 5: Top: lin-lin representation of the absolute intrinsic efficiency curves (grade 2 and 3) obtained by OPLS fitting (ROOT software) of discrete efficiency points associated with radionuclide peaks. Bottom: same curves in a log-log representation.

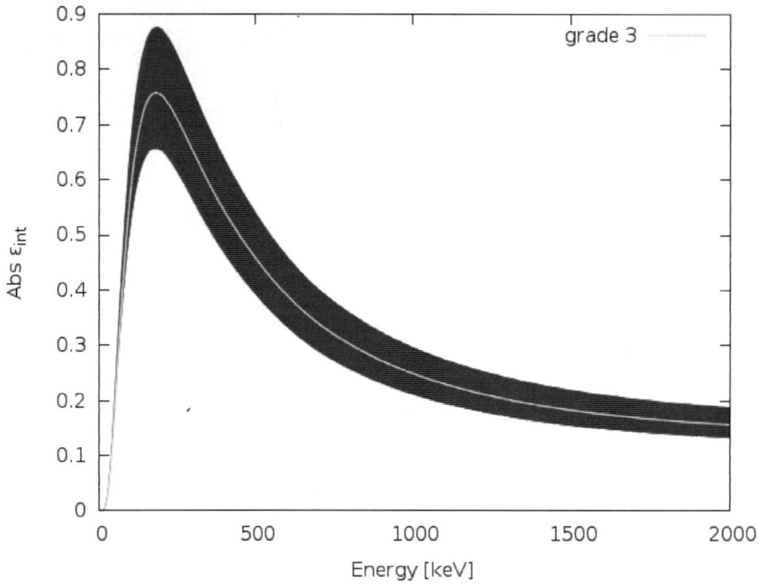

Figure 6: Lin-lin representation of the grade-3 fit for the absolute intrinsic efficiency $\epsilon_{int}(E)$, after normalization to the mean value of the discrete efficiency of barium line at 160.6121 keV - together with its uncertainty region.

4. Discussion

A quick look at Table 2 and Fig. 5 suggests that, in spite of the grade-3 fit showing an unphysical growth of the intrinsic efficiency curve in the high-energy region (where absorption cross-sections tend to decrease [7]), this same curve fits best the low-energy portion of the spectrum.

Consequently, we have opted for saving this grade-3 low-energy "patch", with the purpose to link it - by an appropriate continuity condition imposed on $\epsilon_{int}(E)$ at the matching point - to a high-energy complement obtained by future measurements on other non-calibrated sources, and/or ongoing Monte Carlo simulations of the detector efficiency, and/or similar data from the scientific literature.

5. Conclusions

The absolute efficiency of an Ortec n-type coaxial HPGe detector, used for PGAA and T-PGAA measurements on functional and Cultural Heritage samples at the ISIS pulsed neutron and muon source [2, 3, 17, 18], was determined by the exploitation of a set of calibrated γ-emitting sources following the procedures outlined in Refs. [5, 6]. The results are in agreement with the values reported in the literature [5, 19, 20].

This work will allow to carry out quantitative T-PGAA measurements on multicomponent samples, primarily addressing functional materials and Cultural Heritage artifacts, as well as quantitative estimates of the absolute γ background at ISIS beamlines.

Acknowledgements

This work was supported within the CNR-STFC Agreement 20142020 (No. 3420 20142020) concerning collaboration in scientific research at ISIS Spallation Neutron Source, and partially supported within the SMART Campus Project (Regione Lazio 2015-2017).

IOP Conf. Series: Journal of Physics: Conf. Series **1055** (2018) 012010 doi:10.1088/1742-6596/1055/1/012010

References

[1] Toh Y, Ebihara M, Kimura A, Nakamura S, Harada H, Hara K Y, Koizumi M, Kitatani F and Furutaka K 2014 *Anal. Chem.* **86** 12030
[2] Festa G, Arcidiacono L, Pappalardo A, Minniti T, Cazzaniga C, Scherillo A, Andreani A and Senesi R 2016 *JINST* **11** C03060
[3] Tian K V, Festa G, Szentmiklósi L, Maróti B, Arcidiacono L, Laganà G, Andreani C, Licoccia S, Senesi R and Cozza P 2017 *J. Anal. At. Spectrom.* **32** 1420
[4] Knoll G F 2000 *Radiation Detection and Measurement* 3rd ed (John Wiley & Sons, Inc.)
[5] Molnár G L, Révay Z and Belgya T 2002 *Nucl. Instrum. Methods Phys. Res. A* **489** 140
[6] *IEEE Std 325-1996 - IEEE Standard Test Procedures for Germanium Gamma-Ray Detectors* (New York, U.S.A.: The Institute of Electrical and Electronics Engineers, Inc.)
[7] Molnár G L (ed) 2004 *Handbook of Prompt Gamma Activation Analysis with Neutron Beams* (Kluwer Academic Publishers)
[8] Fazekas B, Révay Z, Östör J, Molnár G L and Simonits A 1999 *Nucl. Instrum. Methods Phys. Res. A* **422** 469
[9] (Last accessed on February 2018) `http://www.lnhb.fr/nuclear-data/nuclear-data-table/`
[10] (Last accessed on February 2018) `http://www.isis.stfc.ac.uk`
[11] *Hypermet-PC, Version 5.0 - User's Manual* (Budapest, Hungary: Institute of Isotopes)
[12] Zahn G S, Genezini F A and Moralles M *2009 International Nuclear Atlantic Conference - INAC 2009*
[13] Debertin K and Helmer R G 1988 *Gamma- and X-Ray Spectrometry with Semiconductor Detectors* (Amsterdam: North Holland)
[14] Kisab Z, Fazekas B, Östör J, Révay Z, Belgya T, Molnár G L and Koltay L 1998 *Nucl. Instrum. Methods Phys. Res. A* **418**(2) 374
[15] Gadjokov V and Jordanova J 1984 *Comput. Phys. Commun.* **31** 53
[16] *ROOT - An Object-Oriented Data Analysis Framework* (Geneva, Switzerland: CERN)
[17] Festa G, Andreani C, Arcidiacono L, Burca G, Kockelmann W, Minniti T and Senesi R 2017 *JINST* **12** P08005
[18] Andreani C *et al.* 2017 *J. Anal. At. Spectrom.* **32** 1342
[19] Wang T K, Mar W Y, Ying T H, Liao C H and Tseng C L 1995 *Appl. Radiat. Isot.* **46**(9) 933
[20] *Profile HPGe Photon Detector - Product Configuration Guide* (Ortec®)

IOP Conf. Series: Journal of Physics: Conf. Series **1055** (2018) 012011 doi:10.1088/1742-6596/1055/1/012011

Nuclear kinetic energies from final-state effects in the harmonic limit

M Krzystyniak[1,2], G Romanelli [1], R Tolchenov[1], M Gigg[1], B Hewer[1], and F Fernandez-Alonso[1,3]

[1] ISIS Facility, Rutherford Appleton Laboratory, Chilton, Didcot, Oxfordshire OX11 0QX, United Kingdom

[2] School of Science and Technology, Nottingham Trent University, Clifton Campus, Nottingham NG11 8NS, United Kingdom

[3] Department of Physics and Astronomy, University College London, Gower Street, London WC1E 6BT, United Kingdom

E-mail: matthew.krzystyniak@stfc.ac.uk

Abstract. This work shows a simple, yet quite powerful, data pre-treatment protocol for mass-resolved neutron spectroscopy, aimed at a better estimation of the main quantum observable linked to a nuclear-momentum distribution, its second moment. From a methodological point of view, the immediate benefit of having such a protocol is twofold. Firstly, a good estimate of a second moment of a nuclear-momentum distribution, and hence nuclear kinetic energy, provided as input for subsequent data fitting, accelerates the convergence of a data fit and minimises the likelihood of it being stuck in a non-physical solution. Secondly, it provides a simple screening tool in the search for quantum systems exhibiting statistically significant departures for a classical behaviour of equipartition of kinetic energy at any given temperature. This second benefit renders the presented protocol an important data screening tool in the search of materials exhibiting exotic properties, possibly attributed to nuclear quantum effects.

1. Introduction

Neutron Compton scattering (NCS), owing to its unique capability to measure nuclear momentum distributions (NMDs) in a mass-resolved manner, has been a method of choice in many material science studies aiming at characterising functional properties related to nuclear quantum effects (NQEs) [1, 2, 3].

The first step on this journey is a robust prediction of the second moment of NMDs, as well as the related values of nuclear mean kinetic energies. These important nuclear quantum observables have been historically linked to nuclear zero-point energy [3], which influences many material properties, starting from thermodynamical potentials [4], through phase boundaries and activation energies for kinetic rate constants and diffusion [5], and all the way to isotope fractionation ratios and pH values, to be mentioned but a few [6].

More recently, a few other 'nuclear quantum yardsticks' have been developed to characterise the degree in which a given material under investigation is affected by NQEs [1, 2]. The first of them is the 'nuclear quantum temperature' which, unlike its thermodynamic counterpart, is a mass-specific property reflecting how the amount of kinetic energy of a given nucleus present in a condensed matter system translates into an equivalent amount of temperature [1, 2]. Recently, another parameter characterising 'nuclear quantumness' of a system has been proposed, a quantum-energy excess (QE), defined as the ratio of the nuclear kinetic energy measured in an NCS experiment to a classical prediction based on energy equipartition theorem [7]. The way the QE is defined is reminiscent of a description provided by Ceriotti when describing the degree in which NQEs play a role in the description of nuclear dynamics of hydrogen and oxygen in liquid ice and water [8, 6]. To this end, a very striking observation is that even at 750 K atoms in water are far from classical behaviour. Oxygen atoms have kinetic energies 10% higher and hydrogen atoms have nearly double the kinetic energy compared to the classical prediction, giving at this temperature a mean kinetic energy per atom of 97 meV [8]. Moreover, even heavier atoms exhibit a considerable amount of QE. In the solid caesium hydrogen sulfide it was found that only the heavier caesium atoms reach the classical limit at room temperature. Contrary to naïve expectation, sulfur exhibited a more pronounced quantum character relative to classical predictions than the lighter oxygen atom, which was interpreted as an NQE arising from the tighter binding environment of this species [7].

The examples of nuclear quantum observables and their links to important material properties, mentioned above, constitute a motivation for a robust NCS data pre-treatment protocol aiming at constraining the fitting of the second moments of NMDs to regions in parameter space lying much closer to physically meaningful solutions. Historically, the first attempt with respect to direct kinetic energy extraction from NMDs was made by Senesi [9] by directly, *i.e.*, without resorting to nonlinear fitting of the scattering spectra, employing the sum rules associated with the scattering functions.

The work presented in this contribution represents an extension of this line of NCS work in two important aspects. Firstly, it is a further contribution to the development of the robust and efficient methodology of the extraction of nuclear kinetic energy from experimental NCS data. Secondly, it extends the methodology towards the fitting-free estimation of nuclear kinetic energies in systems with complicated stoichiometry yielding NCS spectra with partially overlapping peaks. The overarching motivation of the presented work is to provide a simple screening tool in the search for quantum systems exhibiting statistically significant departures from a classical behaviour of equipartition of kinetic energy. Such a screening protocol is a much needed tool for system-tailored *ab initio* modelling based on the degree in which NQEs manifest themselves in the NCS data.

This work is organised in the following manner. First, the Sec. 2 describes the basic theoretical assumptions and approximations outlining the concept of the nuclear momentum distribution and its mathematical connection to the expression describing the overall shape of the signal recorded in the time-of-flight (TOF) domain in the neutron Compton scattering technique. Specifically, the Subsecs. 2.1 and 2.2 describe, respectively, how these systematic departures from the basic approximation can be taken into account in a perturbative manner and how they can be employed to infer the information about the widths of the momentum distributions from the peak positions of a TOF spectrum under an assumption of harmonic lattice vibrations.

2. Methodology

2.1. Nuclear momentum distributions from neutron Compton scattering

Detailed accounts of the NCS method, in terms of both its instrumental and mathematical foundations, can be found elsewhere [1, 3, 2]. In this contribution, we will thus concentrate only on the main physical assumptions and mathematical approximations underlying the way NMDs are obtained in NCS, with special emphasis on the description of departures from them, termed as Final-State Effects (FSEs).

In what follows, throughout the rest of the text, we will use terms nuclear momentum distribution and longitudinal momentum distribution interchangeably. Moreover, in what follows the term 'NMD peak position for a given nucleus' will be defined as the position of the NMD maximum. Mathematically, this means that the NMD peak position is defined for a smooth peak shape as the locus of the place at which the first derivative of the curve describing the peak, calculated with respect to the domain in which the peak is plotted, vanishes. In experimental practice, however, NMD peak positions are rather determined by employing a peak-maximum search algorithm.

We start our description with a a brief account of the Impulse Approximation (IA). In an nutshell, the IA is a way of mathematically describing a relatively simple physical reality of the kinematics of collisions of very fast (epithermal) neutron projectiles with the target being the sample placed in a neutron beam, such as the one available at the VESUVIO spectrometer at the ISIS Pulsed Neutron and Muon Source in the United Kingdom [10, 11]. In such fast collision regime, the energy (E) and momentum (q), transferred from the impinging neutrons to the sample in the scattering process, are sufficiently high (in the range 1–100 eV and 30–200 Å^{-1}, respectively) that the kinematic description boils down essentially to binary collisions between neutrons and individual nuclei present in the sample under investigation. Moreover, such binary collision is, within the IA, governed by energy and momentum conservation [1, 3, 2]. It is this description which underlies the very unique feature of the NCS as a spectroscopic method, that is, its intrinsic mass-selectivity.

Mathematically, the IA is set out using the following chain of approximations. First, the dynamic structure factor $S(q, E)$, for a given atom of mass M, is reduced to a single peak centered at the recoil energy $E_r = \hbar^2 q^2/2M$ with $\hbar = 2.0446$ (meV amu)$^{1/2}$ Å [12, 13, 14, 15, 16, 17, 18, 3]. This first approximation paves the way to the West scaling property of NCS. Namely, a dynamic structure factor, $S(q, E)$, of a sample consisting of N different atomic species with different nuclear masses M_N, can be decomposed into N individual peaks in the domain of the longitudinal-momentum space $J(y)$ of the respective masses, $S(q, E) = \sum_M (M/^{-}\hbar q) J_M(y_M)$. In this decomposition, y_M is the initial radial momentum p of the nucleus with mass M projected onto the scattering vector q, $J_M(y_M)$ is the so called NCS profile [13, 12] and formally corresponds to the NMD of the target nucleus along y_{M_N}.

In this work, motivated by the fact that, in the bulk of present-day NCS work, NQEs are simulated using the approximation of harmonically bound isotropic systems [1, 2], we will consider only spherically averaged three-dimensional longitudinal NMDs. In this case, $J_{IA}(y)$ for a given mass M in the IA limit depends only on the magnitude of y but not on $\hat{q}$. Moreover, as we want to find a robust way to approximate the second moment of an NMD by obtaining its standard deviation (STD) value, denoted by σ. In doing so, we will concentrate on the leading, purely Gaussian term only:

$$J_{IA}(y) = \frac{1}{\sigma\sqrt{2\pi}} \exp\left(\frac{-y^2}{2\sigma^2}\right). \tag{1}$$

where σ, with the present choice of units, is expressed in Å^{-1}. For finite values of q, corrections to the IA are known as FSEs [18, 3]. To account for FSEs, the method of Sears [12] is routinely incorporated into standard NCS data treatments [18] by expressing the measured NCS profile $J(y)$ as a series of the form

$$J(y) = J_{IA}(y) + J_{FSE}(y) = J_{IA}(y) - \frac{M\langle\nabla^2 V\rangle}{36\hbar^2 q}\frac{d^3}{dy^3}J_{IA}(y) + ... \tag{2}$$

where $\langle\nabla^2 V\rangle$ is the mean value of the Laplacian of the potential energy of the nucleus expressed in meV Å^{-2} (cf. Refs. [18, 19]). Importantly, an explicit connection between the magnitude of FSEs k and $\sigma =$ can be established as $k = \sqrt{2}\sigma/12$ [20, 21, 22] with the FSE term, given by the Eq. 2, expressed using the third order Hermite polynomial, $H_3(\frac{y}{\sigma\sqrt{2}})$. This results with an

IOP Conf. Series: Journal of Physics: Conf. Series **1055** (2018) 012011 doi:10.1088/1742-6596/1055/1/012011

expression:

$$J(y) = \frac{1}{\sigma\sqrt{2\pi}} \exp\left(\frac{-y^2}{2\sigma^2}\right)\left[1 - \sigma\frac{\sqrt{2}}{12q}H_3\left(\frac{y}{\sigma\sqrt{2}}\right)\right]. \tag{3}$$

The longitudinal momentum distribution of a given nucleus, $J(y)$ is not directly accessible. Most generally, the starting point for the analysis is the expression describing the count rate, $C_\theta(t)$, of a neutron detector for a scattering angle θ in the time-of-flight, t, that takes explicitly into account the resolution of the instrument. This resolution in most generally incorporated into the expression for $C_\theta(t)$ in the form of multiple integrals, performed independently for each value of the TOF, t, over the distributions of geometrical and energy analyser parameters of the spectrometer [23]. In practice, however, a sufficiently good degree of reproduction of experimental spectra is usually achieved by incorporating the instrument resolution in form of sum of convolutions of longitudinal momentum distributions, $J_M(y_M)$, with their respective mass-dependent resolution functions, $R_M(y_M)$, referred to as the Convolution Approximation (CA) [3, 2, 24]:

$$C_\theta(t) = A'\left[\frac{E_0 I(E_0)}{q}\right]_t \sum_M I_M M J_M[y_M(t)] \otimes R_M[y_M(t)] \tag{4}$$

Equation 4 conveys two important messages about the way in which the information about the NMD is hidden in an NCS experimental spectrum. Both of them are responsible for the difficulties which NCS data treatment faces when trying to predict basic NMD properties, especially at the stage of pre-fitting the initial raw data. The first realisation is that two pre-factors need to be applied to NMD for mass M, $J_M(y_M(t))$ given by Eq. (3), to convert it from the domain of the longitudinal momentum y_M into the t domain: (i) A' is a mass–independent experimental constant, and (ii) the mass–independent factor $[E_0 I(E_0)/q]_t$ that depends on the incident neutron spectrum, $I[E_0(t)]$, the initial neutron energy, $E_0(t)$, and the momentum transfer $q(t)$, all explicit functions of t. Moreover and most importantly in the context of a robust NMD shape-prediction protocol, the second important feature of the count rate in the time-of-flight domain is that it contains a mass-dependent resolution function which can also be expressed in the domain of the longitudinal momentum, $R_M[y_M(t)]$. In general, this resolution function will smear out the features of the respective NMD profiles associated with the same masses and contribute to decreasing the mass-resolution of the technique by increasing the peak overlap in the spectrum. The resolution function of the instrument is also the reason why one resorts to fitting the time-of-flight spectra using the CA and the models such as the ones described by Eqs.(3) and (4).

The initial stage of NCS data fitting has been described in detail elsewhere [2, 3]. Thus, here only a few aspects of this procedure, relevant for the work presented, will be mentioned. The NCS data pre-treatment relies crucially on the fact that integrated peak intensities, denoted in Eq. 4 as I_M, are proportional to the scattering densities $I_M = AN_M 4\pi b_M^2$ where $4\pi b_M^2$ are the total (bound) neutron-scattering cross sections for masses M. [25, 18]. First, raw TOF data are usually fitted sequentially, detector by detector, using Eq.(4) in order to obtain the initial values of the fitting coefficients, I_M and σ_M. Then, the obtained values of I_M and σ_M are used as input parameters to calculate the sample-dependent forward scattering corrections: (i) multiple scattering (MS) from the sample and container, and (ii) gamma background (GB) due to cross-talk between signals created at the forward detectors by primary and secondary gold foils. It is at these two first steps of data pre-treatment were the initial good assessment of the values of σ_M is crucial. Resorting to fitting the count rates using Eq. 4 will naturally introduce correlations between I_M and σ_M as fitting parameters. These correlations will propagate through the whole data pre-fitting process and introduce systematic uncertainties in the shapes of the final NMD curves fitted to isolated recoil peaks after the whole series of correction have been applied to data.

IOP Conf. Series: Journal of Physics: Conf. Series **1055** (2018) 012011 doi:10.1088/1742-6596/1055/1/012011

It is thus crucial to develop a robust protocol to predict initial values of I_M and σ_M, possibly in a way that does not involve non-linear fitting algorithms. The method to derive I_M and σ_M directly employing the sum rules formulated for $J_M(y_M)$, historically the first and important step in this direction [9], performs robustly in case of non-overlapping peaks in a spectrum described using Eq.(4). In this work, we pave the way to a protocol that can be applied even in the case of partially overlapping peaks, and based on the fact that the information about σ_M is encoded in the magnitudes of the FSEs, as can be seen in Eq.(3). We describe the protocol and its mathematical foundations in the next subsection.

2.2. Nuclear kinetic energies from the positions of the maxima of NMDs in the presence of FSEs

The first important question related to the reconstruction of the NMD properties from experimental NCS data is whether the NMD shape is preserved by the convolution with the experimental resolution function. Mathematically, this question belongs to the group of problems tackled by the so-called 'scale-space methods' in image processing [26]. Specifically, the chief question to this end has been whether the number and the structure of local minima and maxima of a curve is altered by the convolution process. Using scale-space terminology, the resolution function $R_M[y_M(t)]$ in the convolution product, $J_M[y_M(t)] \otimes R_M[y_M(t)]$, is a convolution kernel termed as a 'mollifier'. A perfect, *i.e.*, shape-preserving mollifier can be axiomatically derived by requesting a series of properties to be satisfied. Scale-space axioms led Iijima [27] to derive a Gaussian kernel by requiring linearity, translation invariance, scale invariance, semi-group property, and preservation of positivity. Additionally, the scale-space evolution should not create new level curves when increasing the scale parameter s characterising the mollifier, *i.e.*, $R_M = R_M[y_M(t), s]$. If this is satisfied, iso-intensity linking through the scales is possible and a structure at a coarse scale can (in principle) be traced back to the original image. Koenderink [28] showed that a Gaussian mollifier with standard deviation $\sqrt{s}$ will obey this axiom of 'non-creation of local extrema', *i.e.*, no additional extrema will be created in the convolved image as the width $\sqrt{s}$ of the Gaussian mollifier increases [29].

Moreover, a question arises whether convolution preserves the unimodality of a function. In general, unimodality means there is only a single highest value of a given function, *i.e.*, if a is a mode for f then $f(x) < f(a)$ for all x. Ibragimov [30] showed that if one of the two functions is logarithmic concave (the logarithm of the function is itself a concave function) their convolution is unimodal. Furthermore, if both are logarithmic concave, so is their convolution [31]. To this end, the Gaussian mollifier plays a central role in image processing and physics as it can be shown to preserve unimodality also [30, 31, 29]. For this reason, in what follows the NCS resolution function will be approximated as a Gaussian, despite the fact that, when expressed in the longitudinal momentum space, $R_M[y_M(t), s]$, exhibits a line shape that can be best approximated by a Voigt profile [18, 10, 32]. There is a multitude of reasons for adopting this approximation. Firstly, both in the case of a Gaussian and Voigt profile, the resolution function is symmetric and centred at the zero value of the longitudinal momentum distribution of a given mass, M, $y_M(t) = 0$. In consequence, the locus of the maximum of the recoil peak $J_M[y_M(t)]$ will not be altered by switching between both profiles describing the shape of $R_M[y_M(t)]$. However, the mathematical expression defining this locus will be much simplified when using a Gaussian from of $R_M[y_M(t)]$. Secondly, both Gaussian and Voigt profiles are shape-preserving mollifiers. Both are logarithmic-concave functions and, as such, both preserve the structure of maxima and minima of functions they are convolved with. This ensures that, using both Gaussian and Voigt profiles to describe the resolution function, ensures that no additional minima or maxima of $J_M[y_M(t)]$ are created in the process of the convolution. Finally, it is also worth noting that the so-called double difference technique of NCS data acquisition [33, 34, 35, 36, 37] suppresses the Lorentzian wings of $R_M[y_M(t)]$ to such an extent that it can be described as nearly a

IOP Conf. Series: Journal of Physics: Conf. Series **1055** (2018) 012011 doi:10.1088/1742-6596/1055/1/012011

Gaussian. Moreover, the use of uranium, instead of gold foils, as the final energy analyser on VESUVIO leads to $R_M[y_M(t)]$ best described as a purely Gaussian function [18, 10, 32]. Thus, VESUVIO as an instrument, in the limiting case of vanishing Lorentzian part of the final energy resolution function, becomes and example of a perfect Gaussian blurring device from the point of view of image and pattern recognition theory. In passing, it is also worth noting that the 'non-creation of local extrema' axiom, fulfilled by the VESUVIO resolution function, has got one more very important consequence for the analysis of NMDs in systems exhibiting nuclear quantum tunnelling. Namely as noted by Car [38], in such case a theoretical model describing an NMD of a tunnelling nucleus would exhibit nodes. In an NCS experiment a log-convex VESUVIO resolution function would then preserve the nodes of the measured NCS spectrum.

Encouraged by these important results, we can proceed to the description of the protocol to estimate σ_M values, and hence the values of nuclear kinetic energy, from the positions of the maxima of the measured longitudinal momentum distributions in the presence of FSEs. In what follows, we will describe two limiting cases: (i) one of an instrument described by a purely Gaussian resolution function, and (ii) a perfect-resolution instrument. In both cases, the starting point is the Gaussian longitudinal momentum distribution for a given nucleus with mass M, with FSEs in the harmonic limit, $J_M(y_M)$ described by Eq.(3). Let us start with the most general case of non-vanishing width, s of the instrument resolution function $R_M(y_M)$ and write its convolution with the $J_M(y_M)$ explicitly:

$$J_M(y_M) \otimes R_M(y_M) = \int_{-\infty}^{\infty} \frac{\exp\left(-\frac{(y-u)^2}{2\sigma^2}\right)}{\sqrt{2\pi\sigma^2}} \left(1 - \frac{\sqrt{2}\sigma}{12q} H_3\left(\frac{y-u}{\sigma\sqrt{2}}\right)\right) \frac{\exp\left(-\frac{u^2}{2s^2}\right)}{\sqrt{2\pi s^2}} du. \quad (5)$$

Differentiating Eq(. 5) with respect to the variable y, performing the integral analytically, and equating the resultant expression to zero yields the condition:

$$\frac{-\sqrt{2}}{6} \frac{e^{\frac{-u^2}{2(s^2+\sigma^2)}}}{\pi q (s^2+\sigma^2)^{9/2}} P(\sigma, s, y, q) = 0 \quad (6)$$

where

$$P(\sigma, s, y, q) = \sigma^4 \left[\left(y^2 - \sigma^2 - s^2\right)^2 + 2\left(\sigma^2 + s^2\right)^2\right] + 3qy\left(\sigma^2 + s^2\right)^3 \quad (7)$$

Solving the Eq(. 6) amounts to solving the polynomial equation given by Eq.(7), that is a multiply nested radical equation. Algebraically, the equation has got four solutions. However, the only physically meaningful solution, $y_{max}(\sigma, s, q)$, has the following form:

$$y_{max}(\sigma, s, q) = -\sigma_{tot} \frac{Q(\sigma, s, q) - \sqrt{24\sigma^2 - Q^2(\sigma, s, q) + \frac{12\sqrt{2}q\sigma_{tot}^3}{Q(\sigma,s,q)\sigma}}}{2\sqrt{2}\sigma} \quad (8)$$

where $\sigma_{tot}^2 = \sigma^2 + s^2$, and

$$Q(\sigma, s, q) = \sqrt{\frac{8\sigma^3\Sigma^{1/3} + 16\, 2^{1/3}\sigma^6 + (2\Sigma)^{2/3}}{\sigma\Sigma^{1/3}}}, \quad (9)$$

$$\Sigma(\sigma, s, q) = \sigma\left(9q^2\sigma_{tot}^6 + \sqrt{R(\sigma, s, q)} + 32\sigma^8\right), \quad (10)$$

and

$$R(\sigma, s, q) = 81q^4\sigma_{tot}^{12} + 576q^2\sigma_{tot}^6\sigma^8 - 1024\sigma^{16}. \quad (11)$$

In addition to the exact analytic solution provided by Eq. 8, one can obtain an approximate solution by expressing the polynomial in the form

$$y_{max} = -\frac{\sigma^4\left[\left(y_{max}^2 - \sigma_{tot}^2\right)^2 + 2\sigma_{tot}^4\right]}{3q\sigma_{tot}^6} = F\left(y_{max}\right) \tag{12}$$

Furthermore, we can notice that, based on the fact that an FSE correction only slightly shifts the position that the maximum of an NMD would have otherwise had in the IA limit, $y_{max} = 0$, leads to the condition $y_{max}^2 \ll \sigma_{tot}^2$. One can therefore obtain an approximate solution to be refined through an iterating process. In particular, one has

$$y_{max}^{(0)} = -\frac{\sigma^4}{q\sigma_{tot}^2} \simeq \frac{\sigma^2}{q}, \tag{13}$$

where the last term on the right-hand side is expressed in the limit $s \to 0$. By inserting this result in the right-hand side of equation 12, one can define a recurrence relation

$$y_{max}^{(n)} = F\left(y_{max}^{(n-1)}\right) \tag{14}$$

It is worth noting that the solution of Eq.(8) does not explicitly depend on the recoiling nuclear mass M although the nuclear mass dependence is implicitly included in the definition of the West scaling variable y. Naturally, the solution can always be plotted in region of q and σ relevant for a given nuclear mass under investigation. In case of protons, this region is defined by q stretching from $q \approx 25$ Å^{-1} to $q \approx 125$ Å^{-1} and σ in the region of $\sigma \approx 4$ Å^{-1} to $\sigma \approx 6$ Å^{-1}. It is this region, for which the solution given by the Eq.(8) in the limit of vanishing $s \to 0$ is plotted in Figures 1 and 2 . For lighter masses, e.g. proton, deuteron, or lithium, the NCS scattering is by no means elastic [3, 2] and the position of the maximum starts depending non-linearly on the STD, especially for small values of the STD, i.e. for protons (see Fig. 2, bottom).

The next question that arises in the NCS experimental practice is how the position of the maximum of the experimentally observed NMD depends on the width of the resolution function, s. First thing to note to this end is that the width of the VESUVIO resolution function, $R(y_M)$, when expressed in the longitudinal momentum space of a given nucleus, is proven to vary as a function of the scattering angle, and thus the magnitude of the momentum transfer [18, 10, 32]. This variation is more pronounced for the case of the proton resolution function, where the component responsible for the variation is the angular resolution, whereas in case of heavier masses the dominating contribution, the energy resolution function, is shown to vary relatively less as a function of the momentum transfer [18, 10, 32]. The plot of the position of the maximum of the experimentally observed longitudinal momentum distribution as a function of the width of the resolution function in the longitudinal momentum space is shown in Figure 3. The plot has been generated using Eq(. 8) for the value of the width of the proton longitudinal momentum distribution fixed at, $\sigma = 4.5$ Å^{-1}, a value typical for many condensed matter systems containing protons [2, 3]. Moreover, for the sake of the discussion, the plot is shown for a wide range of the values of the widths of the resolution function. However, a typical value of the width of the instrument resolution function, s, expressed in the proton longitudinal momentum space, is $s =$ 1.7 Å^{-1}.

It can be seen that VESUVIO, due to kinematic constraints imposed on the instrument in order to be able to operate within the IA, presents a lucky case as far as the method described here is concerned. Namely, the main values of the resolution functions of the proton, deuteron and lithium are all located in the high q, low s region, whereas the the resolution functions of the heavyweight nuclei have their main values located in the low q, high s region [18, 10, 32].

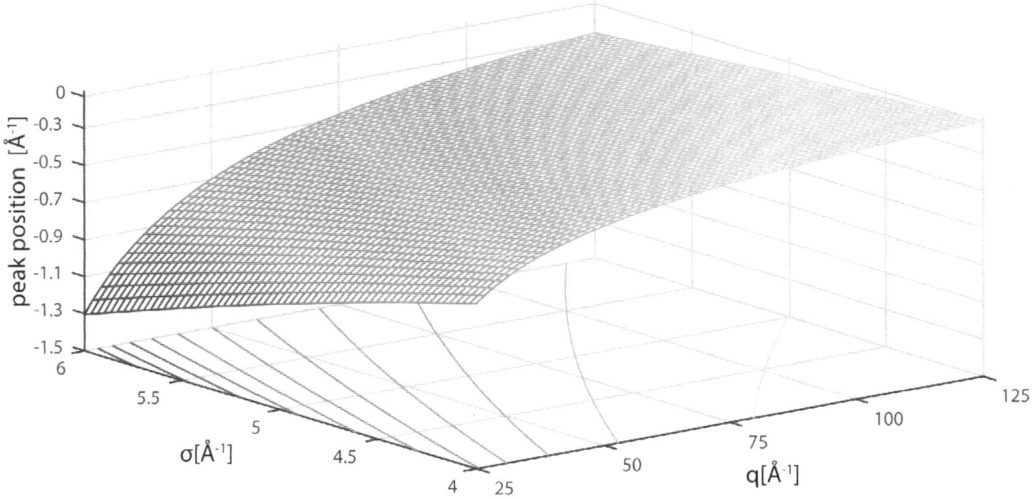

Figure 1. Positions of the maximum of a Gaussian NMD with FSEs calculated within harmonic limit, given by Eq.(8), for the case of perfect instrument resolution $s \to 0$. See text for details.

In both those cases, the position of the maximum of the longitudinal momentum distribution practically does not depend on q and s, *i.e.*, the method proves to be robust. It is also worth mentioning that this lack of sensitivity to the q and s values is observed (plots not shown here) in the entire region of the STDs of NMD of relevance, *i.e.*, σ = 4.5 -30 Å^{-1} [18, 10, 32].

2.3. Experimental Results

All NCS measurements were performed on the VESUVIO spectrometer at the ISIS Pulsed Neutron and Muon Source, United Kingdom [10, 11, 32]. The experimental protocol and NCS data reduction and treatment have been described in great detail elsewhere in the literature [10, 11, 32, 3, 2]. As the method described here differs to some extent from the normally applied NCS data analysis, in what follows we will only concentrate and emphasise those data treatment aspects that are crucial for the understanding of the method.

The starting point for the analysis are raw NCS data obtained in the time-of-flight (t) domain and described by the count rate, $C_\theta(t)$, of a neutron detector for a scattering angle θ [3, 2], given by Eq. 4 with the NMD for mass M, $J_M(y_M(t))$, is given, in the IA limit with harmonic FSEs, by the Eq. (5).

The method, for a given nucleus with the mass M, can be summarised in the following steps:

- raw TOF data, recorded by a given detector (placed at a given scattering angle θ), are transformed into the longitudinal momentum domain y_M;
- a peak search algorithm, implemented in the MantidPlot software [39], is performed in the y_M domain, capitalising on the fact that the expected peak position, to within the algorithm search radius, is expected around $y_M \sim 0$;
- a value of the momentum transfer magnitude $q = q_{max}$ is found numerically for a given value $y_M = y_{max}$, corresponding to the peak centre found by the algorithm in the previous step;
- both values, $[q_{max}, y_{max}]$, are recorded and the procedure is repeated for the next detector;

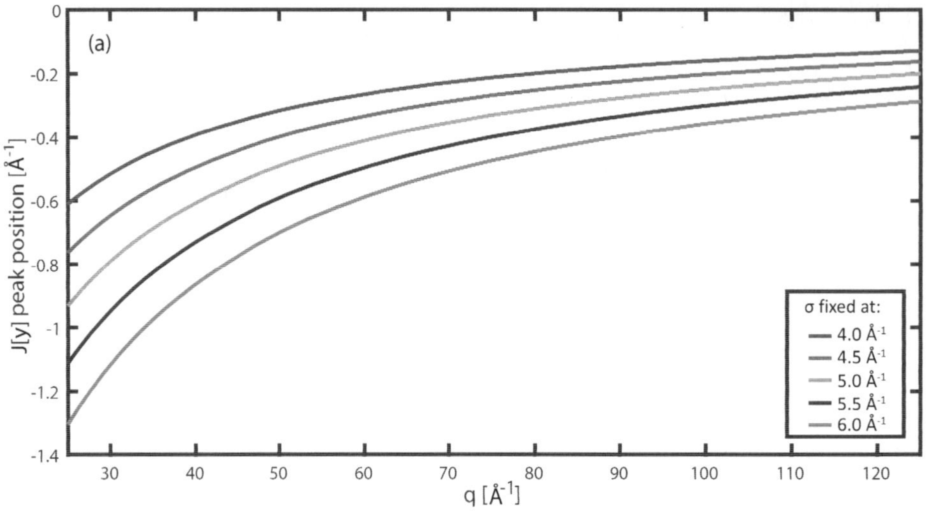

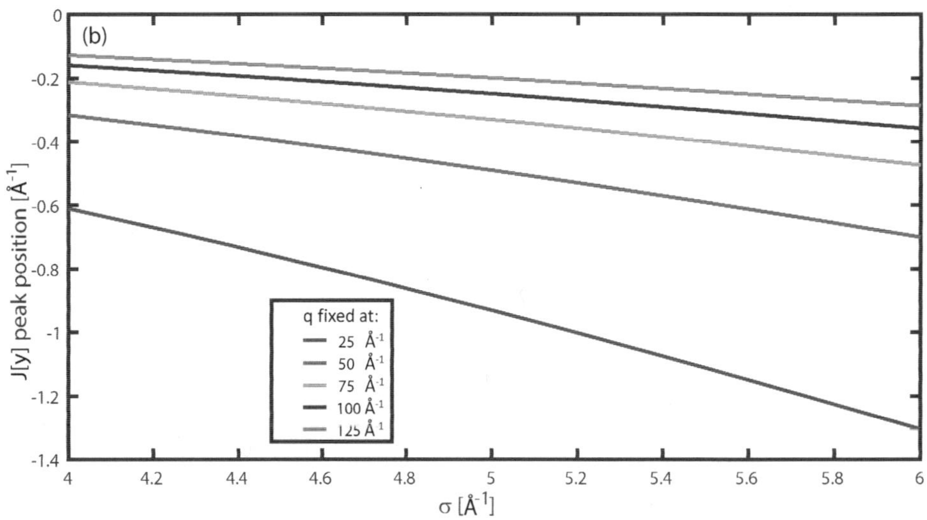

Figure 2. (a) Curves representing the loci of the maxima of the longitudinal momentum distributions, $J(y)$, calculated using Eq.(8), assuming a perfect instrument resolution function with the vanishing width, $s \to 0$. The loci are calculated as functions of the momentum transfer magnitudes at peak maxima, q, calculated keeping the values of the standard deviations of the Gaussian momentum distributions, σ, constant (see figure legend for the values of σ). The values of σ, q, and the loci of the maxima of the longitudinal momentum distributions are all expressed in the units of Å^{-1}.
(b) Curves representing the loci of the maxima of the longitudinal momentum distributions, $J(y)$, calculated using Eq.(8), assuming a perfect instrument resolution function with the vanishing width, $s \to 0$. The loci are calculated as functions of the values of the standard deviations of the Gaussian momentum distributions, σ, calculated keeping the values of the momentum transfer magnitudes at peak maxima, q, constant (see figure legend for the values of q). The values of σ, q, and the loci of the maxima of the longitudinal momentum distributions are all expressed in the units of Å^{-1}. See text for details.

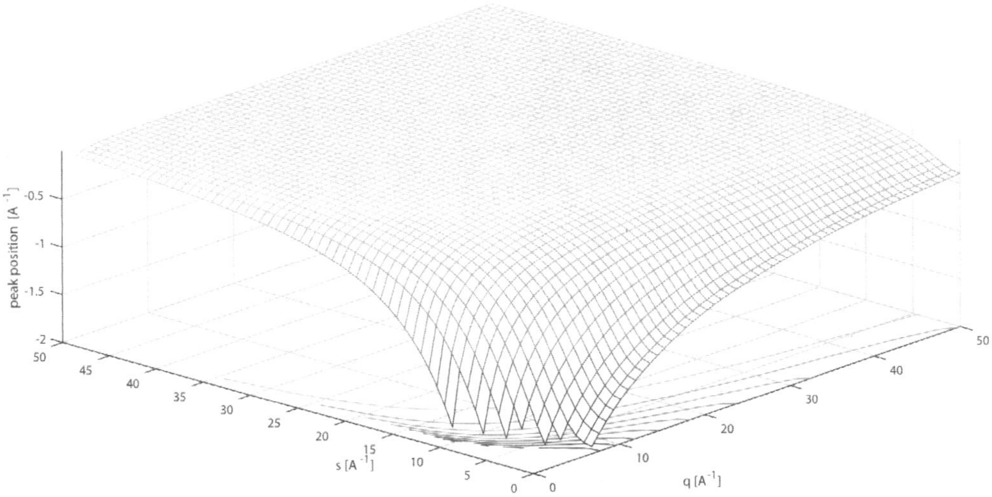

Figure 3. Positions of the maximum of Gaussian NMD with FSEs calculated within harmonic limit as a function of the instrument resolution width s, expressed in the longitudinal momentum space. The plot has been generated using Eq.(8) with $\sigma = 4.5$ Å^{-1}, a value typical for proton, for which a typical value of the resolution function width s is 1.7 Å^{-1} [3, 2]. See text for details.

- the peak maximum loci are fitted with the function given by Eq.(8). In fitting, the parameter, s_M, is fixed at the average of all detectors value of s_M. Two parameters are free in fitting. The first one is the width of the nuclear momentum distribution, σ. The second one is, y_0, accounting for a small albeit systematic shift of the recoil peak positions due to slightly imperfect instrument calibration. A similar parameter is routinely used in fitting NMDs in NCS [3, 2].

In what follows, two different grouping protocols are described: (i) a sequential, detector-by-detector procedure with no detector grouping; and (ii) a procedure whereby signals are first grouped according to detector bank grouping [10, 11, 32], *i.e.*, in groups of 8 detectors for forward scattering, and 6 detectors for backscattering data. All data analysis steps described here have been performed within the MantidPlot software [39, 40, 41, 42].

3. Results and Discussion

The proposed method has been tested on raw proton NCS data from a recent VESUVIO experiment on solid formic acid (FA) at a temperature of 240 K [22]. As a preliminary step of the analysis, the basic assumption of the proposed method were checked. The result of this preliminary assessment are plotted in Figure 4. The figure shows the result of the sequential, applied detector-by-detector, application of the procedure. No detector grouping of the recorded spectra was performed. The only intervention was that a group of few forward scattering detectors, with low counting statistics was excluded from analysis by masking them in MantidPlot data analysis procedure. This preliminary analysis shows that: (i) the positions of the proton recoil peak maxima do vary significantly with the magnitude of the momentum transfer, q from the neutrons, different from detector to detector (Figure 4, top); (ii) the width of the instrument resolution function exhibits very small variations with the detector number and the assumption of using its averaged over all set of detectors value is certainly justified

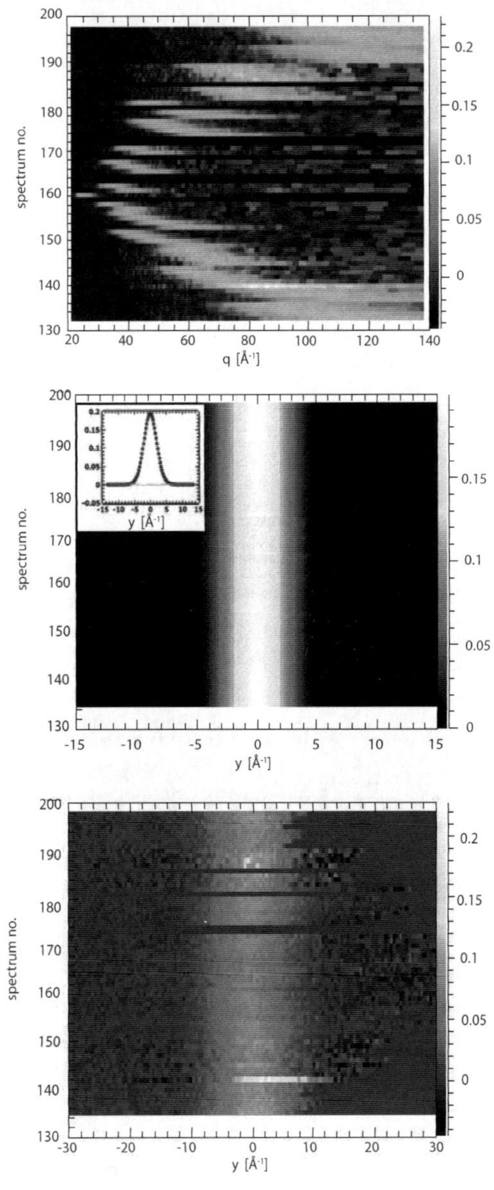

Figure 4. Kinematic range relevant to the NCS experiment on protons in solid formic acid at 240 K [22]. Top: the experimental longitudinal momentum distribution profile, $J(y_H)$ plotted as a function of the the momentum transfer magnitude, q for forward scattering Vesuvio detectors. Middle: the proton resolution function, $R(y_H)$ plotted as a function of the the proton longitudinal momentum, y_H. Bottom: $J(y_H)$ plotted as a function of y_H for forward scattering Vesuvio detectors. Black streaks with no scattering intensity in the top and bottom figure panels correspond to masked detectors. See text for details.

(Figure 4, middle); and (iii) the positions of the proton recoil peak maxima, when plotted in the proton longitudinal momentum domain, do exhibit small albeit pronounced variation with the detector number and thus with the value of the the momentum transfer, q (Figure 4, bottom).

It can be clearly seen in Figure 4 that whereas the position of the maximum of proton NMD in FA proves to depend very sensitively on the magnitude of the momentum transfer q (top panel), the dependence is much less pronounced on the longitudinal momentum variable y_H (bottom panel). Moreover, as shown in the inset in the middle panel of Figure 4, a Gaussian fit to the averaged over the whole set of forward scattering detectors instrument resolution function for the proton is of very good quality, yielding the value of the width of the averaged resolution function, $s = 4.2$ Å^{-1}.

The result of the fit of the position of the maximum of the proton longitudinal momentum

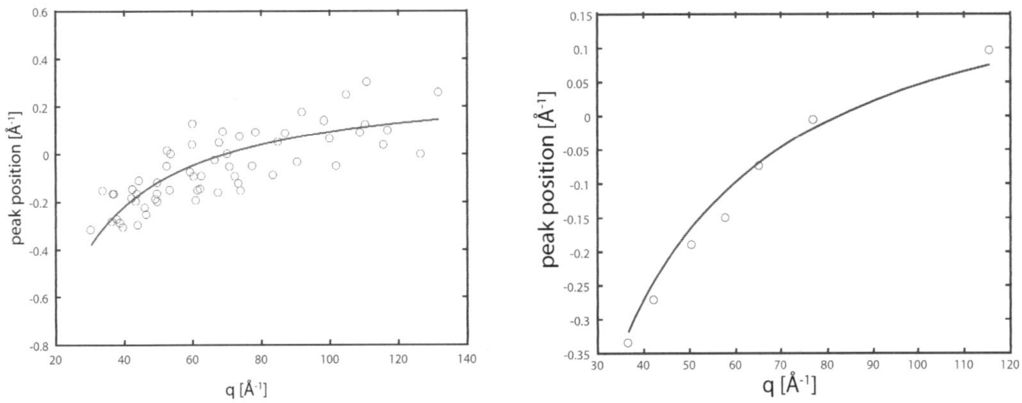

Figure 5. (Left) The result of the fit of the maximum of the proton longitudinal momentum distribution in FA at 240 K. For the proton resolution function, Gaussian approximation was applied with the average resolution function of the proton modelled as a Gaussian with STD, s = 4.2 Å^{-1}. Data not grouped. Individual detector data used for the fit. (Right)The result of the fit of the maximum of the proton longitudinal momentum distribution in FA at 240 K. For the proton resolution function, Gaussian approximation was applied with the average resolution function of the proton modelled as a Gaussian with STD, s = 4.2 Å^{-1}. Data grouped in groups of 8 detectors, corresponding to the VESUVIO detector bank set-up in forward scattering. See text for details.

distribution in FA at 240 K is shown in Figure 5. The STD of the proton NMD, obtained from the fit to the data with the width of the resolution function fixed, $s = 4.2$ Å^{-1}, was $\sigma = 4.5 \pm 0.6$ Å^{-1}, very well within a single standard deviation value in agreement with the value published for FA at 240K, being $\sigma = 4.4 \pm 0.1$ Å^{-1} [22].

The relatively big scatter of the data, giving the relative standard deviation of 10% of the value of the STD of the proton NMD, is the result of the data grouping policy. Thus, an attempt was made, to re-fit the data by grouping data in groups of 8 detectors, corresponding to the VESUVIO detector bank setup in forward scattering. The result of this procedure is shown in the right panel of Figure 5. A much better fit quality is obtained with the value of $\sigma = 4.5 \pm 0.3$ Å^{-1}. Thus, detector grouping improves the fit quality by reducing the error of the STD of the NMD by a factor of two.

4. Summary

On the whole, this work aligns itself closely with the overall direction in the research centered around electron-volt neutron spectroscopy by simultaneously addressing its two mainstream problems. Firstly, it paves the way towards high-throughput NCS data processing by offering a simple yet robust data pre-treatment protocol. Secondly, it provides a solid springboard towards more robust benchmarking of modern *ab initio* code simulation results for solid state systems and molecules.

To this end, we have presented a robust NCS data pre-fitting protocol that aims at obtaining the good initial guess of the second moments of NMDs for further NCS data processing and modelling. The method can be summarised in two steps. These steps are identical for each nuclear mass for which a recoil peak can be clearly resolved in the NCS spectrum recorded by each detector. Moreover, all the steps can be performed either on the detector-by-detector basis, or alternatively, stepping over a set of grouped detectors (e.g., in groups of corresponding to detector banks).

In the first step, a peak-finding algorithm is employed to find the positions of recoil peaks. As a result of this step, a set of pairs of values, q_{max}, y_{max}, is found, corresponding to the values of the recoil peak maxima positions, expressed in the domain of the longitudinal momentum of a chosen nucleus, y_{max} , and the values of the magnitudes of the momentum transfers, q_{max}, from the incident neutrons to the recoiling nucleus that correspond to those recoil peak positions.

Following this, the points on the plot of q_{max} vs. y_{max} are fitted with the function given by Eq.(8), keeping the width of the instrument resolution function fixed and introducing two parameters free in fitting. These free parameters are the width of the nuclear momentum distribution, σ, and, y_0, accounting for a small albeit systematic shift of the recoil peak positions due to slightly imperfect instrument calibration.

The procedure does not require any detailed knowledge of the precise shape of the NMDs as it relies on the approximation of a given NMD by a Gaussian term with FSEs calculated in the harmonic limit. The method can be applied for different detector grouping policies, both in forward and backscattering regimes of the VESUVIO spectrometer. The method owes its robustness to the fact that NMD data are sampled at their maxima where the influence of noise, sample-dependent and independent background signals and the overlap with NMDs of other nuclei are minimal.

Promising results, obtained for proton NMDs in solid FA show that, grouping detectors by banks, in order to reduce the data sparsity, yields results that are of overall good quality and well within the values obtained by much more complicated sequential NCS data fitting protocol, with sample-dependent data corrections to account for multiple scattering and neutron gamma background, followed by data grouping in momentum space. However, the proposed method can only serve as a preliminary step in the NCS data reduction scheme. For precision work, aiming at a detailed characterisation of the shapes of NMDs, a fully-fledged data reduction and fitting protocol, including all the above mentioned corrections, cannot be avoided.

The proposed method can be applied in case of extensive parametric NCS measurements, as a diagnostic tool in search of NMD changes induced by phase transitions or other changes in physical sample properties as a function of external stimuli such as temperature, pressure, or pH value of the environment.

On the *ab initio* materials modelling front, the method proposed and described here can serve as yet another much-desired data screening tool. Namely, the method can capture systematic deviations from the final states effects within the harmonic lattice approximation. To this end, it serves potentially an important role as a filter selecting experimental results potentially requiring time consuming and complicated *ab initio* modelling beyond the harmonic lattice approximation limit. Conversely, when successful in data pre-fitting, it can contribute to avoiding time consuming modelling approach all together, thus providing a much-sought tool for material modelling community. In doing so, the method contributes further to the optimisation of the synergy between neutron Compton scattering and materials modelling.

Moreover, the proposed method can be also applied as a diagnostic tool in the assessment of the overall quality of the instrument calibration. To this end, any trends that will not be captured by the fitting curve with a single offset parameter may signal that a re-assessment of instrument calibration is needed. For such an assessment, a set of well-known calibration samples can be employed with their second moments of nuclear momentum distributions fixed in fitting during the calibration procedure.

5. Outlook

Estimation of the second moment of NMDs with the sheer knowledge of the recoil positions is potentially a very robust and quick data screening tool for both experimental and *ab initio* NCS data modelling. To this end, further method improvements are highly desired.

The first obvious step in this direction is the optimisation of detector signal grouping prior to the analysis of the recoil peak positions. To this end, a Bayesian-like approach may additionally be implemented, to optimise the data grouping protocol given a degree of data sparsity, in a mass-resolved manner [43, 44].

Furthermore, additional work is needed to assess the performance of the method for higher masses, both in forward and backscattering. This part of work is also necessary in case of NCS spectra recorded using samples with complicated stoichiometry with NCS spectra consisting of overlapping peaks. In such case, the above mentioned Bayesian approach together with novel peak search algorithms should be tested, preferably within the MantidPlot data simulation and treatment environment.

Further improvement of the precision of the proposed method, both as NMD data pre-fitting and instrument calibration procedure, can be obtained by relaxing the constraint of the constant instrument resolution function. In principle, a two-dimensional data surface-fit can be performed with both, the magnitude of the momentum transfer at peak position and the width of the resolution function calculated individually.

Acknowledgments

The authors gratefully acknowledge the UK Science and Technology Facilities Council for financial support and for access to beamtime.

[1] Fernandez-Alonso F and Price D L (eds) 2013 *Neutron Scattering – Fundamentals* (New York: Academic Press)
[2] Andreani C, Krzystyniak M, Romanelli G, Senesi R and Fernandez-Alonso F 2017 *Adv. Phys.* **66** 1–73
[3] Andreani C, Colognesi D, Mayers J, Reiter G F and Senesi R 2005 *Adv. Phys.* **55** 377
[4] Fultz B 2010 *Progress in Materials Science* **55** 247 – 352
[5] Cazorla C and Boronat J 2017 *Rev. Mod. Phys.* **89**(3) 035003
[6] Ceriotti M, Fang W, Kusalik P G, McKenzie R H, Michaelides A, Morales M A and Markland T E 2016 *Chem. Revs.* **116** 7529–7550
[7] Krzystyniak M, Druzbicki K and Fernandez-Alonso F 2015 *Phys. Chem. Chem. Phys.* **17**
[8] Ceriotti M 2014 *J. Phys.: Conf. Ser* **571** 012011
[9] Senesi R 2012 *Nucl. Instrum. Meth. A* **661** 70 – 76
[10] Mayers J and Reiter G 2012 *Meas. Sci. Technol.* **23** 045902
[11] `http://www.isis.stfc.ac.uk/instruments/vesuvio/`
[12] Sears V F 1984 *Phys. Rev. B* **30** 44
[13] Watson G I 1996 *J. Phys.: Condens. Matter* **8** 5955
[14] Mayers J, Andreani C and Baciocco G 1989 *Phys. Rev. B* **39** 2022
[15] Mayers J 1990 *Phys. Rev. B* **41** 41
[16] Mayers J 1993 *Phys. Rev. Lett.* **71** 1553
[17] Andreani C, Colognesi D and Pace E 1999 *Phys. Rev. B* **60** 10008
[18] Mayers J and Abdul-Redah T 2004 *J. Phys.: Cond. Matter* **16** 4811
[19] Fielding A L, Timms D N and Mayers J 1998 *Europhysics Letters* **44** 255
[20] Krzystyniak M 2010 *J. Chem. Phys.* **133** 144505
[21] Krzystyniak M and Fernandez-Alonso F 2011 *Phys. Rev. B* **83** 134305
[22] Krzystyniak M, Druzbicki K, Romanelli G, Gutmann M J, Rudic S, Imberti S and Fernandez-Alonso F 2017 *Phys. Chem. Chem. Phys.* **19** 9064–9074
[23] Blostein J J, Dawidowski J and Granada J R 2005 *Phys. Rev. B* **71**(5) 054105
[24] 2017 *Neutron Scattering - Applications in Biology, Chemistry, and Materials Science* (*Experimental Methods in the Physical Sciences* vol 49) ed Fernandez-Alonso F and Price D L (Academic Press) pp 403 – 457
[25] Lovesey S W 1984 *Theory of Neutron Scattering from Condensed Matter* (Oxford: Clarendon)
[26] Weickert J, Ishikawa S and Imiya A 1999 *Journal of Mathematical Imaging and Vision* **10** 237–252
[27] Wada H, Takahashi S, Iijima T, Okumura Y and Imoto K 1959 *Communications of the ACM* **2**
[28] J K J 1984 *Biological Cybernetics* **50** 363–370
[29] Babaud J, Witkin A P, Baudin M and Duda R O 1986 *IEEE Transactions on Pattern Analysis and Machine Intelligence* **PAMI-8** 26–33

[30] Ibragimov I A 1956 *Theory of Probability & Its Applications* **1** 255–260
[31] Dharmadhikari S and Kumar J 1988 *Unimodality, Convexity and Applications* (New York: Academic Press)
[32] Pietropaolo A and Senesi R 2011 *Physics Reports* **508** 45
[33] Seeger A, Taylor A D and Brugger R M 1985 *Nucl. Instrum. Meth. A* **240** 98
[34] Blostein J J, Dawidowski J and Granada J R 2003 *Physica B* **334** 257
[35] Blostein J J, Dawidowski J, Ibanez S A and Granada J R 2003 *Phys. Rev. Lett.* **90** 105302
[36] Andreani C, Colognesi D, Degiorgi E, Filabozzi A, Nardone M, Pace E, Pietropaolo A and Senesi R 2003 *Nuclear Instruments and Methods in Physics Research A* **497** 535
[37] Abdul-Redah T, Krzystyniak M and C-Dreismann C 2005 *Phys. Rev. B* **72** 052202
[38] Carr R personal communication
[39] Oram E and McCann L 2016 Simulation support and further indirect inelastic specific improvements within mantid Tech. Rep. RAL-TR-2016-019 Rutherford Appleton Laboratory URL `http://purl.org/net/epubs/work/30884953`
[40] Oram E 2016 An overview of developments in mantid relating to indirect inelastic spectroscopy july 2016 - july 2017 Tech. Rep. RAL-TR-2016-011 Rutherford Appleton Laboratory URL `http://purl.org/net/epubs/work/29538224`
[41] Nixon D 2015 Simulation support within mantid for molecular spectroscopy Tech. Rep. RAL-TR-2015-012 Rutherford Appleton Laboratory URL `http://purl.org/net/epubs/work/23618523`
[42] Jackson S 2014 Vesuvio data reduction and analysis in mantid Tech. Rep. RAL-TR-2014-009 Rutherford Appleton Laboratory URL `http://purl.org/net/epubs/work/12239830`
[43] Pardo L C, Rovira-Esteva M, Busch S, Ruiz-Martin M D and Tamarit J L 2011 *J. Phys.: Conf. Ser.* **325**
[44] D M, Vispa A, Pardo L C, R Tolchenov R, Mukhopadhyay S and Fernandez-Alonso F 2015 *J. Phys.: Conf. Ser.* **663** 012009

IOP Conf. Series: Journal of Physics: Conf. Series **1055** (2018) 012012 doi:10.1088/1742-6596/1055/1/012012

Model selection in neutron Compton scattering - a Bayesian approach with physical constraints

M Krzystyniak[1,2], G Romanelli [1], K Drużbicki[3], R Tolchenov[1], M Gigg[1], B Hewer[1], and F Fernandez-Alonso[1,4]

[1] ISIS Facility, Rutherford Appleton Laboratory, Chilton, Didcot, Oxfordshire OX11 0QX, United Kingdom

[2] School of Science and Technology, Nottingham Trent University, Clifton Campus, Nottingham NG11 8NS, United Kingdom

[3] Faculty of Physics, Adam Mickiewicz University, Umultowska St. 85, 61-614 Poznan, Poland.

[4] Department of Physics and Astronomy, University College London, Gower Street, London WC1E 6BT, United Kingdom

E-mail: matthew.krzystyniak@stfc.ac.uk

Abstract. This work analyses the performance of the maximum-likelihood estimation approach in fitting Gram-Charlier expansion curves to nuclear momentum distributions with non-negativity constraints. The presented approach guarantees that the most likely model selected to describe the recorded data is also a physically meaningful one, i.e., corresponds to a non-negative probability distribution function. For the case of the most popular momentum distribution model, containing the information about the variance and excess kurtosis of the distribution, we derive a simple and easy to implement non-negativity criterion. We test the performance of the newly developed approach by applying it to interpret proton momentum distribution obtained from neutron Compton scattering from solid phosphoric acid, a system in which nuclear quantum tunnelling was proposed in the limit of low temperature. From a methodological point of view, this work provides a screening tool in the search for systems exhibiting the so-called 'non-trivial nuclear quantum effects'.

1. Introduction

Neutron Compton scattering (NCS), owing to its unique capability to to measure nuclear momentum distributions (NMDs) [1, 2, 3] has been increasingly popular amongst material modelling communities as a benchmarking spectroscopic method-of-choice for a plethora of *ab initio* methodology, including but not limited to, phonon - lattice dynamics (LD) based on Density Functional Theory (DFT), Born Oppenheimer Molecular Dynamics (BOMD), and Path Integral Molecular Dynamics (PIMD) [4, 5, 6, 7, 8, 9, 10, 11, 12, 13, 14, 15]. In practice, however, a hierarchy of approximations, used in treating, analysis and fitting NMD models to NCS data, is confronted with a limited information content that can be extracted from an NCS experiment. More sophisticated NMD models contain a greater number of parameters. Thus, a correspondence naturally establishes itself between the degree of an NMD model sophistication and a level of theory required for the successful modelling of a system under investigation. With the most sophisticated PIMD methodology requiring a lot of computing resources and thus leading to 'heroic simulations' taking months on the most advanced High-Performance Computing (HPC) infrastructure, this benchmarking exercise must be properly optimised.

Thus, a suitable NMD model selection protocol must be established, tested and implemented before one embarks on the materials modelling mission, even with the simplest theoretical tools available. A good example of the need towards establishing such a protocol is the recent NCS work tackling the problem of the phase polymorphism in solid formic acid (FA) [16, 17]. In this work, a series of *ab initio* modelling tools with the systematically increasing level of theory was applied and tested for the agreement with the measured NMD of the proton. LD, BOMD, and the so called quantum-thermostated BOMD (QT-BOMD), making use of the generalised Langevin equation (GLE) [17], were compared. Surprisingly at the first glance, LD-DFT was demonstrated to perform very well, with the QT-BOMD only slightly improving the result [17]. Thus, a preposterous situation arises in which a level of theory requiring at most a day of computing time (LD-DFT) ends up reproducing the experimental result, for all practical intents and purposes, in a same qualitative and quantitative level of agreement, as a 'heroic' QT-BOMD calculation requiring in this case 9 months of HPC time. Moreover, this result fits in very well into a trend, present in the NCS literature, where a simple DFT calculation on an isolated molecule, with additional degrees of freedom (rotational and translational) accounted for in a classical way by resort to the energy equipartition theorem, was able to account very well for the variance of the NMD [18, 19, 9]. The reason behind this lies in the nature of the NCS method. In essence, the second moment of the NMD can be termed as first moment of the atom-projected Vibrational Density of States (apVDOS) [18, 19, 9]. Consequently, its value is most sensitive to the high frequency vibrational excitations chiefly associated with intramolecular vibrations. Thus, as far as the second moment of the NMD is concerned, an isolated-molecule vibrational calculation, using a basic DFT scheme implemented using, e.g., Gaussian software [20], is expected to perform well for molecules in liquids and gases or even condensed matter systems, provided no high-frequency vibrational excitations of inter-atomic nature, with the prominent example of the hydrogen bonding (HB), will be present. Thus, when characterising for instance strongly ionic systems, both solid and liquid, the modelling task can be easily accomplished with a basic molecular DFT level of theory.

Contrary to simple ionic systems, however, systems with HB pose much more challenge from the NMD modelling point of view. Firstly, due to the reasons mentioned above, modelling, even at the level of the second moment of the NMD, will require resorting to LD-DFT methodology, at least. Secondly, when investigating systems with HB at low temperature, nuclear quantum tunnelling (NQT) may show up. In such case, the description of the proton NMD containing the second moment only will not be sufficient. The reason for this is that, protons tunnelling across the HB axis must be described by a coherent superposition of quantum states, naturally leading to bimodal NMDs characterised by non-vanishing excess kurtosis, or even sometimes higher NMD moments [21, 22, 23, 24]. This in turn poses a challenge on both NCS data treatment and modelling fronts. From the data treatment point of view, a model selection protocol is needed to select where to truncate a Gram-Charlier (GC) expansion to describe the observed NMD most accurately but also in a manner which uses only the information present in the data without extending the NMD model overly, not to end up fitting the statistical noise, background or spurious peaks present in recorded NCS data. Motivated by this, a number of Bayesian inference-based model selection protocols for nautron scattering have been established [25, 26, 27, 28, 29], with the recent implementation in the Mantid software, referred to as FABADA, [28, 29]. The second, *ab initio* modelling challenge, is even more formidable. Namely, only one modern *ab initio* modelling approach, the PIMD, is capable of theoretically reproducing NQT in NMD data. With the sophistication and computational cost of the PIMD method sky-rocketing for realistic reproduction of properties of condensed matter systems like quantum liquids or disordered solids, the need for a proper model selection protocol as a gateway to gauge whether PIMD is needed for modelling NMD from NCS data, is rendered even more critical.

Despite the fact that NCS has already witnessed an upsurge of activity in the NMD model selection and establishment of reliable protocols and software tools [25, 26, 27] one crucial aspect of this exercise has been surprisingly omitted. Namely, the whole NMD model selection so far has boiled down to comparing the likelihood levels of different models without constraining them to be positive from the outset. The figure of merit of the NCS, the NMD, is required by the laws of quantum mechanics to be non-negative, as it can be interpreted as the square of the modulus of a nuclear wave function in the momentum space [1, 2, 3]. Towards this end, NCS experiments allow in principle for the model-free reconstruction of the proton wave function together with its underlying effective Born-Oppenheimer (BO) potential-energy surface of a given nuclide [1, 2, 3]. Mathematically, this reconstruction procedure is reminiscent of the way in which crystal structures are obtained from diffraction data. The presented work is largely motivated by these unique features and capabilities of the NCS method. Its ambition is to establish a model-selection gateway as a guidance for tailored *ab initio* modelling of nuclear quantum effects, especially in cases where the most quantum of the quantum mechanics phenomena, the quantum coherence, is hoped to be visible in an experiment.

2. Methodology

2.1. Nuclear momentum distributions from neutron Compton scattering

NCS has been described in great detail elsewhere [1, 2, 3]. Here, for the sake of demonstration of basic properties of NMDs, a brief account only will be made of physical assumptions underlying the way NMDs are obtained in the NCS method. The basic setting for the operation of the VESUVIO spectrometer is that it uses electron-volt (eV) neutrons, exclusively provided by spallation sources, such as the ISIS pulsed neutron and muon source in the United Kingdom [30, 31]. Consequently, the energy (E) and momentum (Q), transferred in the scattering process are sufficiently high (in the range 1-100 eV and 30-200 Å^{-1}, respectively) such that the NCS data can be analysed using the so-called impulse approximation (IA), implying that the neutron scatters from a single atomic centre, with conservation of total kinetic energy and linear momentum.

Within the IA, the dynamic structure factor $S(Q, E)$ for a given atom of mass M reduces to a single peak centered at the recoil energy $E_r = \hbar^2 Q^2/2M$ with $\hbar = 2.0446$ (meV amu)$^{1/2}$ Å [32, 33, 34, 35, 36, 37, 38, 2]. Consequently, $S(Q, E) = (M/^{-}\hbar Q)J(y)$, where y is the initial radial momentum p of the nucleus projected onto the scattering vector Q. $J(y)$ is the so called NCS profile [33, 32] and formally corresponds to the longitudinal nuclear momentum distribution (NMD) of the target nucleus along y. In what follows, we will consider the case of harmonically bound isotropic systems (e.g. liquids or crystalline solid powders), constituting the bulk of present-day NCS work on VESUVIO. In this relevant case, a spherically averaged three-dimensional longitudinal NMD, $J_{IA}(y)$, in the IA limit depends only on the magnitude of y but not on $\hat{Q}$ and can be conveniently expressed via recourse to a Gram-Charlier (GC) expansion of the form [2, 39]

$$J_{IA}(y) = \frac{\exp\left(\frac{-y^2}{2\sigma^2}\right)}{\sigma\sqrt{2\pi}} \sum_n \frac{c_{2n}}{2^{2n} n!} H_{2n}\left(\frac{y}{\sigma\sqrt{2}}\right). \tag{1}$$

where H_{2n} denotes a Hermite polynomial and the standard deviation (STD) of a given NMD is denoted by σ, which, with the present choice of units, is expressed in Å^{-1}. In Eq. (1), the term with $n = 0$ is set to unity due to the requisite normalization of the NMD. Moreover, with c_2 set to zero, the kinetic energy E_k of a nucleus of mass M is related to the STD of its NMD, σ, via $E_k = 3\hbar^2\sigma^2/2M$. This relationship is independent of the precise values of the remaining coefficients c_{2n}. Therefore, the fitting of $J_{IA}(y)$ to the Gram-Charlier expansion amounts to varying both the width of the momentum distribution σ and the coefficients c_{2n} with $n \geq 2$.

For finite values of Q, corrections to the IA are known as 'Final-State Effects' (FSEs) [38, 2]. To account for FSEs, the method of Sears [32] is routinely incorporated into standard NCS data treatments [38] by expressing the measured NCS profile as a series of the form

$$J(y,Q) = J_{IA}(y) + J_{FSE}(y) = J_{IA}(y) - \frac{M\langle\nabla^2 V\rangle}{36\hbar^2 Q}\frac{d^3}{dy^3}J_{IA}(y) + ... \tag{2}$$

where $J_{IA}(y)$ is the IA result given by Eq. (1). $\langle\nabla^2 V\rangle$ is the mean value of the Laplacian of the potential energy of the nucleus expressed in meV Å^{-2} (cf. Refs. [38, 40]).

All three Cartesian direction components of the STD of a given NMD, σ_x, σ_y, σ_z, as well as its associated Laplacian of the effective BO potential felt by a nucleus and underlying the NMD under investigation, $\langle\nabla^2 V\rangle$, may be calculated from the underlying apVDOS under the assumption that the NMD for a given nucleus in a crystal along the $\widehat{Q}$ direction assumes a purely multivariate Gaussian functional form. [18, 2, 41, 42]. Importantly, within this assumption an explicit connection between the FSE magnitude, k and the average value of of the STD, $\sigma = \sqrt{\frac{1}{3}(\sigma_x^2+\sigma_y^2+\sigma_z^2)}$, can be established as $k = \sigma\frac{\sqrt{2}}{12}$ [41, 42, 16]. Moreover, a general connection between $\sigma_x, \sigma_y, \sigma_z$ and the GC expansion coefficients, c_{2n}, can be established [43]. In the bulk of the NCS studies published so far on isotropic harmonic systems the expression given by Eq. (1) is truncated to include the purely Gaussian term and the term proportional to the Gaussian multiplied by the Hermite polynomial term describing the excess kurtosis of the NMD, c_4, as the next term, c_6, often proves to be fitted to experimental NMD data with errors of the order of 100 % [24]. Additionally, the FSE term, given by the Eq. (2 is expressed using the third order Hermite polynomial, $H_3(\frac{y}{\sigma\sqrt{2}})$. This results in an expression:

$$J_{IA}(y) = \frac{\exp\left(\frac{-y^2}{2\sigma^2}\right)}{\sigma\sqrt{2\pi}}\left[1 + \frac{c_4}{32}H_4\left(\frac{y}{\sigma\sqrt{2}}\right) - \sigma\frac{\frac{\sqrt{2}}{12}}{Q}H_3\left(\frac{y}{\sigma\sqrt{2}}\right)\right]. \tag{3}$$

Additionally, in most of the NCS data treatment protocols used, before the final NMD fit is performed, the FSE contributions are subtracted from data on a detector-by-detector basis in order to separate out the Q-dependent terms from the Q-independent terms in Eq. (3) [7, 8] (see, Sect. 2.2). Thus in what follows, we will discuss the model selection protocol, based on the Bayesian inference principle, implemented using the FABADA algorithm in Mantid software [28, 29], using the Eq. (3) in the form:

$$J_{IA}(y) = \frac{\exp(\frac{-y^2}{2\sigma^2})}{(\sigma\sqrt{2\pi})}\left[1 + \frac{c_4}{32}H_4\left(\frac{y}{\sigma\sqrt{2}}\right)\right]. \tag{4}$$

with an additional non-negativity condition imposed, $\forall x \epsilon \Re\, J_{IA}(x) \geq 0$ with x defined as $x = \left(\frac{y}{\sigma\sqrt{2}}\right)$. This condition is equivalent to a condition involving the Hermite polynomial expansion part of Eq. (4) only:

$$\forall x \epsilon \Re, 1 + \frac{c_4}{32}H_4(x) \geq 0. \tag{5}$$

Moreover, a general connection between $\sigma_x, \sigma_y, \sigma_z$ and the GC expansion coefficients, c_{2n}, can be established [43]. In the specific case of the c_4 coefficient this relation can be written in the following explicit manner:

$$c_4 = \frac{2}{15}\left(\frac{\sigma_x^4+\sigma_y^4+\sigma_z^4}{\sigma^4} - 3\right) \tag{6}$$

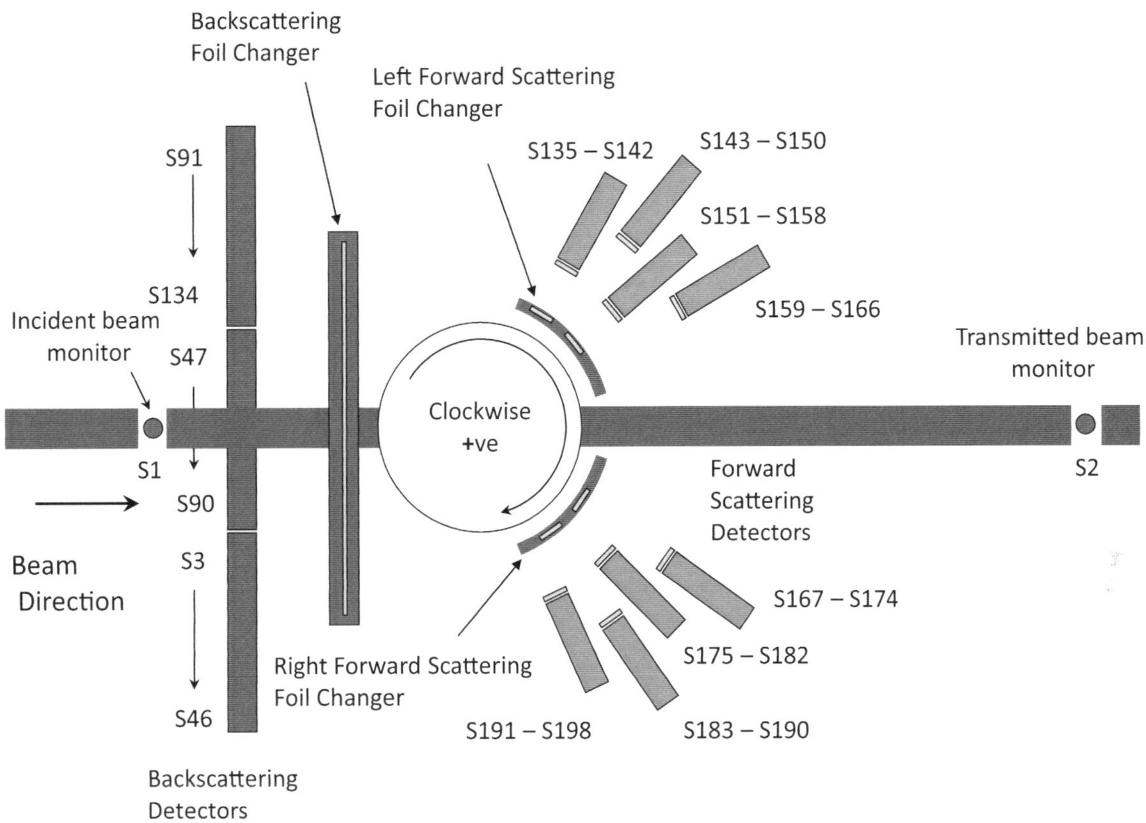

Figure 1. Schematic diagram of the VESUVIO spectrometer. For further details, see, e.g., Refs. [2, 3]

In what follows, we will analyse which values an NMD based on an underlying multivariate Gaussian distribution may take and how it relates to the above mentioned non-negativity constraints.

2.2. Experimental Protocol

All NCS measurements were performed on the latest incarnation of the VESUVIO spectrometer(see, Fig. 1) at the ISIS pulsed neutron and muon source, United Kingdom [30, 31, 44]. The instrument, both in its operational principles and neutronic parameters, has been extensively described elsewhere [30, 31, 44, 2, 3]. Here, we will only concentrate on the main principles underlying the NMD model selection protocol.

In the vast majority of NCS studies tackling the problem of the visibility in the recorded data of signatures of nuclear quantum coherence, possibly leading to nuclear quantum tunnelling, the studies involve the proton in isotropic condensed matter systems. Due to the conservation of momentum and energy in the Compton scattering within the IA, the NCS scattering off protons, specifically the single scattering process, can occur only for forward scattering angles. Thus, in order to test the newly proposed NMD model selection protocol, we have concentrated in forward scattering NCS data.

The starting point for the analysis of forward scattering NCS data is the expression describing the count rate, $C_\theta(t)$, of a neutron detector for a scattering angle θ in the time-of-flight (t) domain [2, 3]:

$$C_\theta(t) = A' \left[\frac{E_0 I(E_0)}{Q}\right]_t \times \sum_{n=1}^{N} I_n M_n J_n[y_n(t)] \otimes R_n[y_n(t)] \tag{7}$$

where A' is a mass–independent experimental constant and the mass–independent factor $[\frac{E_0 I(E_0)}{Q}]_t$ depends on the incident neutron spectrum, $I[E_0(t)]$, the initial neutron energy, $E_0(t)$, and the momentum transfer $Q(t)$, all explicit functions of t. In Eq. (7), the NMD for mass M, $J_M(y_M(t))$, is given by Eq. (2) with the expression for the NMD in the IA limit, $J_{IA}(y)$, given by Eq. (4). Integrated peak intensities I_M for a given scattering mass M are proportional to the scattering density $I_M = AN_M 4\pi b_M^2$ where $4\pi b_M^2$ is the total (bound) neutron-scattering cross section [45, 38].

The NCS forward scattering data reduction performed in this work follows the literature protocol [7, 8] and can be summarised in few steps:

- raw TOF data were fitted sequentially detector by detector using Eq. (7) in order to obtain the initial values of the fitting coefficients, I_M and σ_M;
- the obtained values of the fitting coefficients, I_M and σ_M were used as input parameters to calculate the sample-dependent forward scattering corrections: (i) multiple scattering (MS) from the sample and container, and (ii) gamma background (GB) due to cross-talk between signals created at the forward detectors by primary and secondary gold foils (see Figure 1);
- single scattering (SS) raw TOF spectra were generated by subtracting from raw TOF data a linear combination of corrections: (i) MS multiplied by the ratio MS/SS obtained from the measured sample transmission, (ii) GB multiplied by unit factor, and (iii) empty sample container spectrum multiplied by the measured ratio of sample plus container and empty container transmission;
- a second fit to raw TOF data containing SS contributions from different nuclei present in the sample was performed using Eq. (7) and the second iteration of I_M and σ_M parameter values was obtained. In the second iteration of thc fitting algorithm, the values of the I_M parameters were constrained using the known sample composition (and hence the stoichiometric ratios of numbers of moles of different masses M present in the sample) and tabulated values of the total (bound) neutron-scattering cross sections $4\pi b_M^2$;
- peaks in the TOF spectra representing neutron recoil of protons (H) were isolated by subtracting, detector by detector, from the raw TOF data corrected in the limit of SS of two independent contributions: (i) the fitting curves of recoil peaks from the other nuclei, and (ii) the fitting curves representing the FSE contributions to NMDs of H/D, given by the term $\sigma \frac{\frac{\sqrt{2}}{12}}{Q} H_3\left(\frac{y}{\sigma\sqrt{2}}\right)$;
- the isolated H recoil peaks were transformed, detector by detector, into respective H longitudinal momentum transfer domains, y_H, a transformation termed as 'y-space data focussing'. The focused data with greatly improved (by summing 64 spectra) signal-to-noise ratios were finally subject to fitting H NMDs using Eq. (4).

Using the NMD curves obtained after the data pre-treatment steps, described above, model selection was performed using the Bayesian inference tool, FABADA [28, 29]. The selection was performed between an NMD model containing purely Gaussian shape, and a model cotaining a sum of a pure Gaussian term and a Gaussian term multiplied by the Hermite polynomial accounting for non-zero excess kurtosis. Importantly, throughout the whole data reduction, both in the data pre-treatment steps and in the final analysis of the obtained H NMDs, positivity constraints, given by Eq. (5) were imposed. All data analysis steps were performed within the Mantid software [46, 47, 48, 49].

For the demonstration of the performance of the proposed model selection protocol, a recent example of VESUVIO work, a sample of concentrated, 85wt% phosphoric acid (PA) was chosen. The specimens were enclosed in a 6x6 cm^2 flat aluminum (Al) cell resulting in a total transmission of 92% of the sample in the Al container, and 2% transmission level of the empty container, respectively. The data used in the analysis were recorded in the PA solution with the experiment were carried out at 80 K with the sample and container temperature stabilised using a dedicated helium closed-cycle refrigerator.

3. Results and Discussion

3.1. NMD positivity constraint

We begin the description of the non-negativity constrained NMD model selection protocol by noting that the positivity constraint, given by Eq. (5), can be solved analytically in a simple manner by applying a substitution, $x^2 = u$, and thus converting it to a quadratic equation. A quadratic equation withrealcoefficients can have either one or two distinct real roots, or two distinct complex roots. In this case the discriminant determines the number and nature of the roots. Specifically in our case, if the discriminant is negative, then there are no real roots. Moreover, if the coefficient multiplying the u^2 term is positive, the quadratic function is always positive too. Thus, we have:

$$\forall u \epsilon \Re, \frac{1}{2}c_4 u^2 - \frac{3}{2}c_4 u + \frac{3}{8}c_4 + 1 \geq 0. \tag{8}$$

and the equation for the determinant:

$$\Delta = \left(\frac{3}{2}c_4\right)^2 - 2c_4\left(\frac{3}{8}c_4 + 1\right) = \frac{3}{2}c_4^2 - 2c_4 \leq 0. \tag{9}$$

With the additional condition that c_4 be positive we finally get $c_4(3c_4 - 4) \leq 0$. The final solution for the c_4 is then that $0 \leq c_4 \leq \frac{4}{3}$. Thus, any NMD model selection protocol that ensures a non-negative, physically meaningful solution, should include the condition $0 \leq c_4 \leq \frac{4}{3}$ as an additional constraint imposed in fitting.

It is worth noting that, in case of a most general positivity constraint, imposed on a polynomial expansion of the type of the Gram-Charlier series given by Eq. (1), no closed-form analytical solution exists [50] and for each individual series truncated at a different maximal Hermite polynomial order, n, the search for specific-case solution is a complicated task, oftentimes motivating special case studies dedicated to finding a solution, mostly using numerical methods [50]. Interestingly for our specific case, a study exists in the literature by Jondeau and Rockinger [50] concerning a positivity constraint for a GC expansion of the type $\forall x \epsilon \Re, 1 + cH_4(x) + kH_3(x) \geq 0$. Using sophisticated methods of analytical geometry and numerical analysis, the authors have found a region in parameters space of the excess kurtosis, c, and skewness, k, that corresponds to a positive solution (see Figure 2 taken from Ref. [50]). Importantly, this region is not rectangular in shape which has an important consequence that for any pair of parameters (c, k) giving a positive solution, one of them is not independent on the other.

Interestingly, a specific solution within the class given by Jondeau and Rockinger exists for the case of vanishing skewness of the distribution. This corresponds to the zero value abscissa in Figure 2 from Ref. [50]. The zero-skewness abscissa defining a positive GC expansion value region is bounded from below by zero and from above by a value of 4. Using the definition of

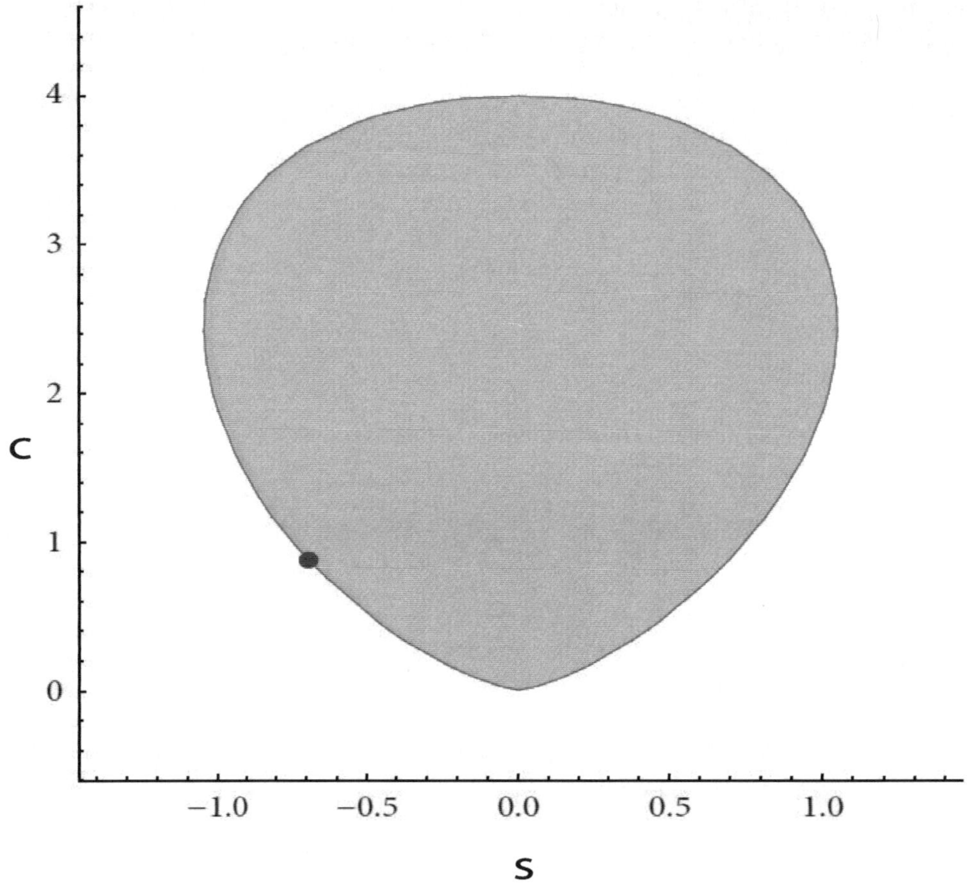

Figure 2. Feasible region of excess kurtosis, c, and skewness, s, defining the parameter region for which positive GC expansion values exist together with a maximum likelihood estimator (denoted by a filled circle symbol), taken from Jondeau and Rockinger, Ref. [50]

the excess kurtosis, $c = \frac{M_4}{\sigma^4} - 3$, with M_4 being the fourth moment of the distribution, defined as, $M_4 = 3(1 + c_4)\sigma^4$, we get $c = 3c_4$ and $0 \leq 3c_4 \leq 4$. Thus, the specific case of the solution given by Jondeau and Rockinger coincides with our solution given above based on the simple consideration of a quadratic equation.

Beyond the special, albeit very popular in the NCS literature, case of the zero-skewness NMD models containing the Gaussian term and the term with excess kurtosis only, one is tempted to formulate a more general NMD positivity criterion beyond this special, relatively simple case. A first step in that direction is to consider an NMD of a type given by Jondeau and Rockinger, i.e. with non-zero skewness. Here, already a first difficulty arises. Namely, the solution given by Jondeau and Rockinger cannot be directly employed as a positivity constraint in NCS as the skewness parameter they consider is independent on an external variable, like the magnitude of the momentum transfer Q in NCS. In case of NCS the problem is even more complicated as, due to the kinematics of momentum and energy conservation within the limits of the IA, Q depends in a parametric manner on the TOF, differently for scattering angle θ. This is exactly the case why, in the NCS data pre-treatment protocol described above, the FSE subtraction is performed before the NCS data for the isolated recoil peak is focused in its y-space and fitted with an NMD model. At present, the non-negativity constraint is not explicitly imposed at the pre-treatment stage. All fitting curves, obtained in a detector-by-detector manner, are simply visually inspected and raw spectra yielding negative solutions in fitting are simply pruned out and excluded from the final selection of spectra for the focusing and final NMD fit. Work is in progress, within the joint effort of the Mantid software development team, to implement a general positivity constraint for any Gram Charlier series truncation scheme, both at the pre-fit (with different detector-grouping schemes) and final NMD fit stage, using numerical methods employing Chebychev polynomial expansion.

3.2. Multivariate Gaussian NMD and the positivity constraint

A second very popular NMD model, exploited extensively beside the GC expansions to describe nuclear chemical dynamics in liquids and solids within the harmonic approximation, is the model in which NMDs are described using a multivariate Gaussian (MVG) [2, 3]:

$$J_{MVG}(y) = \frac{1}{\sqrt{8\pi^3\sigma_x^2\sigma_y^2\sigma_z^2}}\langle \exp(\frac{-y^2}{2\sigma_x^2})\exp(\frac{-y^2}{2\sigma_y^2})\exp(\frac{-y^2}{2\sigma_z^2})\rangle_\Omega \tag{10}$$

where Ω denotes the angular average. It is thus of importance to establish mathematical connection between the MVG and GC expansions, especially in the light of the positivity constraint discussed in Section 3.1. In this work, we will concentrate only on the MVG connection to the truncated GC expansion considered above by noting first that, the connection between the two GC parameters, σ and c_4 has already been given above (see, Eq. (6)).

For the specific form of the c_4 coefficient, given by Eq. (6), one may inspect the connection by plotting the values of c_4 as a function of σ_x, σ_y, and σ_z in a relevant range of those MVG second moment values. Such a plot is shown in Figure 3.

Already a first glance at the Figure 3 reveals that, c_4 values, plotted within a cube of values of the $(\sigma_x, \sigma_y, \sigma_z)$ parameters of the MVG model ranging from 0 to 50 Å^{-1}, seems to be bounded from below by the zero value and from above by the value of 0.8. Indeed, a further function maximum search using an unconstrained optimisation scheme by means of the `fminsearch` algorithm, implemented in Matlab 2017a software [51], reveals that the function is constrained in the region 0–0.8, with the maximum of 0.8 obtained always for the solution having the symmetry, $\sigma_i = \sigma_j = 0, \sigma_k > 0$. Within the domain plotted in Figure 3 the non-linear optimisation algorithm finds two maxima, $c_4 = 0.8$ at $(0, 0, 23.245)$ and $(0, 0, 49.6)$.

A special, albeit very important and popular in the literature [2, 3, 18] case exists when two of the three values of the second moment of the MVG become identical, by virtue of the symmetry of the NMD, i.e. $\sigma_x = \sigma_y = \sigma_t$. In this special case, $c_4 = \frac{4}{45}\frac{(\sigma_t^2-\sigma_z^2)^2}{\sigma^4}$ with $\sigma^2 = \frac{2\sigma_t^2+\sigma_z^2}{3}$ [24]. The plot of the c_4 as a function of σ_t and σ_z is shown in Figure 4.

In this important case, the MVG becomes a relatively simple two dimensional function and its properties can be investigated using simple methods of mathematical analysis. Specifically, it can be proven that the c_4 function in this case does not have a proper minimum nor a maximum. Instead, c_4 attains a minimal valuc at $\sigma_t, \sigma_z = 0, 0$ which is a saddle point. Moreover, c_4 reaches its maximal value, $c_4 = 0.8$, asymptotically as a limit $\sigma_t = 0$, $\sigma_z \to \infty$. At another important case, $\sigma_z = 0$, $\sigma_t \to \infty$, the function reaches its second limit $c_4 = 0.2$.

On the whole, the main conclusion from the analysis of the MVG connection to the GC expansion of the NMD is that, any sensible choice of values of the second moment of the MVG will always lead to an NMD which, when expanded in the GC expansion containing only the

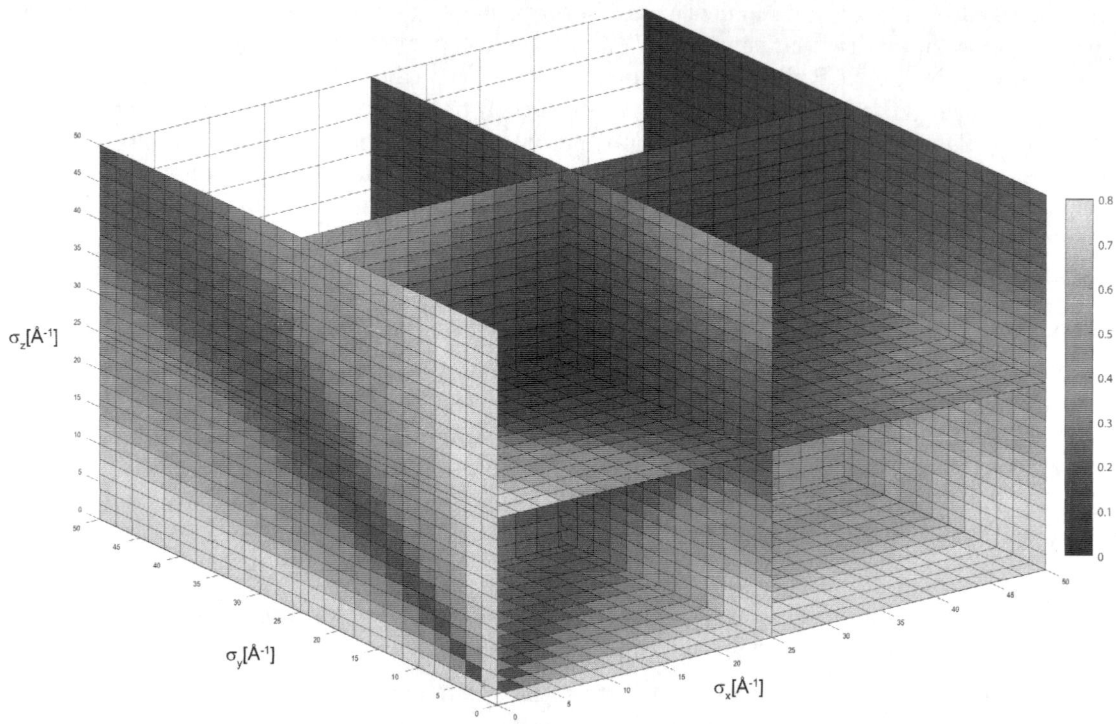

Figure 3. Values of c_4, given by Eq. (6), as a function of σ_x, σ_y, and σ_z. See text for details.

Gaussian term and the terms responsible for non-vanishing excess kurtosis, will always lead to values of c_4 bounded from below by zero and from above by the value of 0.8.

Moreover, this result has another very important consequence for screening the NMDs obtained from NCS experiments for the signs of coherence of a nuclear wave packet and thus possibly nuclear quantum tunnelling. Namely, when a GC expansion fit of the type discussed here, performed on an experimentally obtained NMD form a NQT-candidate NCS data, converges with a c_4 value above 0.8 (taking into account the error bar on the value of c_4), a conclusion must be reached that the underlying BO effective potential cannot correspond to an MVG distribution. In other words, no three-dimensional harmonic potential, leading to an MVG NMD can possibly yield, in a GC expansion, c_4 values above 0.8. This condition is, however, only a necessary but not a sufficient one, i.e., bi-modal BO effective potentials of a type of a double-well, which naturally lead to solutions of nuclear Schroedinger equations exhibiting nuclear quantum coherence across both wells, may still have c_4 values below 0.8. For instance, in a study comparing proton NMDs in liquid water, ice, supercritical water, and water under pressure confined in C60, Reiter et al. [24] fitted an NMD model in which, in a frame of reference where an individual bond is taken to lie along the z axis, the motion transverse to the bond is harmonic and along the bond given by a distribution that corresponds in real space to two Gaussians separated by a distance d. The fitting of this model leads to a result that values of the excess kurtosis above ca 0.3 (with the values of the coefficient c6 vanishing within the experimental error) are associated with a bimodal proton wave function across the hydrogen bond with the two maxima separated by ca 0.3 Å. For the values of c_4 below ca 0.3 and values of c6 equal to zero to within the experimental accuracy the model leads to a wave function which is a single modal bivariate Gaussian ($d = 0$). Thus, within the model by Reiter et al., the values of c_4 above 0.8 will certainly always lead to a double-well effective BO potential with nuclear

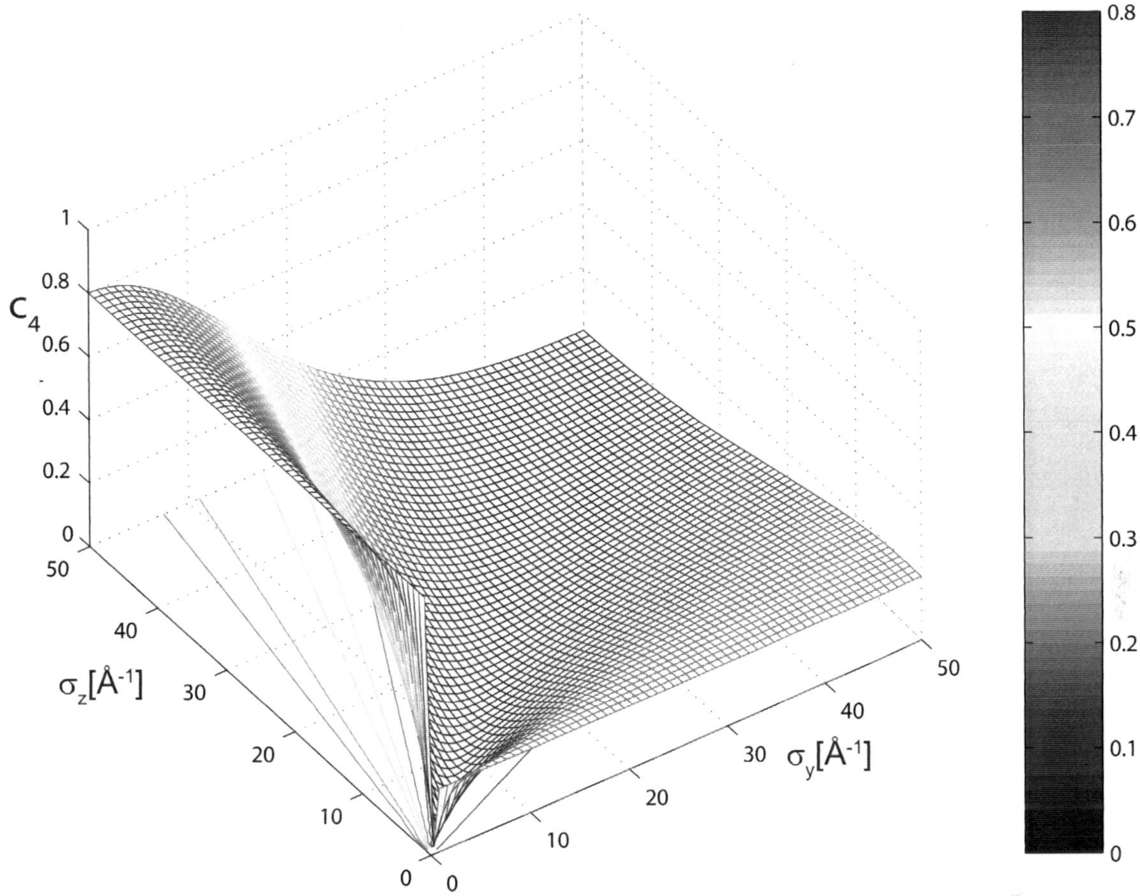

Figure 4. Values of c_4, given by Eq. (6), for the specific case of $\sigma_x = \sigma_y = \sigma_t$. Contour lines plotted are the isolines of constant c_4 magnitude. See text for details.

quantum coherent wave packet as a solution, albeit values of c_4 above ca 0.3 and still below 0.8 already start producing such solutions.

3.3. Bayesian inference with positivity constraint in action: experimental NMD study of an NQT-candidate structure

In order to demonstrate the performance of the Bayesian-inference principle with positivity constraints in fitting NMDs from an NCS experiment, we show here an example from a recent VESUVIO work [52]: a proton NMD obtained from an NCS experiment on frozen disordered concentrated (85 wt%) solution of the phosphoric acid (PA) at T=80K. The following protocol was adopted. The FABADA tool, implemented in Mantid software [46, 47, 48, 49] was employed to compare four models: (i) a model with purely Gaussian NMD; (ii) a model with a sum of two terms, a Gaussian and a Gaussian multiplied by the excess kurtosis term in the GC expansion without positivity constraints, (iii) the same model with positivity constraints imposed in fitting, (iv) a model with a Gaussian term, the excess kurtosis term and an additional term proportional to $c_6 H_6(x)$ without positivity constraints imposed in fitting.

The result of the NMD model selection protocol for the proton NMD obtained from NCS data recorded from the PA sample is illustrated in Figure 5. The first observation is that, to within the

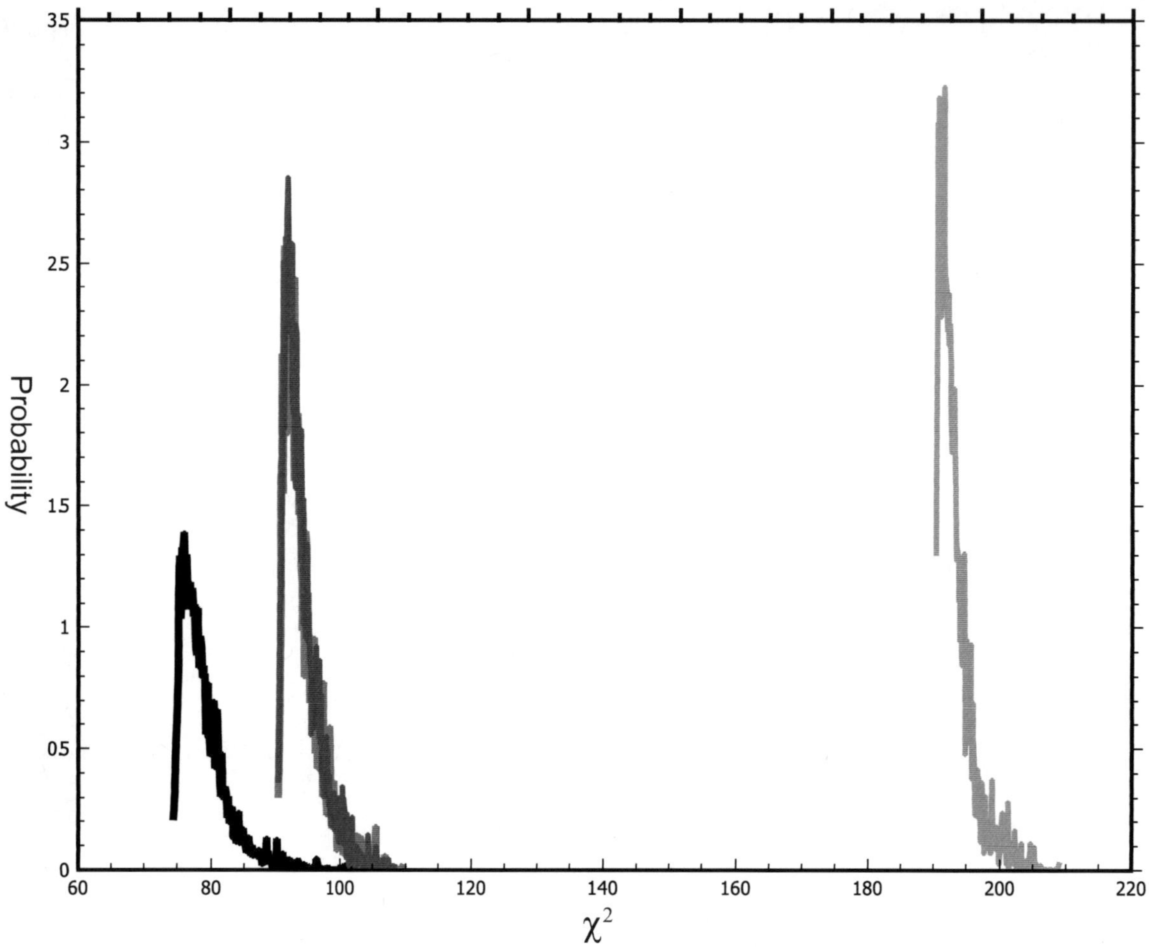

Figure 5. The result of the NMD model selection protocol for the proton NMD obtained from NCS data recorded from frozen disordered concentrated (85 wt%) PA solution at T=80K . (i) green line – a model with purely Gaussian NMD; (ii) red line – a model with a sum of two terms, a Gaussian and a Gaussian multiplied by the excess kurtosis term in the GC expansion without positivity constraints, (iii) blue line – the same model with positivity constraints imposed in fitting, (iv) black line – a model with a Gaussian term, the excess kurtosis term and an additional term proportional to $c_6 H_6(x)$ without positivity constraints imposed in fitting. See text for details.

experimental accuracy, a model with a sum of two terms, a Gaussian and a Gaussian multiplied by the excess kurtosis term in the GC expansion, yields the same probability distribution function (PDF) no matter the positivity constraints are applied or not. Moreover, in both cases the the values of the most likely parameters, c_4 and σ, are identical to within the experimental error and yield, 0.84 ± 0.07 and 6.7 ± 0.2 $\AA^{-1}$ respectively. The proton NMD curve fitted to experimental data using this model is shown in Figure 6. The inset shows the polynomial part of the GC expansion plotted for the optimal c_4 values of 0.84 ± 0.07. It clearly shows that the solution is positive for all values of the reduced variable x, as predicted by the positivity constraint condition developed in this work. Importantly, in the context of the NQT, the solution provided by this model strongly suggests the existence of the coherent supersposition of the proton wave

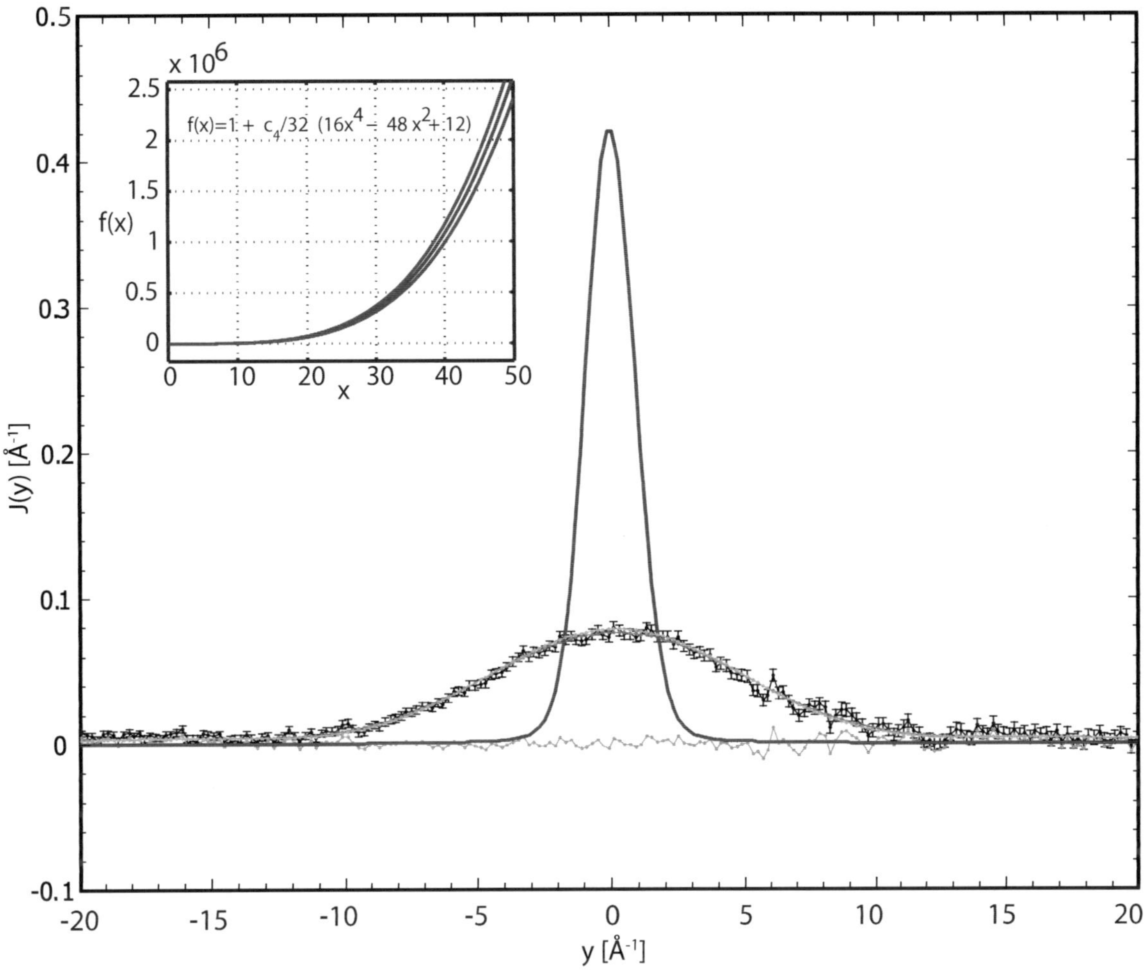

Figure 6. Main: The proton NMD curve fitted to experimental data recorded from frozen disordered concentrated (85 wt%) PA solution at T=80K . The model used involves an NMD containing a sum of two terms, a Gaussian and a Gaussian multiplied by the excess kurtosis term in the GC expansion with positivity constraints: the raw data is shown in black, the best fit NMD curve is reported light blue together with the instrument resolution function for protons in navy blue. The convolution of the NMD curve with the resolution function, which was subject to direct fitting to the experimental data, is shown as red trace and the fit residual is shown in green. Inset: the polynomial part of the GC expansion plotted for the optimal c_4 values of 0.84 $\pm$ 0.07. See text for details.

packet in the PA system under investigation, which may mark the onset of the temperature regime at which tunneling may be possible of protons across the HBs in the PA. It is worth noting that, mathematically similar solution, i.e. c_4= 0.720 $\pm$ 0.158, was found by Reiter et al. in fitting a GC to NCS data recorded in KH_2PO_4 (KDP) at T=90K and taken as a sign of the coherence over two locations for the proton observed in the paraelectric phase. [23]

The second observation is that the Bayesian inference prefers the model with a Gaussian term, the excess kurtosis term and an additional term proportional to $c_6 H_6(x)$. However, without any positivity constraints imposed on this model in fitting FABADA minimiser converges to a non-positive solution, as evidenced by the the polynomial part of the GC expansion plotted for the optimal c_4 and c_6 values found, i.e. 0.605 $\pm$ 0.06 and -0.6 $\pm$ 0.1, respectively, as shown in the inset of Figure 7.

Two important ramifications emerge from the case study illustrated in Figure 7. First of all, unlike suggested by Jondeau and Rockinger [50], the maximum likelihood estimation does not naturally find a solution which is a non-negative GC expansion (as denoted by the closed circle symbol in Figure 2). At least, it does not seem to be an inherent feature of this type of the estimation. Secondly, drawing from the example of the solution for the GC expansion containing the skewness and excess kurtosis terms by Jondeau and Rockinger [50], we can already say at this point that a general solution most likely will contain a high degree of correlation between the c_4 and c_6 values. The shape of the region in (c_4, c_6) parameter space may, in general, be very complicated, most likely to be found by a numerical solution. Moreover, an analytical solution that would describe the $c_4 - c_6$ correlation, even if it exists, would be very difficult to implement in fitting software.

An important realisation from the analysis of the performance of the positivity constraint implementation in the Bayesian NMD model selection, as applied to NMD data from the NCS experiment described above, is that the unconstrained solutions lead to NMD curves with negative values far away from the peak centre, in the regions of the longitudinal momentum distributions which oftentimes escape the scrutiny of the NCS data analysis. However, it is important to note in that respect that the mere fact that a numerical solution of the fitting of a given NMD model adopts positive values only in the kinematic domain accessible by the instrument does not signify that the obtained solution is physically meaningful. Owing to the unique nature of NCS, it is a basic postulate of quantum mechanics, namely the Born interpretation linking the square of the wave function to the probability distribution in the *whole* domain of the parameters on which this function depends, that intervenes here and provides a much-sought physical meaningfulness criterion.

4. Outlook

In this work we have presented a much-needed case study of the implementation of the NMD model ranking protocol, based on Bayesian inference with additional NMD positivity constraints. The protocol ensures that the NMD solutions obtained from the NCS data analysis are physically meaningful. Moreover, owing to the mass-selective nature of the NCS technique [53, 54, 55, 3, 56], this criterion can be implemented for NMDs of all masses in a given system under investigation, which strengthens the protocol performance in pruning out unphysical solutions from the investigation of any molecular and condensed matter nuclear quantum tunnelling-candidate systems.

The results presented herein provide a suitable platform for a future development of a robust and easy to implement NCS data screening procedure that would constitute the first crucial step in the understanding of nuclear quantum effects, a prerequisite for a suitable choice of the level of *ab initio* theory employed for system modelling. The ultimate goal to this end is the development of a hierarchy of the applicability of subsequent *ab initio* theory levels and their practical implementations in the quantum chemistry software that would match the degree of the information about the nuclear quantum effects present in the recorded NCS data, as categorised by Bayesian inference. In this way, 'heroic quantum simulations' would be proposed and applied to the characterisation of *only* those NCS data sets that would positively pass the Bayesian screening protocol. Conversely, systems exhibiting no marked 'non-trivial' nuclear quantum effects (e.g. no signatures of tunnelling in the NCS data) could be tackled by relatively low-cost lattice dynamics DFT-based methods. The overarching goal of the strategy described in this work is to further increase the level of much-needed synergy between NCS experiments and computational materials modelling.

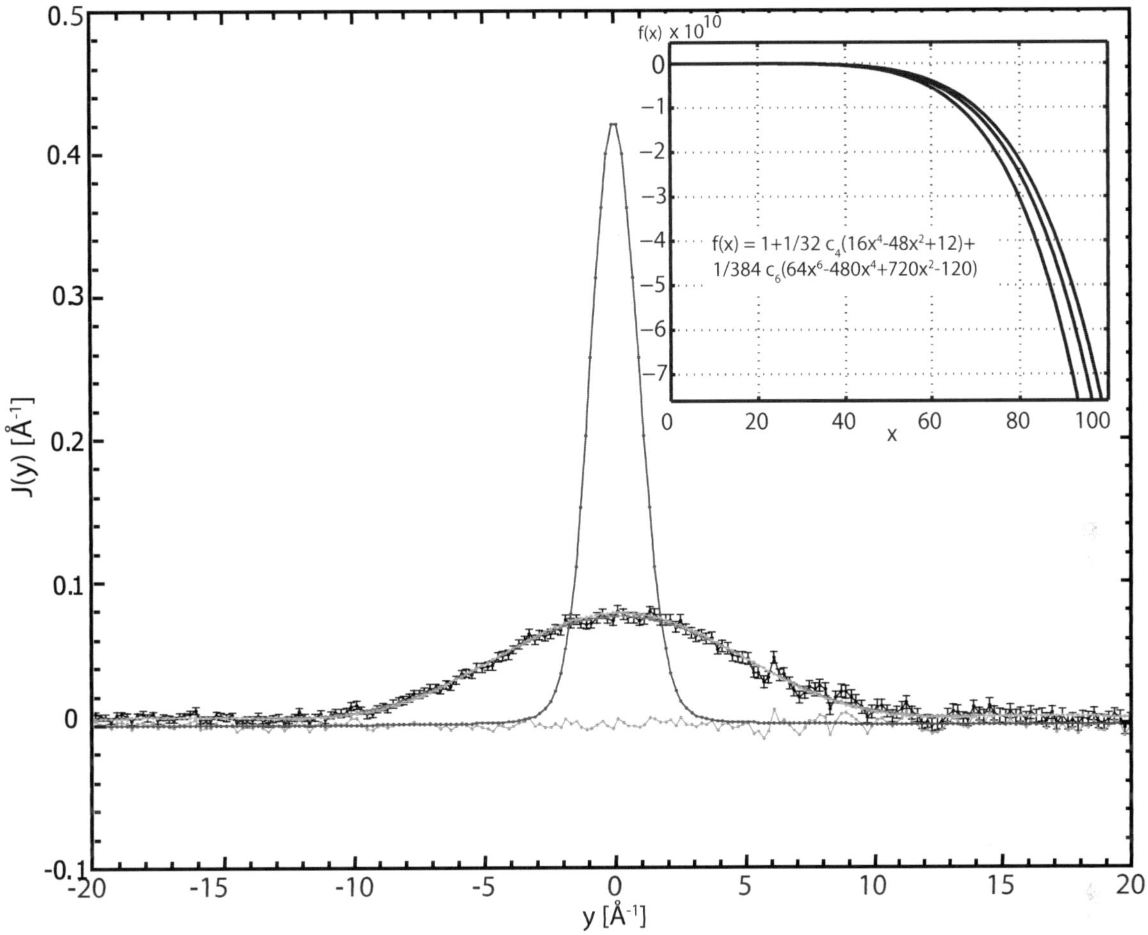

Figure 7. Main: The proton NMD curve fitted to experimental data recorded from frozen disordered concentrated (85 wt%) PA solution at T=80K. The model used involves a Gaussian term, the excess kurtosis term and an additional term proportional to $c_6 H_6(x)$ without positivity constraints imposed in fitting: the raw data is shown in black, the best fit NMD curve is reported light blue together with the instrument resolution function for protons in navy blue. The convolution of the NMD curve with the resolution function, which was subject to direct fitting to the experimental data, is shown as red trace and the fit residual is shown in green. Inset: the polynomial part of the GC expansion plotted for the optimal c_4, c_6 values of 0.605 $\pm$ 0.06,-0.6 $\pm$ 0.1 respectively. See text for details.

5. Acknowledgments

The authors gratefully acknowledge the UK Science and Technology Facilities Council for financial support and for access to beamtime.

[1] Fernandez-Alonso F and Price D L (eds) 2013 *Neutron Scattering – Fundamentals* (New York: Academic Press)
[2] Andreani C, Colognesi D, Mayers J, Reiter G F and Senesi R 2005 *Adv. Phys.* **55** 377
[3] Andreani C, Krzystyniak M, Romanelli G, Senesi R and Fernandez-Alonso F 2017 *Adv. Phys.* **66** 1–73
[4] Ceriotti M, Bussi G and Parrinello M 2009 *Phys. Rev. Lett.* **103**
[5] Ceriotti M, Miceli G, Pietropaolo A, Colognesi D, Nale A, Catti M, Bernasconi M and Parrinello M 2010 *Phys. Rev. B* **82**
[6] Lin L, Morrone J A, Car R and Parrinello M 2010 *Phys. Rev. Lett.* **105** 110602
[7] Krzystyniak M 2010 *J. Chem. Phys.* **133**
[8] Krzystyniak M and Fernandez-Alonso F 2011 *Phys. Rev. B* **83**
[9] Krzystyniak M, Adams M A, Lovell A, Skipper N T, Bennington S M, Mayers J and Fernandez-Alonso F 2011 *Farady Discussions* **151**
[10] Romanelli G, Ceriotti M, Manolopoulos D E, Pantalei C and Senesi R 2013 *J. Phys. Chem. Lett.* **4** 3251
[11] Wikfeldt K T and Michaelides A 2014 *J. Chem. Phys.* **140** 041103
[12] Andreani C, Ceriotti M, Fernandez-Alonso F, Gorini G, Krzystyniak M, Parmentier A, Parker S, Seel A, Senesi R and Walewski L 2014 *J. Phys.: Conf. Ser* **571**
[13] Andreani C, Ceriotti M, Chass G, Drechsel-Grau C, Fernandez-Alonso F, Greaves N, Gidopoous N, Krzystyniak M, Parmentier A, Refson K, Reiter G F, Romanelli G, Seel A, Walewski L and Wiles T 2014 *J. Phys.: Conf. Ser* **571**
[14] Ceriotti M, Drechsel-Grau C, Fernandez-Alonso F, Greaves N, Krzystyniak M, Major D, Parmentier A, Salzmann C, Senesi R, Sturniolo S, Walewski L and Wikfeltd T 2014 *J. Phys.: Conf. Ser* **571**
[15] Ceriotti M, Fang W, Kusalik P G, McKenzie R H, Michaelides A, Morales M A and Markland T E 2016 *Chem. Revs.* **116** 7529–7550
[16] Krzystyniak M, Druzbicki K, Romanelli G, Gutmann M J, Rudic S, Imberti S and Fernandez-Alonso F 2017 *Phys. Chem. Chem. Phys.* **19** 9064–9074
[17] Druzbicki K, Krzystyniak M, Romanelli G, Gutmann M J, Rudic S, Imberti S and Fernandez-Alonso F 2018 Hydrogen dynamics in solid formic acid: Insights from simulations with quantum colored-noise thermostats, submitted.
[18] Senesi R, Colognesi D, Pietropaolo A and Abdul-Redah T 2005 *Phys. Rev. B* **732** 054119
[19] Colognesi D, Ramirez-Cuesta A J, Zoppi M, Senesi R and Abdul-Redah T 2004 *Physica B* **350** E983
[20] Frisch M J *et al.* 2016 Gaussian16 Revision A.03 gaussian Inc. Wallingford CT
[21] Reiter G, Mayers J and Platzman P 2002 *Phys. Rev. Lett.* **89**
[22] Reiter G, Mayers J and Platzman P 2002 *Ferroelectrics* **268** 283–288
[23] Reiter G, Shukla A, Platzman P M and Mayers J 2008 *New J. Phys.* **10**
[24] Reiter G, Li J, Mayers J, Platzman P and Stillinger F 2004 *Braz. J. Phys.* **34** 142–147
[25] Sivia D S a D W, Knight K and Gull S 1993 *Physica D* **66** 234–242
[26] Sivia D S 2006 *Advance Mathematical and Computational Tools in Metrology VII* (*Series on Advances in Mathematics for Applied Sciences* vol 72) ed Ciarlini P, Filipe E, Pavese F, Perruchet C and Siebert B pp 108–118
[27] Sivia D S and Silver R 1989 *Momentum Distributions* ed Silver R and Sokol P pp 377–380
[28] Pardo L C, Rovira-Esteva M, Busch S, Ruiz-Martin M D and Tamarit J L 2011 *J. Phys.: Conf. Ser.* **325**
[29] D M, Vispa A, Pardo L C, R Tolchenov R, Mukhopadhyay S and Fernandez-Alonso F 2015 *J. Phys.: Conf. Ser.* **663** 012009
[30] Mayers J and Reiter G 2012 *Meas. Sci. Technol.* **23** 045902
[31] `http://www.isis.stfc.ac.uk/instruments/vesuvio/`
[32] Sears V F 1984 *Phys. Rev. B* **30** 44
[33] Watson G I 1996 *J. Phys.: Condens. Matter* **8** 5955
[34] Mayers J, Andreani C and Baciocco G 1989 *Phys. Rev. B* **39** 2022
[35] Mayers J 1990 *Phys. Rev. B* **41** 41
[36] Mayers J 1993 *Phys. Rev. Lett.* **71** 1553
[37] Andreani C, Colognesi D and Pace E 1999 *Phys. Rev. B* **60** 10008
[38] Mayers J and Abdul-Redah T 2004 *J. Phys.: Cond. Matter* **16** 4811
[39] Flammini D, Pietropaolo A, Senesi R, Andreani C, McBride F and Hodgson A 2012 *J. Chem. Phys* **136**

024504
[40] Fielding A L, Timms D N and Mayers J 1998 *Europhysics Letters* **44** 255
[41] Krzystyniak M 2010 *J. Chem. Phys.* **133** 144505
[42] Krzystyniak M and Fernandez-Alonso F 2011 *Phys. Rev. B* **83** 134305
[43] Romanelli G and Krzystyniak M 2016 *Nucl. Instrum. Meth. A* **819** 84 – 88
[44] Pietropaolo A and Senesi R 2011 *Physics Reports* **508** 45
[45] Lovesey S W 1984 *Theory of Neutron Scattering from Condensed Matter* (Oxford: Clarendon)
[46] Oram E and McCann L 2016 Simulation support and further indirect inelastic specific improvements within mantid Tech. Rep. RAL-TR-2016-019 Rutherford Appleton Laboratory URL `http://purl.org/net/epubs/work/30884953`
[47] Oram E 2016 An overview of developments in mantid relating to indirect inelastic spectroscopy july 2016 - july 2017 Tech. Rep. RAL-TR-2016-011 Rutherford Appleton Laboratory URL `http://purl.org/net/epubs/work/29538224`
[48] Nixon D 2015 Simulation support within mantid for molecular spectroscopy Tech. Rep. RAL-TR-2015-012 Rutherford Appleton Laboratory URL `http://purl.org/net/epubs/work/23618523`
[49] Jackson S 2014 Vesuvio data reduction and analysis in mantid Tech. Rep. RAL-TR-2014-009 Rutherford Appleton Laboratory URL `http://purl.org/net/epubs/work/12239830`
[50] Jondeau E and Rockinger M 2003 *Journal of Economic Dynamics and Control* **27** 1699–1737
[51] 2017 Matlab optimization toolbox the MathWorks, Natick, MA, USA
[52] DAnna F, Billeci F, Ferrero Vallana F M, Krzystyniak M, Plechkova N V and Seddon K 2018 An anomaly in proton conductivity of phosphoric acid, in preparation.
[53] Krzystyniak M, Richards S, Seel A and Fernandez-Alonso F 2013 *Phys. Rev. B* **88** 184304
[54] Krzystyniak M, Druzbicki K and Fernandez-Alonso F 2015 *Phys. Chem. Chem. Phys.* **17**
[55] Krzystyniak M, Seel A G, Richards S E, Gutmann M J and Fernandez-Alonso F 2014 *J. Phys. COnf. Ser.* **571** 012002
[56] 2017 *Neutron Scattering - Applications in Biology, Chemistry, and Materials Science* (*Experimental Methods in the Physical Sciences* vol 49) ed Fernandez-Alonso F and Price D L (Academic Press) pp 403 – 457

IOP Conf. Series: Journal of Physics: Conf. Series **1055** (2018) 012013 doi:10.1088/1742-6596/1055/1/012013

Procedure for the determination of effective temperatures employing VESUVIO spectrometer

J Dawidowski[1], L A Rodríguez Palomino[1], G Romanelli[2], and M Krzystyniak[2]

[1]Consejo Nacional de Investigaciones Científicas y Técnicas (CONICET), Comisión Nacional de Enegía Atómica-Universidad Nacional de Cuyo, Argentina
[2]ISIS Facility, Rutherford Appleton Laboratory, Chilton, Didcot, Oxfordshire OX11 0QX, United Kingdom

E-mail: jdawidowski@gmail.com

Abstract. We describe the procedure to determine effective temperatures employing the data collected by the detectors of the VESUVIO spectrometer (ISIS pulsed neutron source), employing the Deep Inelastic Neutron Scattering technique. A scheme to group the detectors is proposed, based on their positions and angular ranges, in order to add their spectra and improve the statistics. We show the details of all the corrections that must be made on the raw experimental data to reach the spectra on which the desired effective temperatures can be fitted. Most of the correction stages require ad-hoc auxiliary simulations by the Monte Carlo method, that were deviced and described in previous papers. The corrections, that involve background subtraction, empty cell subtraction, and multiple scattering corrections, are described in detail. The efficiencies of the defined detector groups are determined, and employed in the expressions used in data fitting. Data treatment steps performed in hydrogenated and deuterated alcohols, are shown as examples of application of this procedure.

1. Introduction

The knowledge of the kinetic energy of the atoms in condensed matter is a subject of primary interest in Neutron Physics. This is because this observable determines properties of interest, such as the width of nuclear neutron resonances (through the Doppler broadening) [1], or the spectrum shape produced by neutron moderators [2], among others often related to Nuclear Engineering calculations. In all the mentioned cases, a basic parameter that is necessary to know is the mean value of the kinetic energy, which is determined by the thermodynamic temperature as well as the dynamic properties of the system. The mean kinetic energy is often referred to as the effective temperature, depending upon the scope of a given application. While dependent on the thermodynamic temperature, the effective temperature of the atoms that compose a system is also a function of the vibrational densities of states [3, 4].

In Nuclear Engineering it is usual to determine the effective temperatures indirectly from models of frequency spectra [5]. With Deep Inelastic Neutron Scattering (DINS) it is possible to measure directly the effective temperatures of each atomic species composing a system, since it is a mass-resolved technique [6]. This provides a unique tool to access experimentally this observable.

IOP Conf. Series: Journal of Physics: Conf. Series **1055** (2018) 012013 doi:10.1088/1742-6596/1055/1/012013

The VESUVIO spectrometer (ISIS, Rutherford Appleton Laboratory, UK) has the possibility to cover a wide range of scattering angles, with 64 detectors in forward scattering and 132 detectors in backscattering directions. Originally deviced to study the detail of the momentum distributions profiles [7], it also proved to be a useful tool to determine neutron total cross sections as a function of energy by employing its transmission monitor [9]. In this paper, we address the situation when no specific details of the Compton profile are sought, and only the mean value of the kinetic energy needs to be obtained. In such cases, reliable results can be obtained by adding the signal of groups of detectors, with the advantage of improving the statistics, a common practice among the various neutron scattering spectrometers. A different approach to the problem is to transform the detector signals to the variable y (West scaling [6]), and add the spectra afterwards. However as shown in Ref. [10] the y-scaling (associated with the Convolution Approximation) is an approximation, that consists in considering that the filter has a single well-defined energy in the passage from time of flight to y, while the uncertainties associated with the energy width of the filter and others are packed in the resolution function. We showed that there are discrepancies of this approximation from the exact expression noticeable for the case of light nuclei. In this paper we decided to avoid such intermediate uncertainty, by using the exact expression, adding the signals of the individual detectors in the time-of-flight variable at the expense of reducing the angular resolution.

As the purpose of this paper is mainly methodological, we will focus our attention on the procedure to treat the data of the detector groups, from the raw data to the final effective temperature fitted from the profiles, along the lines traced in Ref. [5]. The systems studied are a set of normal and deuterated alcohols, whose total cross sections were previously determined [9]. The systematic study of the effective temperatures of these systems will be left for a forthcoming paper.

2. Experimental

The experiments described in this work were carried out at the VESUVIO spectrometer, placed at target station 1 (TS1) of the ISIS pulsed neutron and muon source at the Rutherford Appleton Laboratory (United Kingdom). A general sketch of VESUVIO is shown in Fig. 1. The forward scattering detectors are yttrium aluminum perovskite (YAP) γ -ray detectors (S135 to S198), and the backscattering detectors (S3 to S134) are ^{6}Li-doped glass scintillators. The analyzer filters are movable ^{197}Au foils that allow filter-in and filter-out measurements, employed to perform the single-difference (SD) technique. A thorough description of the system and its operation can be found in Ref. [7].

In this paper we will refer to measurements performed on hydrogenated and deuterated alcohols in two batches. The samples were liquid high-purity 1-propanol ($CH_3CH_2CH_2OH$), 2-propanol ($CH_3CHOHCH_3$), n-butanol ($CH_3CH_2CH_2CH_2OH$), 1-propanol-OD ($CH_3CH_2CH_2OD$), 2-propanol-D8 ($CD_3CDODCD_3$) and n-butanol-D10 ($CD_3CD_2CD_2CD_2OD$), bought from Sigma Aldrich [8] , with 99% deuteration in the case of the last three [9]. The samples were in the liquid state kept at room temperature, held into aluminum flat cells of square sections 65 mm inner side and 90 mm outer side. The thicknesses were 1.5 mm in the case of normal and 2 mm in the case of the deuterated samples.

The spectra were recorded as a function of the time of flight, and treated with programs based on the Open Genie display and data analysis suite, developed for the neutron scattering instruments at the ISIS facility [11].

3. Data treatment

3.1. Fundamentals

One of the purposes of this work is to integrate the spectra of several detectors to improve the statistics. The difference spectra is the usual working magnitude in DINS experiments and

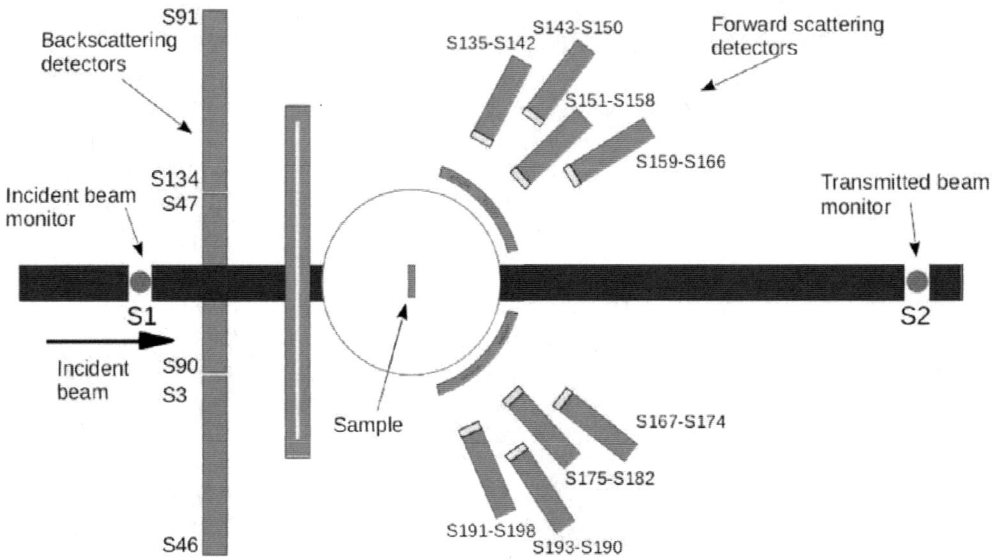

Figure 1. VESUVIO general setup, showing the forward scattering and backscattering detectors.

consists in the 'filter-out' minus 'filter-in' spectra. We will consider the sum of the spectra of detectors at scattering angles $\theta_{\rm i}$, in an angular range such that $\theta_{\rm inf} < \theta_{\rm i} < \theta_{\rm sup}$. Then, the expression of the recorded spectra is [12]

$$c(t) = \sum_{\theta_{\rm i}=\theta_{\rm inf}}^{\theta_{\rm sup}} \int_{\substack{\rm E_{inf} \\ \rm t=const}}^{\infty} dE_0 \, \Phi(E_0) \frac{d^2\sigma(E_0, E, \theta_{\rm i})}{d\Omega dE} \, \varepsilon(E) \left[1 - e^{-nT\sigma_F(E)}\right] \left|\frac{\partial E}{\partial t}\right| \Delta\Omega. \tag{1}$$

In Eq. (1), $\Phi(E_0)$ is the energy dependent incident neutron flux, $\frac{d^2\sigma(E_0,E,\theta)}{d\Omega dE}$ is the double-differential cross section of the sample, and $\varepsilon(E)$ is the detector group efficiency. The expression $\left[1 - e^{-nT\sigma_F(E)}\right]$ corresponds to the absorption probability of the resonant filter, characterized by a number density n, a thickness T, and a total cross section $\sigma_F(E)$. In the present work, gold foils were employed with a thickness of 0.0125 mm for the forward scattering and 0.05 mm for the backward scattering detectors. Finally, $\Delta\Omega$ is the solid angle subtended by a detector. The lower limit of integration $E_{\rm inf}$ is determined by kinematic conditions, while the sum spans over the signals of all the detectors in the considered angular range. It is worth noting that Eq. (1) is strictly valid for single scattering. For a real sample of finite dimensions, the total count rate recorded by the detectors is composed by singly and multiply scattered neutrons. The main route of calculation is by Monte Carlo simulations [12], as it will be shown in Sect. 4.

Group	Number of detectors	θ_{inf} (°)	θ_{sup} (°)
1	4	31.7	35.0
2	6	36.9	39.4
3	6	41.0	45.7
4	6	41.9	45.9
5	4	47.0	47.7
6	8	50.7	54.1
7	6	50.7	54.7
8	8	55.4	60.1
9	6	60.6	63.9
10	10	65.0	69.7

Table 1. Definition of the 10 detector groups in the forward scattering direction. The number of detectors in each group and the lower and upper angular limits (θ_{inf} and θ_{sup} of Eq. (1) in degrees) are indicated.

Group	Number of detectors	θ_{inf} (°)	θ_{sup} (°)
1	21	152.5	165.9
2	51	139.5	151.2
3	60	131.0	139.6

Table 2. Definition of the 3 detector groups in backscattering direction. The number of detectors in each group and the lower and upper angular limits (θ_{inf} and θ_{sup} of Eq. (1) in degrees) are indicated.

Eq. (1) will serve as the basic expression to fit the effective temperatures to the spectra, after the experimental corrections below described are applied.

3.2. Detectors grouping

As mentioned above we defined detector groups in order to add their signals together. The criteria followed to include detectors in a group were to chose

- detectors placed at neighboring scattering angles,
- detectors placed at similar distances to the sample.

For the forward scattering detectors in Fig 1 we observe that there are basically two typical sample-detector distances. Fig. 2 shows the detectors layout as seen from the sample position along the beam direction. The angular range covered by the detectors is roughly from 31° to 70 °. Applying the above mentioned criteria, 10 detector groups are defined and shown in different colors. Table 1 shows the angular limits of each group.

The backscattering detectors are arranged on a plane (as shown in Fig. 1). Fig. 3 shows the detectors layout as seen from the sample position in the beam direction. The angular range covered by the detectors is roughly from 131° to 166 °. In this case 3 detector groups are defined, which are shown in different colors. Table 2 shows the angular limits of each group.

Fig. 4 illustrates the resulting difference spectra ('filter-out' minus 'filter-in') for the defined groups in forward and backward scattering, taking as examples the cases of 2-propanol(H) and 2-propanol(D) respectively.

IOP Conf. Series: Journal of Physics: Conf. Series **1055** (2018) 012013 doi:10.1088/1742-6596/1055/1/012013

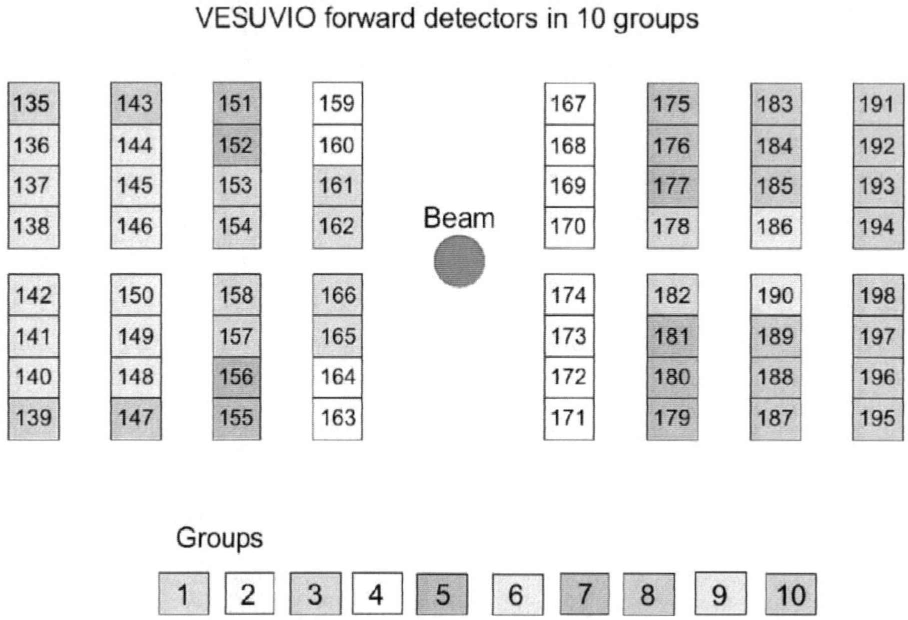

Figure 2. VESUVIO forward scattering detectors as seen from the sample position in the beam direction, showing their numbering and the 10 groups employed for the analysis in this work. The colors indicate the group numbers.

3.3. Determination of the detector groups' efficiencies

In Eq. (1), the efficiency of the detector group is a required quantity to be known when calculating the time-of-flight spectra $c(t)$. In practice, this determination is made by comparing a spectrum measured with the detector whose efficiency we want to know, with the same spectrum measured with a detector of known efficiency. In this case, we firstly measured the incident spectrum $\Phi(E)$, with the transmitted beam monitor S2, shown in Fig. 1, a ^{6}Li-doped glass scintillator of known efficiency. In the case of a detector group, $\Phi(E)$ was determined by measuring the spectrum of a lead target (placed in the sample position), that was considered an elastic and isotropic scatterer as commented in Ref. [12]. If $\mathcal{D}_{\rm Pb}(E)$ is the measured Pb spectrum (after subtracting the background, measured by leaving empty the sample position), the efficiency $\varepsilon(E)$ can be determined from the relation

$$\mathcal{D}_{\rm Pb}(E) = \varepsilon(E)\Phi(E). \tag{2}$$

Since this method to determine the efficiency is based on the measurement of two spectra, and none of them are absolutely measured, we obtain the shape of the efficiency curve and not its absolute value (as in the case of the spectrum), therefore leaving an undetermined multiplicative constant in Eq. (1).

In Fig. 5 we illustrate the determination of the efficiency for the specific case of detector

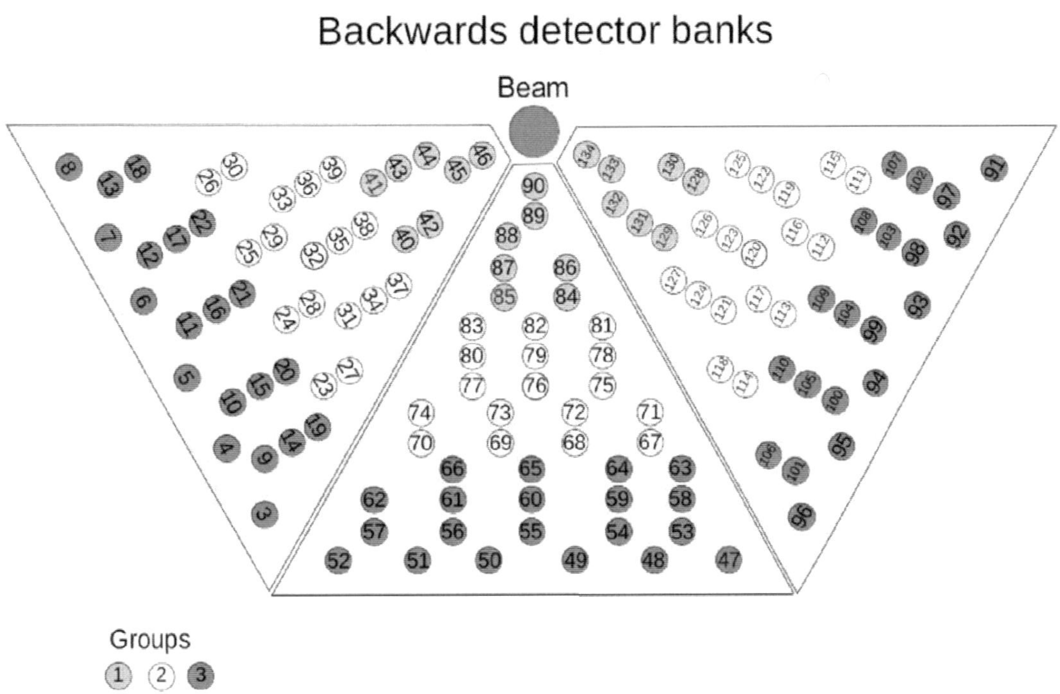

Figure 3. VESUVIO backward scattering detectors as seen from the sample position in the direction opposed to the beam, showing their numbering and the proposed 3 groups employed for the analysis in this work. The colors indicate the group numbers.

group 10 in the forward scattering direction. In the left frame we show the incident neutron spectrum measured with the transmission detector S2 and corrected by its efficiency, together with a Pb sample and background spectra determined with the sum of the signals of detector group 10. All the spectra have been transformed from time-of-flight to energy units with Open Genie routines considering the total flight path of the neutron (incident and scattered paths) and applying the Jacobian of the transformation. In the right frame we show the function $\varepsilon(E)$ (black line) calculated from Eq. (2). The peaks above 1 eV correspond to gold resonances, for which the detection system has a higher neutron cross section that is converted into photons that are counted by the YAP scintillators [13]. Below 1 eV we observe the Bragg peaks corresponding to the lead sample, for which the considerations leading to Eq. (2) are no longer valid. Therefore in that region, we adopt a constant value (dotted red lines) to describe the efficiency. This regime will have only marginal interest in our data treatment, where the epithermal range is the main focus. This procedure was employed to determine all the detector-group efficiencies.

3.4. Background corrections

The residual filter-dependent background in the difference spectra observed on VESUVIO, that was described and treated by Mayers [14], and previously examined in [5], must be subtracted as a first processing step. The treatment we employ for the background is that followed by

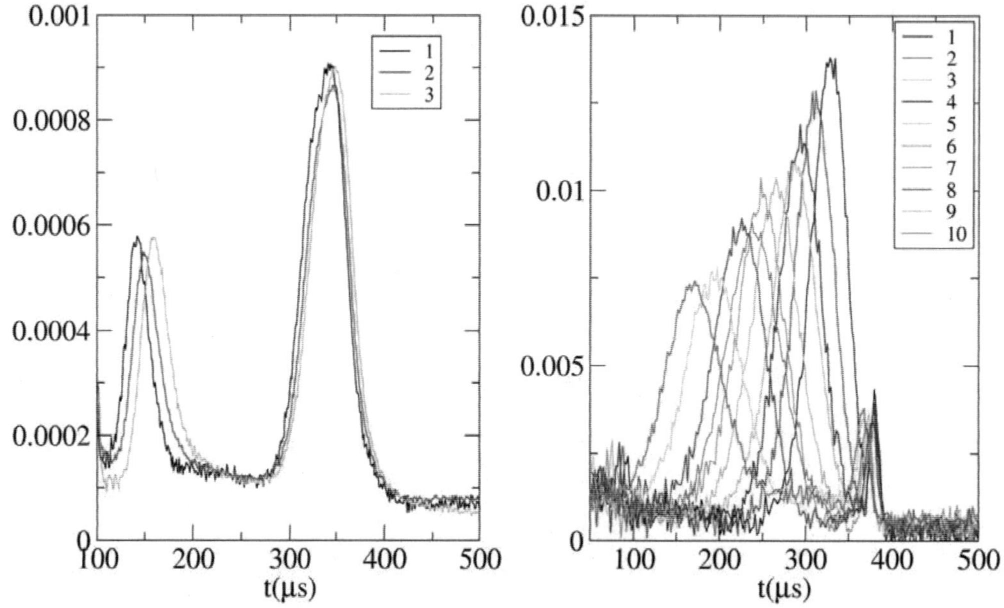

Figure 4. Difference spectra for the three groups of backscattering detectors (left, 2-propanol (D)), and the 10 groups of forward scattering detectors (right, 2-propanol(H).

Perelli-Cippo et al. [15] by means of polynomial functions. After treating different kinds of samples, two main conclusions can be drawn on the background on the difference spectra:

(i) It is filter-dependent. It means that the 'filter-out' minus 'filter-in' spectrum, leaves a remaining background that is not canceled by the difference.

(ii) It is sample dependent. The intensity and shape of the background depends on the scattering power of the sample. Thus, we distinguished two different behaviors depending on whether the samples are hydrogenated or deuterated.

In practice, we proceeded as follows. The hydrogenated and the deuterated samples data sets were treated separately, searching a common background spectrum for each one. The measured difference spectra were compared with the total Monte Carlo calculated profiles that will be described in Sect. 4. The difference between the measured and simulated spectra was ascribed to the background.

The process is depicted in Fig. 6, where we focus on the hydrogenated alcohols spectra of group 2 of backscattering detectors (see Fig 3). In its right frame we show the experimental difference spectra (in full lines), together with the total Monte Carlo simulations. The apparent differences between measured and simulated spectra are due to the backgrounds, that are shown in the right frame, for each sample. Their great similitude allows us to represent them by a single fourth degree polynomial (drawn in blue line).

Fig. 7 depicts the process of background corrections for the hydrogenated alcohols in detector group 2 (backscattering). The upper row shows the measured spectra, together with the polynomial background and the corrected spectra after background subtraction. The lower

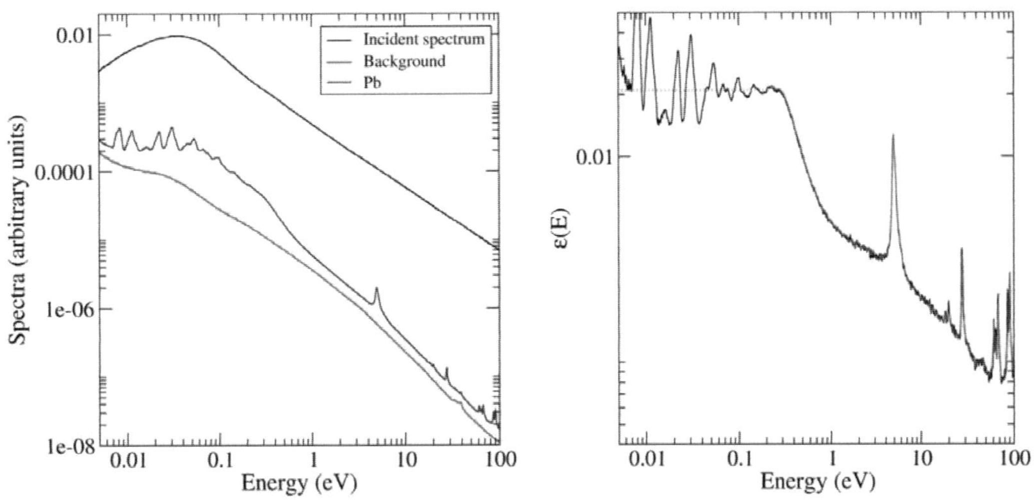

Figure 5. Left: Incident neutron spectrum (measured with transmission detector S2 and corrected by its efficiency), and Pb sample and background spectra (measured with forward scattering detector group 10). Right: the function $\varepsilon(E)$ (black) calculated from the spectra in the right frame, and the adopted efficiency function (red continuous and dotted lines, see text for details).

row shows the corrected data compared with the Monte Carlo simulations. The good agreement shown serves as a consistency check of the procedure.

3.5. Empty cell subtractions

When working with samples that require the use of a container such as for liquids, gases or powders, empty cell measurements are normally made, in order to subtract their signal. This can not be done directly due to the attenuation of the beam in the sample and in the cell, and the multiple scattering, which makes the contribution of the full cell different from that of the empty cell. Paalman and Pings formulated analytical expressions [16] for empty-cell subtraction in the case of X-ray scattering, and well-defined sample geometries. As commented in Ref. [17], the method is not well suited for the case of neutron scattering (although its use is common practice), because only elastic scattering is considered, (and therefore the cross sections are constants). This is usually far from the real case with neutrons, for which multiple inelastic scattering may be important.

The empty-cell subtraction we consider here, relies on Monte Carlo simulations (described in Sect. 4), of the sample-cell (SC), empty cell (C) and sample-only (S) systems. Thus, we calculate the spectra $C_{\rm SC}(t)$, $C_{\rm S}(t)$, $C_{\rm C}(t)$, and define the *cell subtraction factor*

$$f_{\rm C}(t) = \frac{C_{\rm SC}(t) - C_{\rm S}(t)}{C_{\rm C}(t)}, \tag{3}$$

that establishes the relation

$$C_{\rm S}(t) = C_{\rm SC}(t) - f_{\rm C}(t) C_{\rm C}(t). \tag{4}$$

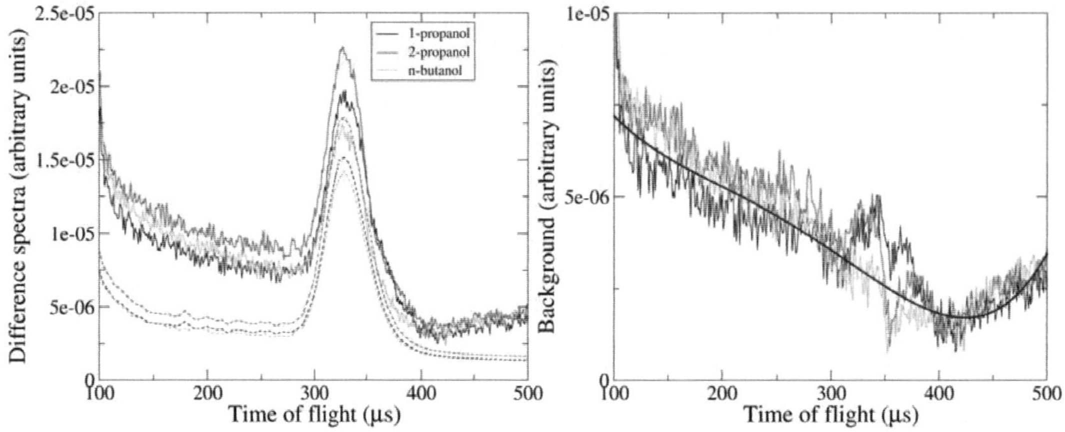

Figure 6. Left: In full lines the experimental difference spectra of the Hydrogenated alcohols for detector group 2 in backscattering. In dashed lines the corresponding total Monte Carlo simulations (with the same color convention as for the experimental). Right: The differences between the experimental and the calculated spectra, showing that they can be approximately described by a common background (with the same color convention indicating the samples) The blue thick line is the fitted polynomial function employed to represent the background.

The next step is to apply the same relation to the experimental data, $\mathcal{E}_{SC}(t)$ and $\mathcal{E}_{C}(t)$, which provides the sought way to subtract the cell

$$\mathcal{E}_{S}(t) = \mathcal{E}_{SC}(t) - f_{C}(t)\mathcal{E}_{C}(t), \tag{5}$$

where f_C is calculated from (3), and $\mathcal{E}_{SC}$ and $\mathcal{E}_{C}$ are measured. The precision of the method depends on the accuracy of the description of the scattering cross sections and the mean free paths as a function of the energy in the Monte Carlo code.

We found that the factors $f_C(t)$ can be represented by one function for the hydrogenated and another for the deuterated samples, regardless the detector group. In Fig. 8 we illustrate such results for the case of a forward and a backward scattering group. We also show the 7th degree polynomials employed to represent $f_C(t)$, together with values of $f_C(t)$ calculated from Eq. (3) with Monte Carlo results in some of the studied systems. Although noticeable statistical fluctuations are evident, the calculations allow a proper definition of the polynomials. In the insets of Fig. 8, we show two representative Monte Carlo results for sample-cell, empty cell and sample-only systems in hydrogenated n-butanol, for forward (left) and backward scattering (right). The proximity of the polynomial to a value about 1 for for the deuterated samples, is a consequence of the transparency of the sample, in which case Eq. (5) is a direct subtraction of the empty cell.

Fig. 9 shows an application of the proposed cell subtraction procedure, for the cases of hydrogenated and deuterated 2-propanol for detector group 3 in backscattering. Rather than the raw experimental empty cell spectra, we fitted Lorentzian functions (shown in red) to avoid amplifying the noise in the subtraction. The blue line indicates the term $f_C(t)\mathcal{E}_C(t)$ of Eq. (5), and the orange line shows the cell-subtracted spectra.

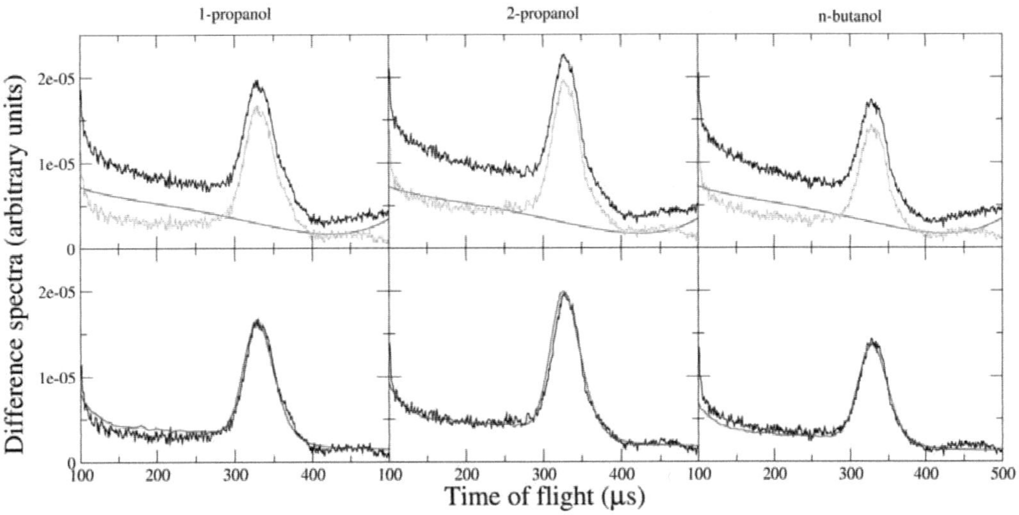

Figure 7. Background corrections for the hydrogenated alcohols detector group 2 in backscattering. In the upper row of frames we show the measured spectra (black lines), the background fitted with a single fourth-degree polynomial (red lines), and the resulting spectra after background subtraction (green lines). In the lower row of frames we show the spectra after background subtraction (black lines) compared with the total Monte Carlo simulations (red lines).

4. Monte Carlo simulations

Monte Carlo simulations have the dual purpose of correcting the experimental data for multiple scattering, attenuation, and detector efficiency effects, and also helping in the analysis of the background as shown in Sects. 3.4 and 5. A detailed account of the Monte Carlo procedure employed can be found in Ref. [18], where the reader is referred for further details. As above referred, the Monte Carlo procedure requires the knowledge of the incident spectrum and the detector efficiency, together with instrumental parameters such as the sample and filter geometries and their total cross sections as a function of the energy. An essential input for the Monte Carlo simulations are the total cross sections of the alcohols and their synthetic frequency spectra that governs the dynamics of the modeled systems. Both subjects were extensively studied for the alcohols presented in this work in Ref. [9]. Finally, the filter analyzer absorption cross section as a function of energy is read from absorption cross section tables, that can be obtained from Ref. [19].

In Fig. 10 we show the comparison of the experimental and Monte Carlo data of normal and deuterated 2-propanol for the group number 10 in forward position (left) and number 3 in backward position (right), respectively. The good agreement observed guarantees the suitability of the Monte Carlo simulations for the aforementioned uses.

5. Multiple scattering corrections

As extensively treated in Ref. [9], we correct the experimental data by beam attenuation and multiple scattering effects, by Monte Carlo simulations (described in Sect. 4). In the present case we made such corrections after subtracting the cell effects as described in Sect

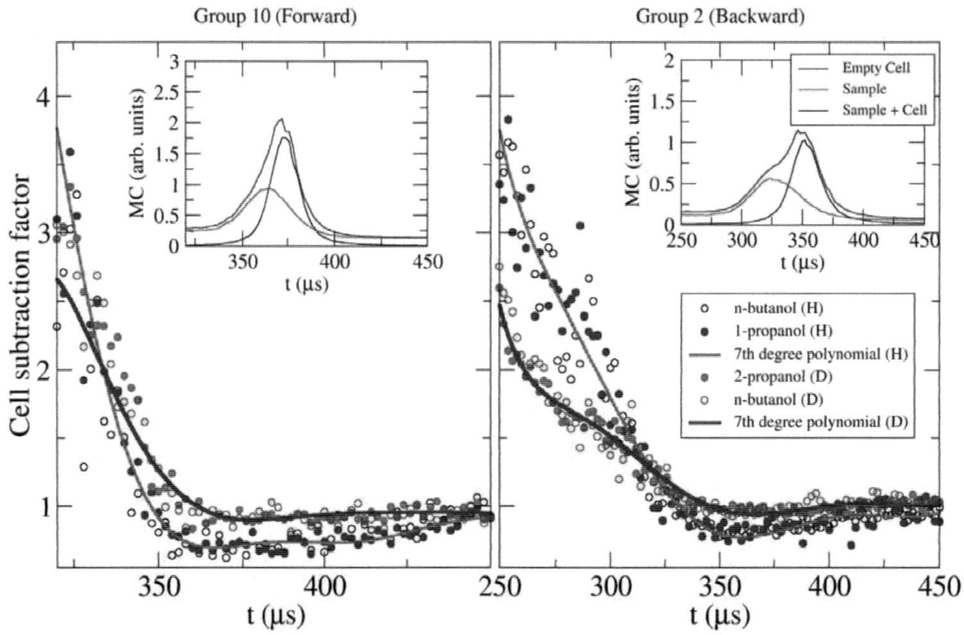

Figure 8. The cell subtraction factor $f_C(t)$ calculated from Monte Carlo results for hydrogenated and deuterated alcohols and different detector groups, in forward scattering (left) and backward scattering (right). The results are fitted with 7th degree polynomials for each type of sample. Insets: Monte Carlo results for n-butanol (H) used to determine the factors.

3.5. Therefore, the Monte Carlo simulations for this corrections were made only on the sample system (without considering the cell). The attenuation correction factor $H(t)$ is defined as ratio of the detected singly-scattered neutrons to the total singly-scattered neutrons spectra, and the multiple scattering correction factor $M(t)$ is defined as the ratio the detected singly-scattered neutrons to the total scattered neutrons spectra [9]. The experimental data $\mathcal{E}(t)$ are then corrected by

$$\mathcal{E}_{\rm corr}(t) = \frac{M(t)\mathcal{E}(t)}{H(t)} = \mathcal{F}(t)\mathcal{E}(t), \tag{6}$$

where we defined a single factor $\mathcal{F}(t) = \frac{M(t)}{H(t)}$ that we call experimental correction factor.

As examples, in Fig. 11 we show the experimental data for the hydrogenated samples for group 10 in forward scattering (frame (a)) and the deuterated samples for group 3 in backscattering (frame (b)), and their corresponding correction factors determined by simulations (insets). It is worth to notice that these factors are very similar for each family of samples, so only one of them was shown. The right frames show the data corrected by these effects.

Upon reaching this instance we summarize the corrections made on the raw data:

- Background subtraction
- Empty cell effects
- Beam attenuation
- Multiple scattering.

IOP Conf. Series: Journal of Physics: Conf. Series **1055** (2018) 012013 doi:10.1088/1742-6596/1055/1/012013

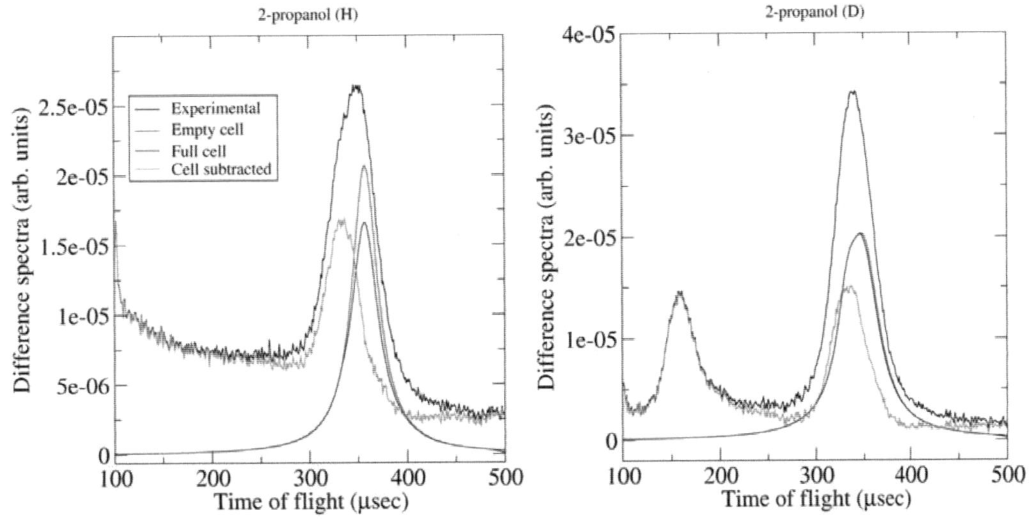

Figure 9. Empty cell spectra subtraction for hydrogenated and deuterated 2-propanol for detector group 3 in backscattering. The empty cell spectra are measured and fitted with a Lorentzian function (red line). The blue line indicates the term $f_C(t)\mathcal{E}_C(t)$ of Eq. (5).

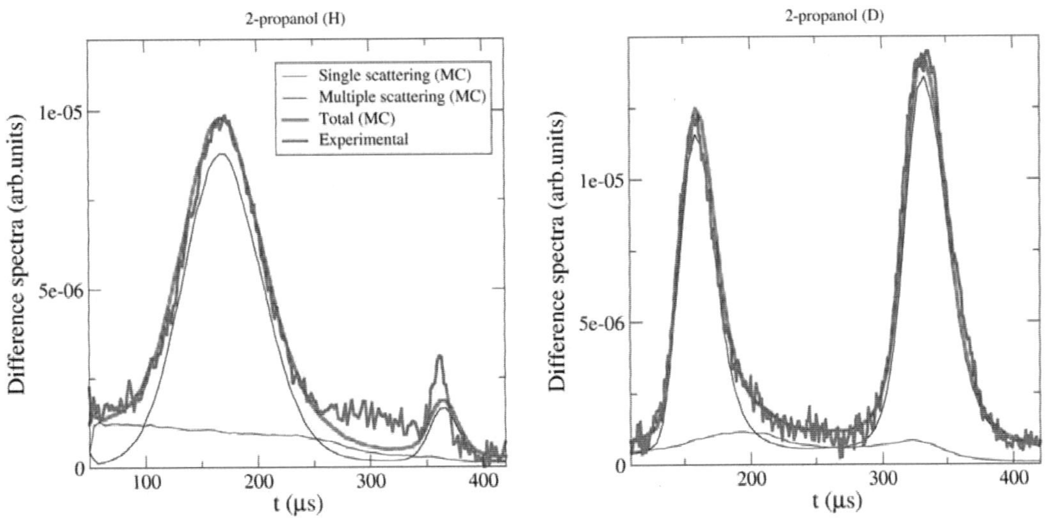

Figure 10. Comparison of experimental and Monte Carlo data of hydrogenated 2-propanol for the detector group number 10 in forward scattering (left) and deuterated 2-propanol for the detector group number 3 in backward scattering (right). The detail of the single scattering and multiple scattering components is shown.

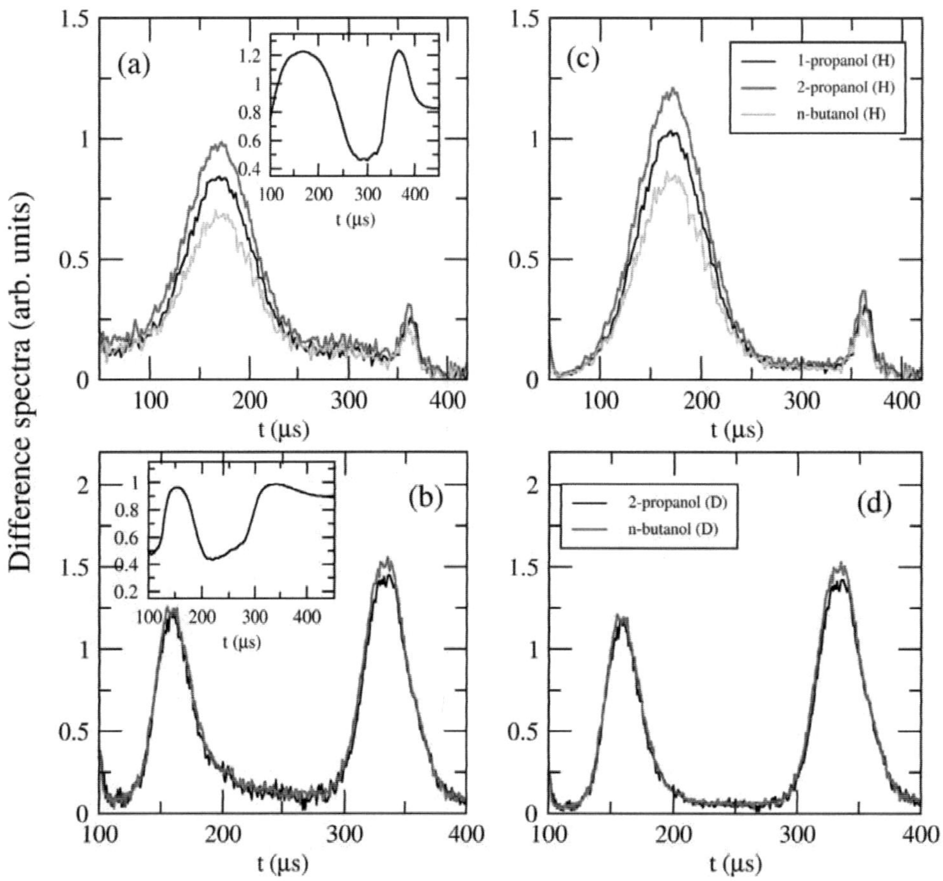

Figure 11. The frames (a) and (b) show the experimental data after background and can subtraction for detector-group 10 (forward scattering, hydrogenated samples) and group 3 (backscattering, deuterated samples), respectively. Their corresponding correction factors defined in Eq. (6) and determined by Monte Carlo simulations are shown in the insets. The frames (c) and (d) show the data corrected by multiple scattering, attenuation and efficiency.

Therefore, the data on the right frames of Fig. 11 are already corrected by the aforementioned effects, and are ready to be used in the analysis of effective temperatures.

6. Data fitting

Fig. 12 shows representative measured profiles of deuterated 2-propanol (group 1 and 3 of backscattering detectors) together with the corresponding fitted curves and the separate contributions of D, C and O, employing Eq. (1). The double-differential cross section of the

IOP Conf. Series: Journal of Physics: Conf. Series **1055** (2018) 012013 doi:10.1088/1742-6596/1055/1/012013

system was modeled by a mixture of gases with effective temperatures, as described in Ref. [5].

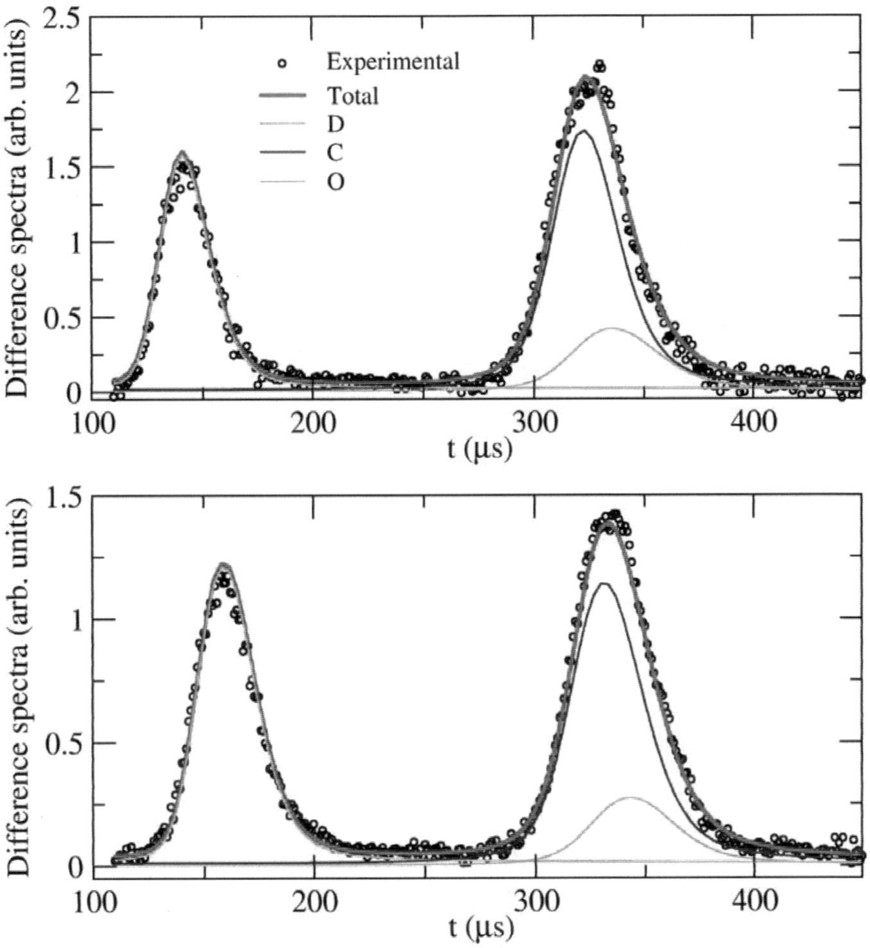

Figure 12. Profiles of deuterated 2-propanol, at group 1 (left) and group 3 (right) of backscattering detectors corrected by the experimental effects described in this work together with the corresponding fitted curves and the separate contributions of D, C and O.

A good precision in the determination of the deuterium effective temperature is found (60.2 ± 0.1 meV for the group-1 spectrum and 60.3 ± 0.1 meV for the group-3 spectrum), whereas the superposition of peaks prevented us from achieving acceptable values for C and O. Further results of effective temperatures of alcohols will be shown elsewhere.

7. Conclusions

In this work we have shown a procedure to treat VESUVIO data from groups of detectors, as a way to improve the statistics for the determination of effective temperatures of the atoms that compose the sample. A scheme was proposed in which the 64 forward detectors were grouped into 10 groups, and the 132 backward detectors were grouped into 3 groups. The knowledge of effective temperatures is relevant in Nuclear Engineering and the method shown here allows a direct experimental way to obtain these parameters.

Raw experimental data must go through an instrumental characterization such as the detector groups efficiency determination, and several stages of corrections, before being ready to be analyzed. To show such stages was the purpose of this work.

All the correction steps were facilitated by Monte Carlo simulations. The determination of the residual background characteristic of the instrument was obtained through the comparison of spectra of similar samples and their corresponding Monte Carlo simulations. To suitably subtract the empty cell spectrum, Monte Carlo simulations were carried out to determine the cell attenuation factor to be used in the subtraction. Finally the Monte Carlo simulations were used in their well-known role to determine the effects of multiple scattering and attenuation, to correct the data.

To be able to perform adequate Monte Carlo simulations, a prior knowledge of the system to be modeled and the experimental parameters of the instrument are required. Key parameters to be included in the simulations are the total cross sections of the studied systems as a function of the energy, also available as output of a VESUVIO experiment through the information given by transmission detector S2, provided the incident spectrum is also measured.

The procedure shown allowed us to obtain the profiles ready to be analyzed. The fitted effective temperatures shown as examples evidence satisfactory precision in the results. A systematic study of the effective temperatures of the studied hydrogenated and deuterated alcohols will be shown in a forthcoming paper.

References

[1] Moreh R, Shahal O and Volterra V 1976 *Nuclear Physics, Section A, Volume 262, Issue 2, p. 221-230* **262** 221–230
[2] Utsuro M 1973 *Journal of Nuclear Science and Technology* **10** 428–434
[3] Granada J R 1984 *Zeitschrift Naturforschung Teil A* **39** 1160–1167
[4] Moreh R, Sellyey W C, Sutton D and Vodhanel R 1985 *Phys. Rev. C* **31**(6) 2314–2316
[5] Dawidowski J, Rodríguez Palomino L A, Márquez Damián J I, Blostein J J and Cuello G J 2016 *Annals of Nuclear Energy* **90** 247 – 255 ISSN 0306-4549
[6] Andreani C, Senesi R, Krzystyniak M, Romanelli G and Fernndez-Alonso F 2017 *Atomic Quantum Dynamics in Materials Research* vol 49 (Amsterdam: Elsevier) p 403 ISBN 9780128053249
[7] Andreani C, Krzystyniak M, Romanelli G, Senesi R and Fernandez-Alonso F 2017 *Advances in Physics* **66** 1–73
[8] URL `https://www.sigmaaldrich.com/`
[9] Rodríguez Palomino L A, Dawidowski J, Márquez Damián J I, Cuello G J, Romanelli G and Krzystyniak M 2017 *Nuclear Instruments and Methods in Physics Research Section A: Accelerators, Spectrometers, Detectors and Associated Equipment* **870** 84 – 89 ISSN 0168-9002
[10] Blostein J J, Dawidowski J and Granada J R 2005 *Phys. Rev. B* **71**(5) 054105
[11] Akeroyd F A, Ashworth R L, Johnston S D, Martin J, Moreton-Smith C M and Sivia D S 2000 Open GENIE User Manual Technical Report RAL-TR-2000-002 Rutherford-Appleton Laboratory
[12] Palomino L A R, Blostein J J and Dawidowski J 2013 *Journal of Instrumentation* **8** P08016 URL `http://stacks.iop.org/1748-0221/8/i=08/a=P08016`
[13] Tardocchi M, Gorini G, Pietropaolo A, Andreani C, Senesi R, Rhodes N and Schooneveld E M 2004 *Review of Scientific Instruments* **75** 4880–4890
[14] Mayers J 2011 *Measurement Science and Technology* **22** 015903
[15] Perelli-Cippo E, Gorini G, Tardocchi M, Andreani C, Pietropaolo A, Senesi R, Rhodes N and Schooneveld E 2006 *Applied Physics A* **83** 453–460 ISSN 0947-8396
[16] Paalman H H and Pings C J 1962 *Journal of Applied Physics* **33** 2635–2639
[17] Palomino L R, Dawidowski J, Blostein J and Cuello G 2007 *Nuclear Instruments and Methods in Physics Research Section B: Beam Interactions with Materials and Atoms* **258** 453 – 470 ISSN 0168-583X
[18] Dawidowski J, Cuello G and Rodríguez Palomino L 2012 *Data Processing Steps in Neutron Diffraction:*

From the Raw Data to the Differential Cross Section (InTech, Prof. Irisali Khidirov (Ed.)) pp 77–100 ISBN 978-953-51-0307-3
[19] 1997 ENDF/B-VI, IAEA Nuclear Data Section

IOP Conf. Series: Journal of Physics: Conf. Series **1055** (2018) 012014 doi:10.1088/1742-6596/1055/1/012014

A McStas simulation of the incident neutron beam on the VESUVIO spectrometer

A Di Giulio[1], M Zanetti[2], G Romanelli[3], M Krzystyniak[3,4], R Senesi[1,5,6] and F Fernandez-Alonso[3,7]

[1]Università degli Studi di Roma "Tor Vergata", Department of Physics and NAST Centre, Via della Ricerca Scientifica 1, Roma, 00133 Italy
[2]CNR, Department of Physical Sciences and Technologies of Matter, Piazzale Aldo Moro 7, 00185 Rome, Italy
[3]ISIS Facility, Rutherford Appleton Laboratory, Chilton, Didcot, Oxfordshire OX11 0QX, UK
[4]School of Science and Technology, Nottingham Trent University, Clifton Campus, Nottingham NG11 8NS, UK
[5]Centro Fermi - Museo Storico della Fisica e Centro Studi e Ricerche "Enrico Fermi", Piazza del Viminale 1, Rome 00184, Italy
[6]CNR-IPCF Sezione di Messina, Viale Ferdinando Stagno d'Alcontres 37, Messina 98158, Italy
[7]Department of Physics and Astronomy, University College London, Gower Street, London WC1E 6BT, UK

E-mail: alexdigiulio@gmail.com

Abstract. We present a Monte Carlo simulation of the incident neutron beam on the VESUVIO spectrometer at the ISIS Facility using the McStas code. As VESUVIO allows for concurrent measurements of neutron diffraction, neutron transmission, and deep inelastic neutron scattering, both incident and transmitted beams are characterized by a broad energy range, spanning over several orders of magnitude from fractions of meV to tens of keV. A transport simulation in the case of the VESUVIO spectrometer is a challenging task, for the McStas code has been traditionally applied to cold and thermal neutrons, and never used in the modelling of electron-volt neutron spectrometers, to the best of our knowledge. In this simulation study, we discuss the modelling of the collimation stages along the primary flight path so as to reproduce the absolute intensity of the incident neutron beam and its shape, both recently characterized experimentally. Finally, we show some preliminary results employing incoherent scattering samples so as to compare the epithermal component of the simulated backscattering spectra to experimental results from Pb.

1. Introduction

The design of new neutron instruments and the upgrade of existing facilities strongly depend upon simulation codes for neutron transport [1, 2]. A quick development of available software has enabled higher simulation speed and an increase of tools and features. One of the most used pieces of software in neutron-scattering-instrument simulations is the McStas code [3]. The package is based upon a special meta-language designed for Monte-Carlo ray-tracing calculations, and includes a library of highly-adaptable "components", with geometrical and physical properties that can be adjusted to the instrument design.

The development of these tools was supported with validation with experiments, however, only on a restricted dataset based on the most common kind of instruments. In many cases,

IOP Conf. Series: Journal of Physics: Conf. Series **1055** (2018) 012014 doi:10.1088/1742-6596/1055/1/012014

neutron instruments are designed so as to operate in a narrow range of energies, as for example diffractometers using cold neutrons and traditional spectrometers using thermal neutrons. As the operational energies of the instrument are fixed, the modelling of the incident and scattered neutron spectra can be verified over limited energy ranges, where reliable experimental data from the real instrument can be provided. In this scenario, the VESUVIO spectrometer [4, 5] at the ISIS pulsed neutron and muon source [6], represents a challenging instrument to be modelled. Owing to the absence of permanent energy filters on both the primary and the secondary paths, VESUVIO allows for concurrent measurements over a broad energy range from the fraction of meV to tens of keV. In recent years, VESUVIO has been referred to as an *epithermal and thermal neutron analysis station* [7, 8], as the same sample can be investigated using deep inelastic neutron scattering [9, 10], neutron diffraction [11, 12], neutron transmission [13, 14], and gamma dopplerimetry [8, 15].

In this article, we present some preliminary comparisons of the simulated incident neutron spectrum and beam profile against recent experimental characterisations [5]. Moreover, we show simulated backscattering time-of-flight spectra including both coherent and incoherent scattering from Pb. The development of McStas components able to reproduce VESUVIO spectra over a broad energy range is the first step in order to successfully simulate deep inelastic neutron scattering spectra within McStas.

2. McStas Model

Monte-Carlo calculations have been performed using McStas 2.4.1 [16]. Simulations were run on a local computer with Windows 10 operating system, as well as on the SCARF cluster [17]. The VESUVIO spectrometer, for which a detailed description can be found in Reference [5], was simplified in McStas with a moderator acting as the neutron source, a series of collimation slits along the incident neutron path, a sample at about 11 m from the moderator, and a monitor representing a ^{6}Li-doped scintillating detector in backscattering, as schematically showed in Figure 1. In particular,

- The moderator component used was Commodus_I [18] (the most recent version of viewModISIS [19]), in order to take into consideration the recent upgrade of the water moderator serving the VESUVIO instrument at the ISIS Target Station 1 [5]. This component can simulate the emission of neutrons from the moderator face with a proton current of $1\mu A$.
- The set of B_4C collimation slits in the range of distances 1.71 m to 9.66 m from the moderator were modelled by ideal "Slit" elements placed at the positions 1.71, 8.46 and 9.66 m. In particular, as the collimation stages are octagonal in transverse cross section, two square slits titled by 45 degrees with respect to each other were used.
- Images of the beam shape and intensity were obtained using virtual Position Sensitive Detector (PSD) components and virtual monitors "PSD_monitor" before and after the first slit, and at the sample position.
- A "Filter-gen" component at the sample position was used to take into account discrepancies between the simulated and experimental beam profiles prior to the scattering from the sample. The Filter_gen component reads a table generated from the left setup in figure 1, and corrects it to match the experimental one. The corrected flux is then back-traced to the last slit position and used to illuminate the sample. The table is an ASCII file where in the first column are listed the energy values and in the second column are reported correcting factors corresponding to the ratio of experimental-to-simulated intensities, so as to correct for the discrepancies in the two intensities.
- The components used to reproduce the backscattering data from a Pb sample were "PowderN" [20] and "Isotropic_Sqw" [21].

- The simulated spectra corresponding to a backscattering detector were recorded using a "Monitor_nD" component, and the count rate was corrected by the known efficiency of ^{6}Li-detector efficiency $\eta(E)$ according to

$$\eta(E) = 1 - e^{-n\sigma(E)d}. \quad (1)$$

where n is the atom density, $\sigma(E)$ cross section and d detector thickness.

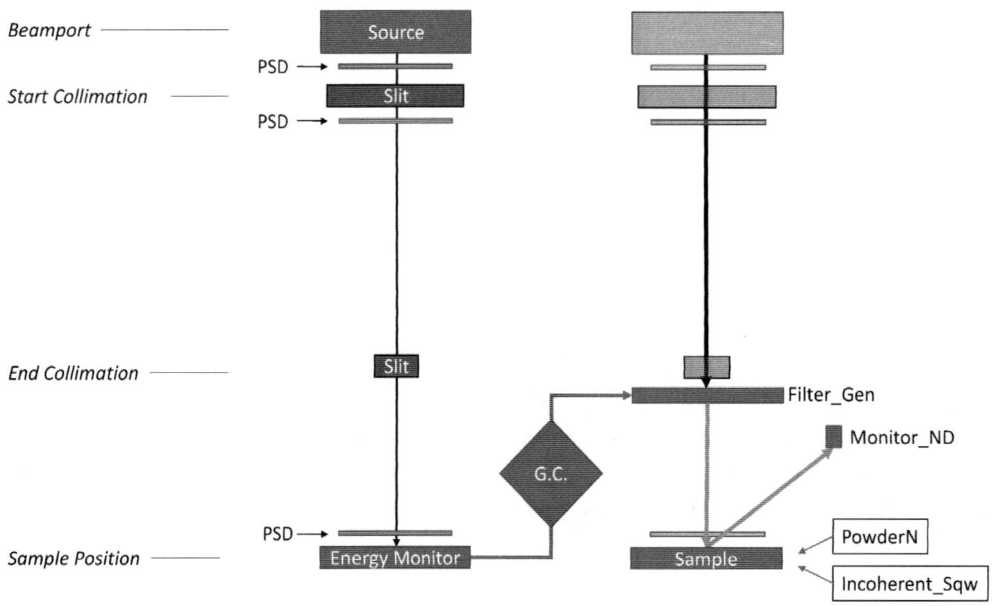

Figure 1. Simulation setup for collimation analysis (left) and backscattering analysis (right) after the inclusion of the "generating correction" (GC) component, "Filter_Gen".

3. Results and Discussion

Inspection of Figure 2 shows how the collimation stages modify the shape of the beam from a large square of about 8 cm side to a circle of diameter about 4 cm. The sole reduction of the size of the beam corresponds to a decrease of about one order of magnitude in the beam intensity at the sample position compared to that before the collimation. Indeed, the ratio of simulated beam fluxes at the two positions is found constant, with no visible dependence upon the incident neutron energy. Moreover, the original beam, homogeneous in the (x, y) plane transverse to the sample-moderator direction, is modified into a Gaussian-like profile. The file representing the brightness of the upstream water moderator was provided by the ISIS Neutronics Group and was created using current MCNP-X model of ISIS Target Station 1 [2].

The simulated results for the energy spectrum and the beam profile of incident neutrons at the sample position were compared with recent experimental measurements [5]. As the result from the simulation are expressed for unit proton current, the simulated spectra were multiplied by the actual proton current ca. 155 μA to TS1 during the measurement. The experiment was performed with the detector and about 1.5 m of the beamline in air, causing an attenuation of the neutron beam of about 0.895 with respect to standard measurement in vacuum. Moreover, the neutron beam needs to cross about 1.5 cm of Al windows between the moderator and the

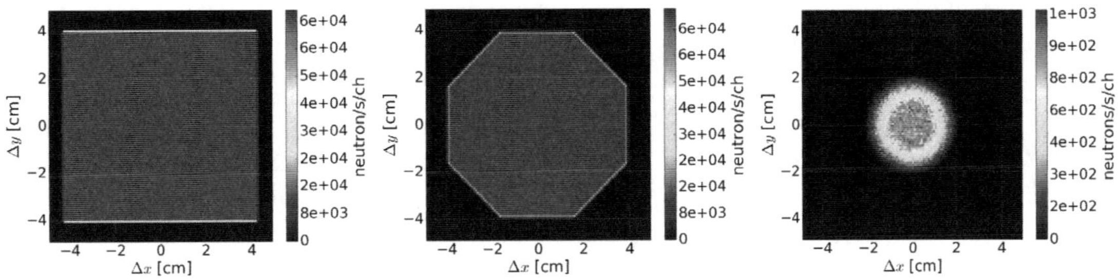

Figure 2. Transverse shape of the neutron beam before (left) and after (center) the beam port slit. (right) Shape for the neutron beam at sample position. The PSDs consist on a 120x120 pixel square matrix with 10 cm side length.

detector, therefore requiring an additional attenuation factor of about 0.855. Figure 3 shows how the present model gives an absolute intensity of the neutron beam between 1.5 and 3 times larger than the experimental data [5]. The agreement between experiment and simulation improves as the incident neutron energy approaches the epithermal region. Similar discrepancies have been previously discussed in Ref[22] about the modelling of the TOSCA spectrometer, where a ratio of simulated-to-experimental intensities ca. 1.77 was reported between 0.28 $\AA$ and 4.65 $\AA$, that favourably compares to the factor 2.39 in our case and in the same energy window.

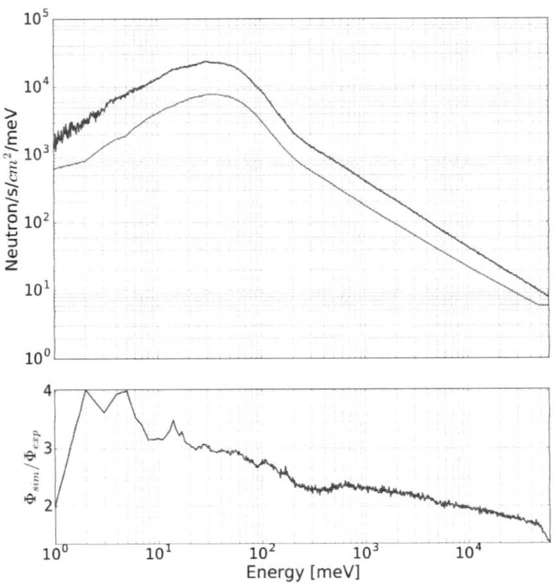

Figure 3. Simulated (blue) and experimental (green) fluxes at the sample position, and the simulated-to-experimental ratio in the bottom panel.

In particular, the flux in the epithermal region is expected to behave as $\propto E^{\alpha-1}$ for real moderators with α leakage exponent. By fitting this model to the experimental and simulated data, we obtained $\alpha = 0.04$ and $\alpha = 0.08$ for simulation and experiment, respectively. Such a small discrepancy could be related to the uncertainties present in the MCNP model which the table of emission TS1_S02_Vesuvio.mcstas that describes the moderator in McStas is based on. It is interesting to notice how, despite the discrepancy in the absolute value of the flux intensity, the simulated shape of the beam compares very favourably to the experimental results. The slightly wider experimental profile, shown in Figure 4, can be partially explained by the broadening due to the spatial resolution component from the nGEM detector used in Ref. [5].

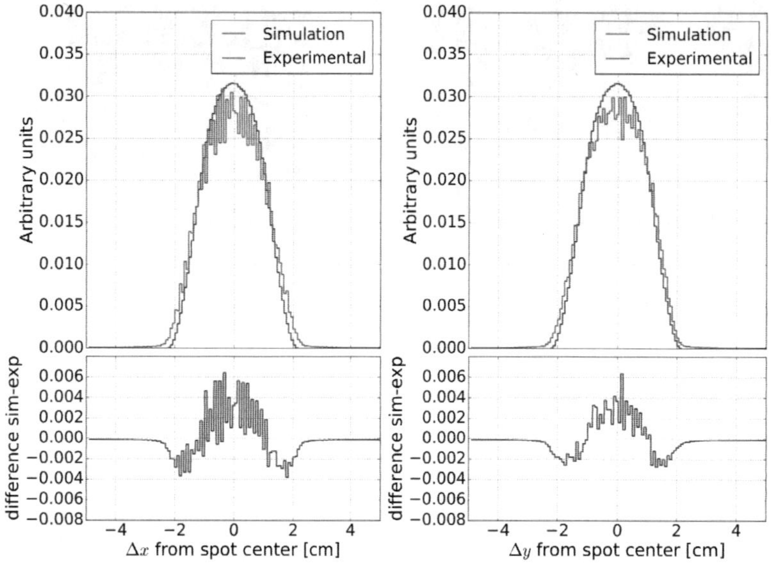

Figure 4. Simulated (blue) and experimental (green) along the two directions in the transverse plane, as well as their difference in the bottom panel. The experimental beam profiles from Ref. [5] have been shifted in order to share the same axis as in the simulation. Data is normalized to unit area.

Finally, the simulation of the backscattering from a 2-mm-thick lead foil was attempted, to be compared to the standard sample available on VESUVIO. Backscattering detectors on the instrument have distances from the sample between 0.45 m and 0.70 m, dimensions of 4×2×0.6 cm^3, and they cover a range of scattering angles between 130 degrees and 163 degrees. Comparisons were drawn with the detector S44 at the Cartesian coordinates $(x, y, z) = (-0.158 \text{ m}, -0.048 \text{ m}, 0.460 \text{ m})$ [24], with the origin at the sample position, the direction z along the incident neutron beam pointing from the sample to the moderator, and the directions (x, y) in the transverse plane. A table with positions of the detectors is available within the MANTID software [25, 26, 27]. In order to properly compare the spectra from the simulation and the experiment, a series of normalisations and corrections were applied. First, the output of the McStas simulation expressed as a count rate in neutrons per second was converted into neutrons per microsecond, generally provided within MANTID. Then, the simulation spectra were multiplied by detector efficiency described in Equation 1. Simulation results generated for a proton beam current of 1 μA were normalised to the nominal current of ca. 160 μA in TS1, and finally multiplied by the duration of the measurements expressed in seconds. In order to reproduce the experimental time-of-flight spectra, we used both a "PowderN" virtual sample, and an "Incoherent_SQW" virtual sample. The former component uses the "Pb.laz" file where the scattering properties of the sample are stored, including the information about Bragg peaks. Additionally, the incoherent cross section was manually set to 11.118 barn [28]. In the case of the "Incoherent_SQW" component, a table was generated including the scattering law as predicted in the framework of the impulse approximation [29]. Figure 5 shows the comparison between simulation and experimental data in the case of detector S44, after the correction by the "Filter_gen" component. The shape and intensity of the spectra in the energy region of epithermal neutrons, for time-of-flight values lower than 2000 μs, are in satisfactorily agreement between the experiment and the simulation. Moreover, in the case of cold and thermal neutrons, the position of the Bragg peaks is very well reproduced by the simulation. Discrepancies in the intensity of the Bragg peaks, in particular the excess intensity in the case of the simulated peak, is completely ascribed to the fact that the actual sample was a foil rather than a powder, and lacking of the correction by the Debye-Waller factor in the simulation.

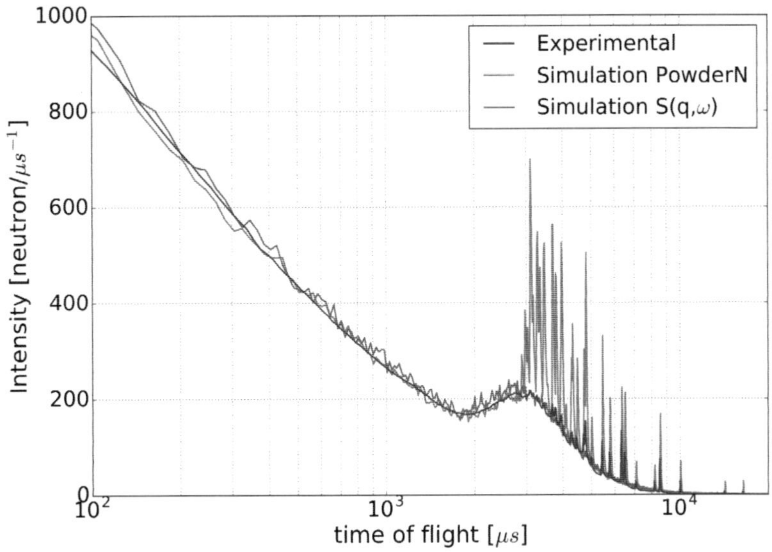

Figure 5. Raw experimental data (blue) for detector S44 and simulation results with PowderN component (green) and custom S(q,ω) from Pb (red). Data are normalized to $1\mu Ah$

4. Outlook and Conclusions

We presented preliminary results of the modelling VESUVIO spectrometer using McStas. Due to the geometry of the instrument, a broad-range white neutron beam flows in both the primary and secondary paths, making the modelling of VESUVIO challenging with respect to other instruments using cold and thermal neutrons. We have shown how the incident neutron beam intensity and shape are overall satisfactorily reproduced by the simulation, and that the backscattering time-of-flight data can be reproduced for cold-to-epithermal neutron energies. Having assessed the capabilities of McStas to cope with the broad energy band of VESUVIO, this work paves the way for the simulation of electron-Volt spectrometers and deep-inelastic-neutron-scattering virtual-experiments.

Acknowledgement

This work was partially supported within the CNR-STFC Agreement (2014-2020) concerning collaboration in scientific research at the ISIS pulsed neutron and muon source. The authors gratefully acknowledge the UK Science & Technology Facilities Council for financial support and access to beam time at the ISIS Facility. Computing resources were provided by the STFC Scientific Computing Department's SCARF cluster. Finally, we thank Dr G Škoro for useful and constructive discussions.

References

[1] Pinna, Roberto S, RudiÄĞ, Svemir, Parker, Stewart F, Gorini, Giuseppe and Fernandez-Alonso, Felix 2015 *EPJ Web of Conferences* **83** 03013 URL `https://doi.org/10.1051/epjconf/20158303013`

[2] Škoro G, Lilley S and Bewley R 2017 *Physica B: Condensed Matter* URL `https://doi.org/10.1016/j.physb.2017.12.060`

[3] Lefmann K and Nielsen K 1999 *Neutron News* **10** 20–23

[4] Last accessed on September 2017 Vesuvio URL `http://www.isis.stfc.ac.uk/Pages/VESUVIO.aspx`

[5] Romanelli G, Krzystyniak M, Senesi R, Raspino D, Boxall J, Pooley D, Moorby S, Schooneveld E, Rhodes N J, Andreani C and Fernandez-Alonso F 2017 *Measurement Science and Technology* **28** 095501 URL `http://stacks.iop.org/0957-0233/28/i=9/a=095501`

[6] Last accessed on November 2017 Iris webpage URL `http://www.isis.stfc.ac.uk/Pages/IRIS.aspx`

[7] Andreani C, Krzystyniak M, Romanelli G, Senesi R and Fernandez-Alonso F 2017 *Advances in Physics* **66** 1–73 URL `http://dx.doi.org/10.1080/00018732.2017.1317963`

[8] Krzystyniak M and *et al* in press *Journal of Physics Conference Series*

[9] Romanelli G, Liscio A, Senesi R, Zamboni R, Treossi E, Liscio F, Giambastiani G, Palermo V, Fernandez-Alonso F and Andreani C 2016 *Carbon* **108** 199–203
[10] Syrykh G F, Stolyarov A A, Krzystyniak M, Romanelli G and Sadykov R A 2017 *JETP Letters* **105** 591–594 URL `https://doi.org/10.1134/S0021364017090041`
[11] Andreani C, Romanelli G and Senesi R 2016 *The Journal of Physical Chemistry Letters* **7** 2216–2220
[12] Krzystyniak M, Druzbicki K, Romanelli G, Gutmann M J, Rudic S, Imberti S and Fernandez-Alonso F 2017 *Physical Chemistry Chemical Physics* **19**(13) 9064–9074 URL `http://dx.doi.org/10.1039/C7CP00997F`
[13] Rodríguez Palomino L A, Dawidowski J, Márquez Damián J I, Cuello G J, Romanelli G and Krzystyniak M 2017 *Nuclear Instruments and Methods in Physics Research Section A* **870** 84 – 89 ISSN 0168-9002 URL `http://www.sciencedirect.com/science/article/pii/S0168900217307696`
[14] Romanelli G, Rudic S, Zanetti M, Andreani C, Fernandex-Alonso F, Gorini G, Krzystyniak M and Skoro G 2018 *Nuclear Instruments and Methods in Physics Research Section A* **in press**
[15] G Romanelli *et al* in press *Journal of Physics Conference Series*
[16] Last accessed on February 2018 Mcstas homepage URL `http://mcstas.org/`
[17] Last accessed on February 2018 Scarf homepage URL `http://www.scarf.rl.ac.uk/`
[18] Last accessed on April 2018 URL `https://isisneutronmuon.github.io/mcstas/How-to-Use-the-ISIS-moderator-component`
[19] Ansell S S G 2015 The viewmodisis component, mcstas simulation package (accessed on the 11 january 2018). URL `http://www.mcstas.org/download/components/contrib/ViewModISIS.pure.html`
[20] Willendrup P, Filges U, Keller L, Farhi E and Lefmann K 2006 *Physica B: Condensed Matter* **385** 1032–1034
[21] Farhi E, Hugouvieux V, Johnson M and Kob W 2009 *Journal of Computational Physics* **228** 5251–5261
[22] Pinna R S, Rudić S, Capstick M J, McPhail D J, Pooley D E, Howells G D, Gorini G and Fernandez-Alonso F 2017 *Nuclear Instruments and Methods in Physics Research Section A: Accelerators, Spectrometers, Detectors and Associated Equipment* **870** 79–83
[23] Bewley R I Ansell S S G 2016 *Proc. of the 21st Meeting of the International Collaboration on Advanced Neutron Sources ICANS XXI*
[24] Mayers J and Adams M A 2011 *Nuclear Instruments and Methods in Physics Research Section A* **625** 47–56
[25] Arnold O, Bilheux J, Borreguero J, Buts A, Campbell S, Chapon L, Doucet M, Draper N, Leal R F, Gigg M, Lynch V, Markvardsen A, Mikkelson D, Mikkelson R, Miller R, Palmen K, Parker P, Passos G, Perring T, Peterson P, Ren S, Reuter M, Savici A, Taylor J, Taylor R, Tolchenov R, Zhou W and Zikovsky J 2014 *Nuclear Instruments and Methods in Physics Research Section A* **764** 156 – 166 ISSN 0168-9002 URL `http://www.sciencedirect.com/science/article/pii/S0168900214008729`
[26] Jackson S, Krzystyniak M, Seel A G, Gigg M, Richards S E and Fernandez-Alonso F 2014 *Journal of Physics: Conference Series* **571** 012009 URL `http://stacks.iop.org/1742-6596/571/i=1/a=012009`
[27] Last accessed on September 2017 Mantid website URL `http://download.mantidproject.org/`
[28] Sears V F 1992 *Neutron News* **3** 26–37 (*Preprint* `http://dx.doi.org/10.1080/10448639208218770`) URL `http://dx.doi.org/10.1080/10448639208218770`
[29] Ivanov G K and Sayasov Y S 1964 *Soviet Physics Doklady* **9** 171

IOP Conf. Series: Journal of Physics: Conf. Series **1055** (2018) 012015 doi:10.1088/1742-6596/1055/1/012015

Neutron-resonance capture analysis on the VESUVIO spectrometer: Towards high-throughput material characterisation

G Romanelli[a], **M Krzystyniak**[a,b], **and F Fernandez-Alonso**[a,c]

[a] ISIS Facility, Rutherford Appleton Laboratory, Chilton, Didcot, Oxfordshire, OX11 0QX, UK

[b] School of Science and Technology, Nottingham Trent University, Clifton Campus, Nottingham, NG11 8NS, UK

[c] Department of Physics and Astronomy, University College London, Gower Street, London, WC1E 6BT, UK

E-mail: Giovanni.romanelli@stfc.ac.uk

Abstract. We discuss the possibility to characterise nuclear quantum effects on the dynamics of heavy nuclei by means of neutron-resonance capture analysis on the VESUVIO spectrometer at ISIS. VESUVIO is equipped with yttrium-aluminium-perovskite gamma-sensitive detectors that can be used to record the Doppler-broadened line shape from a neutron-induced resonance in the neutron-energy range between 1 and 100 eV. The measurement of nuclear momentum distributions for heavy atoms using deep inelastic neutron scattering, the traditional technique for light-weight nuclei from hydrogen to fluorine, is currently severely limited by the resolution of the instrument. On the other hand, gamma-Dopplerimetry studies as the one presented here on gold allow for exquisite precision and short data-collection measurements, of the order of one hour, for the measurements, rendering the technique ideal for high-throughput investigations.

1. Introduction

The effect of quantum mechanics on the nuclear dynamics in condensed-matter systems is a topic of increasing interest [1, 2, 3]. Nuclei, especially at low temperatures and in strongly bonded systems, experience zero-point motions and other nuclear quantum effects dictated by the solution of a Schroedinger equation for their local potential. At present, a remarkable effort is invested in reproducing these effects in first-principles simulations [4, 5], so as to replace more standard, and sometimes inadequate, approaches such as the Einstein or Debye models for vibrations in crystals.

From the experimental point of view, deep inelastic neutron scattering (DINS) is a Compton-like technique where the momentum distribution of nuclei can be directly probed, and values of the nuclear mean kinetic energy can be extracted. The flagship among the instruments using DINS is the VESUVIO spectrometer [6, 7] at the ISIS pulsed neutron and muon source [8]. Recent studies using DINS have tackled the role of nuclear quantum effects in nano-confined water [9], caesium hydrogen sulfate [10], and zirconium-beryllium amorphous alloys [11]. Traditionally, DINS has been applied to the investigation of the dynamics of light- to moderate-weight elements, with the recent exception of niobium in radio-frequency superconducting cavities [12]. However, the higher the mass of an element, the more severe is the limitation from the instrument resolution function on the analysis of nuclear momentum distributions [3].

Neutron Resonance Capture Analysis (NRCA) [13, 14] is a technique traditionally employed to investigate the chemical composition of samples. It is based on the capture of a neutron with energies in the epithermal region, and the subsequent emission of a prompt-gamma cascade [15, 16]. The shape of the resonance spectrum is broadened by the motion of the nucleus, accordingly to the Doppler effect, and it is therefore affected by the momentum distribution and mean kinetic energy of the photon-emitting nucleus. In this sense, NRCA represents a powerful and complementary tool to DINS, as it exploits the intense cross sections for neutrons in heavy elements such as indium, caesium, tantalum, and gold, that cannot be tackled without difficulties by DINS studies.

In this article, we show how NRCA can be employed to obtain values of the nuclear mean kinetic energy, therefore gathering information on the lattice dynamics, in the case of the VESUVIO spectrometer. In particular, Section 2 discusses the theory behind NRCA, the way to extract reliable values of the mean kinetic energy from measured spectra, and the link to materials modelling and first-principles simulations. Section 3 presents the experimental set-up and describes the measurements performed at the VESUVIO spectrometer, and the results of the investigation are discussed in Section 4. Finally, conclusions are drawn in Section 5.

2. Methods

2.1. Neutron Resonance Capture Analysis

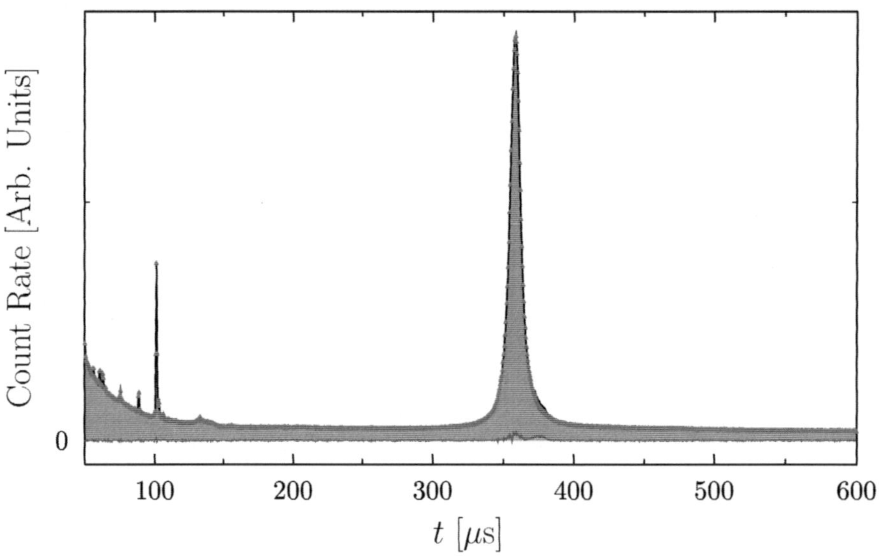

Figure 1. Time-of-flight spectra recorded from one yttrium-aluminium-perovskite detector for one period of data acquisition with the analyser foil in (black continuous line and shadow) and foil out (red triangles). Their difference, corresponding to the foil-cycling procedure, is reported as the tiny green line and shadow.

The neutron scattering cross section in the proximity of a resonance centred at the energy value E_0 and multiplicity g_J experienced by a neutron of incident energy E and wavelength

$\lambda(E)$ can be expressed as the sum of the two contributions [14]

$$\sigma_\gamma(E) = \frac{\lambda^2(E) g_J}{4\pi} \frac{\Gamma_n \Gamma_\gamma}{(E - E_0)^2 + \Gamma^2/4}, \tag{1}$$

related to the processes where a neutron is absorbed and a gamma ray is emitted, and

$$\sigma_n(E) = \frac{\lambda^2(E) g_J}{4\pi} \frac{\Gamma_n^2}{(E - E_0)^2 + \Gamma^2/4} + 2\lambda(E) g_J \frac{R\Gamma_n(E - E_0)}{(E - E_0)^2 + \Gamma^2/4} + 4\pi b^2, \tag{2}$$

defining the probability to have a neutron in the final state as well. Equation 2 is composed of three terms on the right-hand side, of which the first corresponds to the probability for a neutron to be resonantly absorbed and subsequently emitted, the third contribution corresponds to the total scattering term via the Fermi pseudo-potential with b the scattering length [17], and the second term arises from the interference between the two aforementioned channels [18]. The line shape of the resonance is well described by a Lorentzian function, according to the theory by Breit and Wigner [19], with a full width at half maximum (FWHM)

$$\Gamma = \Gamma_n + \Gamma_\gamma. \tag{3}$$

Finally, the total neutron-induced cross section is the sum of the two contributions

$$\sigma_t(E) = \sigma_n(E) + \sigma_\gamma(E). \tag{4}$$

The broadening of the resonance from the Lorentzian line shape is a consequence of the instability of the excited nucleus after the absorption of the incoming neutron, and the parameter Γ is defined from the half-life time of the metastable state [20].

In addition to the Lorentzian broadening, the resonance line shape is furthermore widened by two additional effects that manifest themselves as a Gaussian broadening. The first term, δ, arises from the uncertainty in the time-of-flight measurement because of the finite size of the moderator. At a spallation neutron source such as ISIS, the resolution of the measurement can be expressed as

$$\delta^2 = \left(\frac{\partial E}{\partial L}\right)^2 \Delta L^2 + \left(\frac{\partial E}{\partial t}\right)^2 \Delta t^2, \tag{5}$$

with L the distance between the moderator and the sample, ΔL the associated uncertainty, t the time a neutron takes travelling the distance L, and Δt the width of the neutron pulse. The second contribution to the Gaussian broadening of the resonance comes from the motion of the nucleus induced by its chemical bonding to surrounding atoms. Such Doppler broadening, Δ, can be expressed as proportional to the atom's mean kinetic energy $\langle E_K \rangle$ as [14]

$$\Delta^2 = \frac{4}{3} \frac{mM}{(m + M)^2} E_0 \langle E_K \rangle \simeq \frac{4}{3} E_r \langle E_K \rangle, \tag{6}$$

where the final equality [21] holds when the recoil energy E_r is approximated as

$$E_r \simeq \frac{mM}{(m + M)^2} E_0, \tag{7}$$

with m and M the mass of the neutron and the struck atom, respectively. Reference [21] also discusses the origin of the Doppler broadening from the convolution of the nuclear cross section

$\sigma(E)$ with the dynamic structure factor $S(Q, \Delta E)$, with Q and ΔE the momentum and energy transferred by the neutron to the system of interest. One has

$$S(Q, \Delta E) = \frac{1}{\sqrt{2\pi}\Delta} \exp\left(-\frac{(E - E_r)^2}{2\Delta^2}\right). \tag{8}$$

Finally, one can define the Doppler-broadened experimental cross sections as

$$\sigma^D(E) = \frac{1}{\sqrt{2\pi}\Delta} \int_{-\infty}^{\infty} \sigma(E - \eta) \exp\left(-\frac{\eta^2}{2\Sigma^2}\right) d\eta, \tag{9}$$

where the convolution can be applied to $\sigma_n(E)$, $\sigma_\gamma(E)$, or $\sigma_t(E)$ separately. For additional details, the reader is referred to Ref. [14].

2.2. First-principles calculations

The first-principles vibrational analysis was performed using CASTEP [22], a total energy package based on the plane-wave pseudo-potential density functional theory method. The exchange-correlation functional developed by Perdew, Burke and Ernzerhof in the general gradient approximation was adopted to describe the exchange-correlation energy. The effective interaction between the valence electrons and atom cores were modelled by on-the-fly optimized norm-conserving pseudo-potentials, which allow to choose a relatively small plane-wave basis set without compromising the computational accuracy. The kinetic energy cut-off at 500 eV, and the dense Monkhorst-Pack k-point mesh spanning less than 0.04 Å^{-1} in the Brillouin zone were chosen. To calculate lattice dynamics, a super cell defined by a cut-off radius of 5 Å was built. Before the calculation of vibrational properties, the crystal structure was geometrically optimized to find the energy minimum. The Broyden-Fletcher-Goldfarb-Shanno (BFGS) minimization scheme was employed in the geometry optimisation, in which the convergence criteria for the structure optimisation were set to 5.0×10^{-6} eV per atom, 0.01 eV Å^{-1}, 0.02 GPa and 5.0×10^{-4} Å for energy, maximum force, maximum stress, and maximum displacement, respectively. On the basis of the crystal configuration in the minimal energy, the frequencies of the phonon modes and their dispersion were calculated by the finite-displacement method.

According to the discussion in Refs. [3, 23, 24], the mean kinetic energy of an atom can be deduced from the calculated vibrational density of states $G(\omega)$, within the approximation of harmonic vibrations, as

$$\langle E_K \rangle = 3 \int_0^{\infty} \frac{\hbar\omega}{4} G(\omega) \coth\left(\frac{\hbar\omega}{2k_B T}\right) d\omega, \tag{10}$$

where ω is the frequency of each vibrational mode, and the hyperbolic-cotangent term takes into account the Boltzmann term for the population of excited states for the mode of energy $\hbar\omega$ at every temperature T of the system.

3. Experimental Details

Measurements were performed at the VESUVIO spectrometer [6, 7] on a commercially available Au foil [25] of thickness 10 μm. The sample was inserted in the VESUVIO closed-circuit refrigerator (CCR) for temperature control, and the temperature of the sample was deduced using the sensor at the end of the sample stick, T_2 about 10 cm away from the sample, and the temperature of the CCR, T_1. Measurements were about 30-minute long when a proton current of 160 μA was flowing in the ISIS synchrotron, and each run had a statistics corresponding to 90 μAh of proton charge. After setting each temperature in the range 20 K to 220 K, three runs were acquired, the first one being discarded during the data analysis because of the significant

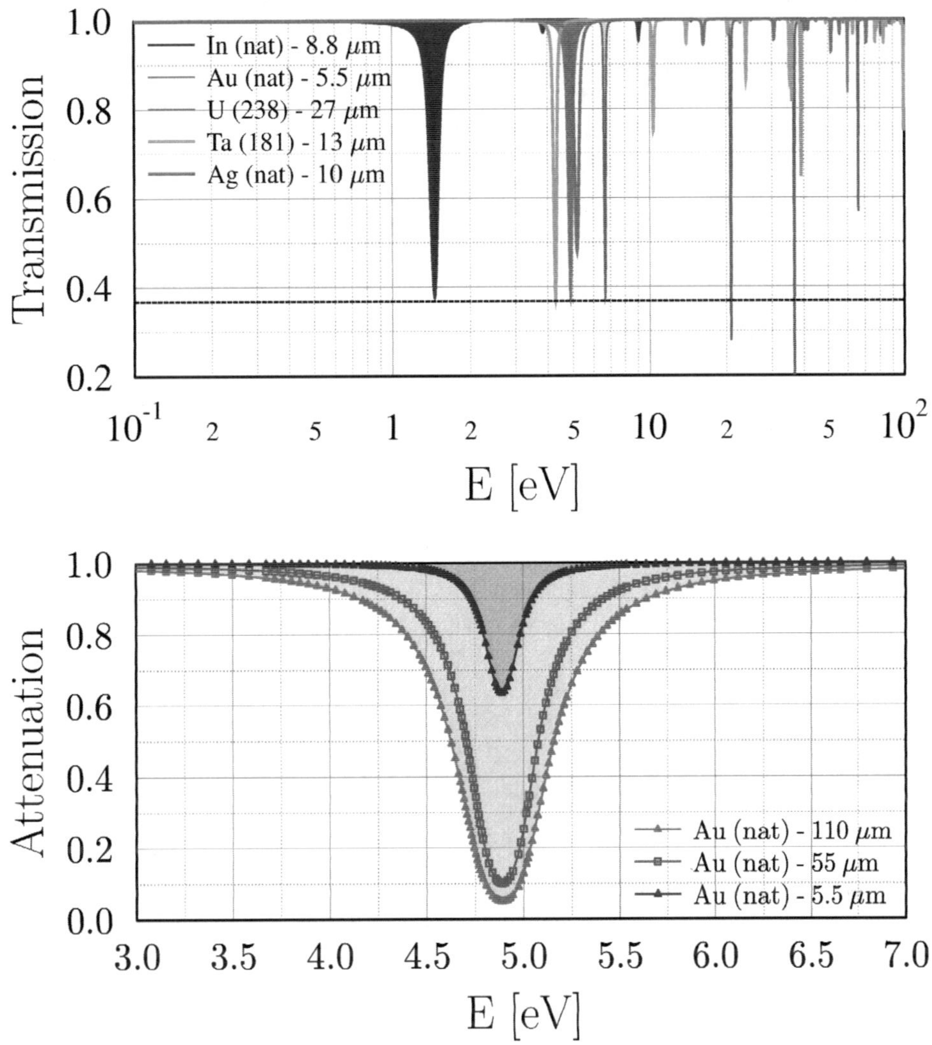

Figure 2. On the top, the transmission of a series of thin metal foils. In particular, the thickness is chosen so as to have a minimum transmission at the centre of the resonance $\simeq e^{-1}$. On the bottom, the attenuation correction in the case of a ^{197}Au foil of thickness 5.5 μm (blue triangles), 55 μm (green squares), and 110 μm (blue triangles).

variation of the sample temperature during the measurement, while the second and third ones have been regrouped in what follows in Dataset 1 and Dataset 2, respectively. Heaters and sensors were not applied on the sample directly.

During a run, six periods of data acquisition are traditionally recorded on VESUVIO, corresponding to different positions of the analyser foils used to select the final energy of neutron scattered by the sample, so as to obtain DINS spectra [26, 27]. This procedure, at the base of DINS experiments [3], was not useful for the present experiment. However, it was decided to mimic normal operations for future concurrent measurements. Therefore, the spectra from the six acquisition periods were summed together in order to increase the statistical quality of the data. Figure 1 shows the spectra corresponding to the first and second periods as recorded by detector 151 in front scattering. The two spectra, namely the solid black line and the red markers, almost perfectly overlap over the entire time-of-flight range from 50 μs to 600 μs. The tiny difference, shown as a green line in the same figure, corresponds to the DINS spectrum. It is composed by the gold neutron Compton profile, on the right, and a residual intensity from the gamma rays emitted by the sample and slightly attenuated when passing through the filter-in period. Such filter-in-filter-out difference is hereafter considered as an experimental background to be subtracted from the data. An additional background from the empty CCR was separately measured and similarly subtracted from the data during the fit.

Experimental data were fitted using the function

$$f(E) = AE^{\alpha}\sigma_{\gamma}^{D}(E)\left(\frac{1-\exp\left(n\sigma_t^D(E)d\right)}{n\sigma_t^D(E)d}\right) + B_{\text{exp}}(E), \tag{11}$$

where the Doppler-broadened gamma cross section needs to be corrected by the self attenuation of the sample, defined by its thickness d, number density n, and the total cross section. The attenuation correction is therefore defined by the thickness of the sample and its scattering power. The neutron transmission from a series of metal foils is shown in Figure 2 using the tabulated values of the total neutron cross section available in the ENDF-B/VII libraries [28]. Here, the thickness of the foils is chosen so as to be in the thin-foil limit, that is the value of the transmission corresponding to the centre of the resonance peak equals e^{-1}. The attenuation correction is then shown in the same Figure 2 for several values of the foil thickness in the case of gold. One can appreciate how the attenuation correction, already in the thin-foil limit, suppresses the intensity at the peak centre at the expenses of the peak wings, therefore apparently broadening the shape of the resonance. Equation 11 shows the explicit dependence upon the incident-beam spectrum through the parameter α, with the expected condition $\alpha > -1$ at spallation neutron sources.

4. Results and Discussion

Data were analysed in three different stages, differing by the number of free parameters in the fit. In the first stage, the value of α was considered as a free parameter. As its average value was found in agreement with recent measurements [7, 29], it was subsequently fixed to the value -0.92. In the second stage, the average of the fitted values of Γ was found to be $\Gamma = 136 \pm 4$ meV, in good agreement with the values reported in a recent review [30] on the neutron-induced cross section from gold ^{197}Au, $\Gamma = 139.2$ meV. Values of $\Gamma_n = 15.2$ meV, $\Gamma_\gamma = 125$ meV from the same reference were fixed in the fitting model. Final results, with both α and Γ fixed, are shown in Table 1 together with the run number, the temperatures measured by the sensors in the stick T_2 and the CCR T_1, the width of the Gaussian broadening Σ prior to the subtraction of the resolution component, and the centre of the resonance peak, E_0. One can notice that the temperature measured by the stick is generally about 1 – 2 K above the temperature measured by the CCR, suggesting a similar uncertainty on the sample temperature. The values of E_0 in the Table are slightly in excess to the value of 4906 meV generally reported in the nuclear libraries [30], for it contains the average contribution of the recoil energies E_r due to the scattering immediately preceding the creation of the compound nucleus in the resonant state [20], as previously described in Equations 6 and 7. Figure 3 (left) shows the quality of the fit using Equation 11 on the spectrum from run 1. One can notice a slight oscillation of the difference of the data and fitting function around the peak centre. Considering the otherwise satisfactory result over a broad energy range, we conclude that such discrepancy at the peak position can be directly ascribed at the statistics corresponding to the short acquisition time of the measurement.

The experimental results for the values of Σ, after subtraction of the resolution component δ, provide estimates for the Doppler broadening Δ and the gold mean kinetic energy $\langle E_K\rangle$ as a

Table 1: Estimated values for the temperature of the sample, as measured using the sensor on the CCR (T_1) and on the sample stick (T_2), and the fitted values of the Gaussian broadening Σ and peak centre (E_0) for each experimental run. Some runs have been discarded as the sample temperature during the data acquisition changed more than 2 K.

Run	T_1 [K]	T_2 [K]	Σ [meV]	E_0 [meV]
1	19.99	20.95	21.55	4935
2	20.00	20.97	21.80	4935
4	39.98	41.61	22.04	4935
5	39.98	41.53	22.27	4935
7	60.00	62.11	23.83	4935
8	60.01	62.57	23.87	4935
10	80.00	82.08	24.86	4935
11	80.00	83.32	25.00	4935
13	100.00	101.56	26.37	4936
14	100.01	103.30	26.17	4935
16	120.00	121.25	27.69	4936
17	120.01	123.26	28.36	4935
19	140.01	141.33	27.94	4936
20	140.01	143.10	29.13	4936
22	160.00	161.30	29.36	4936
23	160.01	162.97	30.61	4936
25	180.01	181.58	30.40	4936
26	180.02	183.09	31.96	4936
28	200.00	201.74	33.23	4936
29	200.00	203.12	33.21	4936
31	220.01	222.12	34.18	4936

function of the sample temperature. The theoretical predictions, together with the experimental values for $\langle E_K \rangle$ are reported in Figure 4. The resolution component, δ, according to Equation 5 provides a value $\delta \simeq 17.8$ meV, with no expected dependence upon the sample temperature. However, a slightly reduced value of $\delta = 15$ meV was used to get the best agreement between the simulation and the experimental data in Figure 4. Of the two datasets, one can notice how Dataset 2 systematically shows the best agreement to the theoretical prediction. This behaviour suggests that the time needed for the sample to reach the set temperature was slightly longer than expected, and only the third measurement for each nominal temperature can be considered having reached the set point. Such behaviour could be easily tested in future experiments by applying a sensor directly on the sample. Figure 4 (right) also shows the difference between two experimental spectra at the temperatures of 20 K and 200 K. One can notice how the difference of the spectra is largely exceeding the dimension of the associated error bars, suggesting the high sensitivity of the technique. Moreover, the difference in the resonance peaks markedly resembles the difference in neutron Compton profiles discussed in Reference [31], therefore suggesting that the fitting procedure discussed in that paper could be readily applied to the case of parametric studies of gamma resonances as well.

5. Outlook and Conclusions

We have presented experimental results from an NRCA study performed at the VESUVIO spectrometer on a gold foil in the range of temperature 20 K to 220 K. We have demonstrated how the instrument can acquire data of exquisite quality with a duration of the measurements

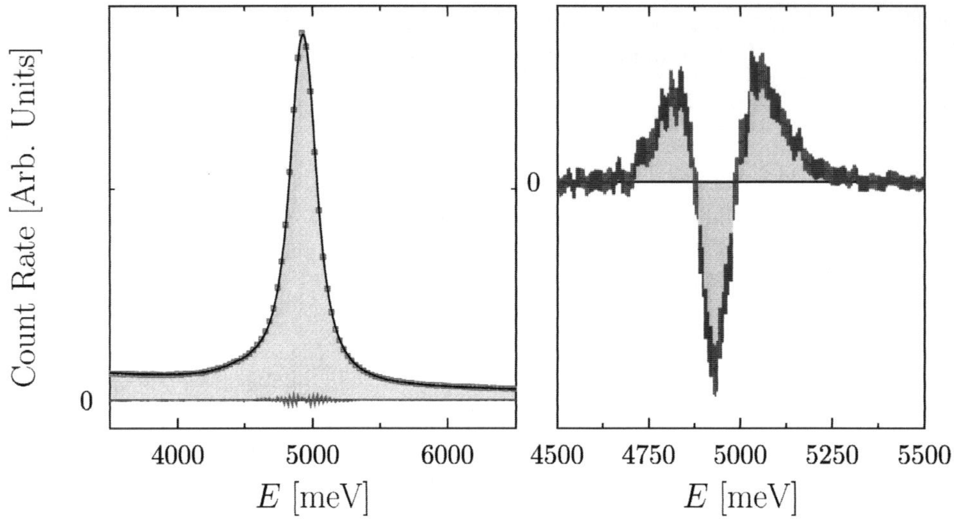

Figure 3. On the left, the fit (black line) of the experimental data (red squares) corresponding to run 1 ($T \simeq 20$ K), together with the difference of experimental data and fit, reported as a green line. On the right, the maximum difference on the experimental data between run 1 ($T \simeq 20$ K) and run 31 ($T \simeq 220$ K).

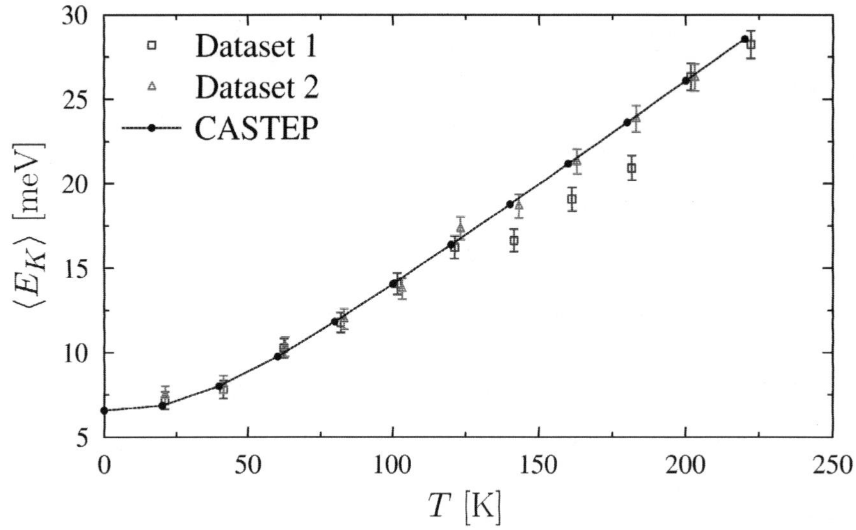

Figure 4. Experimental values of the mean kinetic energy of ^{197}Au as a function of temperature as obtained from the experimental run in Dataset 1 (blue squares) and Dataset 2 (red triangles), as compared to the theoretical prediction based on a CASTEP simulation.

of the order of one hour, an order of magnitude improvement with respect to traditional acquisition times on electron-volt spectrometers. Moreover, VESUVIO proves to be a highly versatile instrument where several neutron-based techniques can be applied concurrently. This assessment paves the way to possible high-throughput parametric investigations, *e.g.*, studying the change of the mean kinetic energy of a heavy nucleus as a function of temperature or time by analysing the change in the shape of a well-resolved resonance. Finally, we note that an additional enhancement of the signal-to-background ratio can be achieved by including a thermal-neutron filter along the incident beam, so as to suppress the background from gamma rays generated within the block house and in the beam stop, as suggested by Onorati *et al* in a separate contribution to this volume [32].

Acknowledgements

The authors gratefully acknowledge the UK Science & Technology Facilities Council for financial support and access to beam time at the ISIS Facility, and are grateful to Dr T Minniti for useful discussions. Computing resources were provided by STFC Scientific Computing Department's SCARF cluster.

References

[1] Ceriotti M, Miceli G, Pietropaolo A, Colognesi D, Nale A, Catti M, Bernasconi M and Parrinello M 2010 *Physical Review B* **82** 174306
[2] Dupuis R, Benoit M, Tuckerman M E and Mheut M 2017 *Accounts of Chemical Research* **50** 1597–1605
[3] Andreani C, Krzystyniak M, Romanelli G, Senesi R and Fernandez-Alonso F 2017 *Advances in Physics* **66** 1–73
[4] Ceriotti M, Fang W, Kusalik P G, McKenzie R H, Michaelides A, Morales M A and Markland T E 2016 *Chemical Reviews* **116** 7529–7550
[5] Cazorla C and Boronat J 2017 *Rev. Mod. Phys.* **89**(3) 035003
[6] Last accessed on June 2018 URL `http://www.isis.stfc.ac.uk/Pages/VESUVIO.aspx`
[7] Romanelli G, Krzystyniak M, Senesi R, Raspino D, Boxall J, Pooley D, Moorby S, Schooneveld E, Rhodes N J, Andreani C and Fernandez-Alonso F 2017 *Measurement Science and Technology* **28** 095501
[8] Last accessed on June 2018 URL `http://www.isis.stfc.ac.uk`
[9] Romanelli G, Liscio A, Senesi R, Zamboni R, Treossi E, Liscio F, Giambastiani G, Palermo V, Fernandez-Alonso F and Andreani C 2016 *Carbon* **108** 199–203
[10] Krzystyniak M, Druzbicki K and Fernandez-Alonso F 2015 *Physical Chemistry Chemical Physics* **17**(46) 31287–31296
[11] Syrykh G F, Stolyarov A A, Krzystyniak M, Romanelli G and Sadykov R A 2017 *JETP Letters* **105** 591–594
[12] Krzystyniak, M *et al* in press *Journal of Physics Conference Series*
[13] Chrien E R and Kane W R 1979 *Neutron Capture Gamma Spectroscopy* (New York: Plenum Press)
[14] Postma H and Schillebeeckx P 2006 *Neutron Resonance Capture and Transmission Analysis* (John Wiley & Sons, Ltd) ISBN 9780470027318
[15] Gorini G, Festa G and Andreani C 2014 *Journal of Physics: Conference Series* **571** 012005
[16] Pietropaolo A, Gorini G, Festa G, Reali E, Grazzi F and Schooneveld E M 2010 *Applied Spectroscopy* **64** 1068–1071
[17] Sears V F 1992 *Neutron News* **3** 26–37
[18] Fano U 1961 *Phys. Rev.* **124**(6) 1866–1878
[19] Breit G and Wigner E 1936 *Phys Rev* **49** 519–531
[20] Bethe H A 1937 *Rev. Mod. Phys.* **9**(2) 69–244
[21] Mayers J, Baciocco G and Hannon A 1989 *Nuclear Instruments and Methods in Physics Research Section A: Accelerators, Spectrometers, Detectors and Associated Equipment* **275** 453 – 459
[22] Clark S, Segall M, Pickard C, Hasnip P, Probert M, Refson K and Payne M 2005 *Zeitschrift fr Kristallographie* **220** 567–570 ISSN 0044-2968 reproduced with permission from the publisher.
[23] Andreani C, Romanelli G and Senesi R 2013 *Chemical Physics* **427** 106–110
[24] Krzystyniak M, Druzbicki K, Romanelli G, Gutmann M J, Rudic S, Imberti S and Fernandez-Alonso F 2017 *Physical Chemistry Chemical Physics* **19**(13) 9064–9074
[25] http://wwwgoodfellowcom/ Last accessed on February 2018
[26] Schooneveld E M, Mayers J, Rhodes N J, Pietropaolo A, Andreani C, Senesi R, Gorini G, Perelli-Cippo E and Tardocchi M 2006 *Review Scientific Instruments* **77** 5103

[27] Andreani C, Colognesi D, Degiorgi E, Filabozzi A, Nardone M, Pace E, Pietropaolo A and Senesi R 2003 *Nuclear Instruments and Methods in Physics Research Section A* **497** 535–549
[28] Chadwick M B *et al* 2006 *Nuclear Data Sheets* **107** 2931 – 3060
[29] Di Giulio A *et al* in press *Journal of Physics Conference Series*
[30] Massimi C *et al* (n_TOF Collaboration) 2010 *Phys. Rev. C* **81**(4) 044616
[31] Romanelli G and Krzystyniak M 2016 *Nuclear Instruments and Methods in Physics Research Section A* **819** 84 – 88
[32] Onorati, D *et al* in press *Journal of Physics Conference Series*

IOP Conf. Series: Journal of Physics: Conf. Series **1055** (2018) 012016 doi:10.1088/1742-6596/1055/1/012016

Data analysis of neutron Compton scattering experiments using MANTID

G Romanelli[a,*], **B Hewer**[a], **M Krzystyniak**[a,b], **M Gigg**[a], **R Tolchenov**[a], **S Mukhopadhyay**[a], **and F Fernandez-Alonso**[a,c]

[a] ISIS Facility, Rutherford Appleton Laboratory, Didcot OX11 OQX, UK
[b] School of Science and Technology, Nottingham Trent University, Clifton Campus, Nottingham, NG11 8NS, UK
[c] Department of Physics and Astronomy, University College London, Gower Street, London, WC1E 6BT, UK

E-mail: `Giovanni.romanelli@stfc.ac.uk`

Abstract. The VESUVIO spectrometer at the ISIS pulsed neutron and muon source is an epithermal and thermal neutron station requiring a software suite able to tackle the analysis of experimental data from several neutron techniques. We present an update on the tools available within the MANTID environment for the treatment of neutron Compton scattering data, and we discuss new strategies for the analysis of neutron Compton profiles and momentum distributions that provide robustness to the scientific studies based on this technique.

1. Introduction

Neutron Compton Scattering [1, 2], also referred to as deep inelastic neutron scattering, is a spectroscopic technique that measures Nuclear Quantum Effects (NQEs) in condensed-matter systems. The main observable available in a NCS experiment is the width of the Nuclear Momentum Distribution (NMD), σ_M, of each element of mass M in the investigated sample. The analysis of the shape of NMDs yields direct information on the nuclear dynamics, underlying, for example, the degree of anharmonicity and anisotropy of the local potential, or the importance of NQEs in different phases of a condensed-matter sample, or at different temperatures. At present, the flagship instrument routinely performing NCS experiments is the VESUVIO spectrometer [3, 4] at the ISIS pulsed neutron and muon source [5]. Several measurements can be performed during the same experiment, including NCS, neutron transmission, neutron diffraction, neutron resonance capture analysis, and prompt-gamma activation analysis [6]

The analysis of NCS data has been recently included within the MANTID software [7, 8], and first results have already been published [9, 10], also in conjunction with the FABADA tool for data analysis and robust model selection, based on Bayesian inference [11]. A preliminary discussion on the implementation of NCS analysis in MANTID was presented in Ref. [12], yet in recent years a considerable effort has been directed towards the improvement of the functionalities and capabilities of the software, as well as the development of new algorithms [13–15].

We report here on the updated capabilities of the MANTID software in visualising, reducing, and analysing data collected on the VESUVIO spectrometer, with particular emphasis on the analysis of NCS spectra in the so-called West-scaling domain.

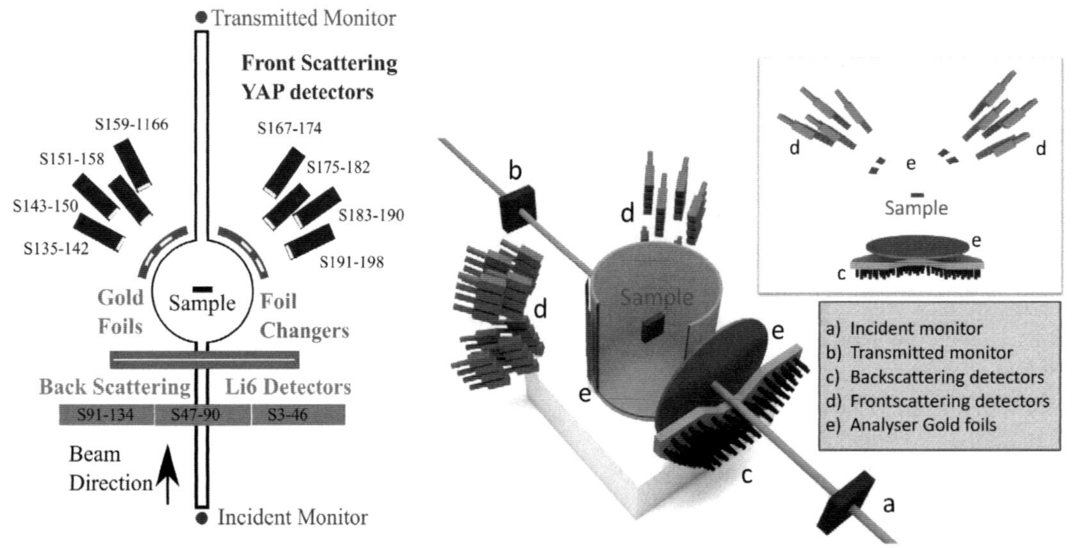

Figure 1. Schematic 2D (left) and 3D (right) layout of the VESUVIO spectrometer at ISIS, reproduced from Ref. [3].

2. Materials and methods

NCS is based on the use of neutron energy transfers in excess of some eV, and momentum transfers in excess of tens of $Å^{-1}$, in a scattering regime referred to as the Impulse Approximation (IA) [16]. VESUVIO is an inverted-geometry spectrometer [4,17], where the energy of scattered neutrons is selected by cycling gold foils. Gold has a very intense absorption cross section at about $E_1 = 4.9$ eV related to a (n,γ) nuclear resonance. When a neutron within a narrow energy range around E_1 passes through the foil, there is a high probability for it to be absorbed and a cascade of prompt γ-rays is emitted. NCS spectra on VESUVIO are obtained by taking the difference of raw spectra obtained with different configurations of the gold foils. In particular, two configurations are available for the photon-sensitive front-scattering detectors [18,19], and three configurations for neutron-sensitive backscattering detectors [20,21]. Additional details on the modes of detection on VESUVIO can be found in Refs. [1,2,22]. As a consequence of such procedures, every data-acquisition *run* is composed of 6 *periods* corresponding to different positions of the energy-selecting gold foils. The differences of raw data from different periods allow the subtraction of the environmental contributions, with the exception of a residual sample-dependent γ-Background (GB) due to the change in the position of the gold foils [23,24].

The VESUVIO experiments discussed in this paper were performed on a polyethylene foil with thickness 0.25 mm in the direction of the neutron beam, and cross-sectional area 8×8 cm^2, larger than the beam cross section [4]. Figure 1 shows a schematic representation of the spectrometer, including the numbering of spectra from different detectors. The foil was attached to a square aluminium frame, not directly hit by the incident neutron beam to avoid unnecessary correction for sample-container backgrounds during the data analysis, and measured at room temperature. In what follows, we shall refer to two measurements, primarily characterised by the accumulated proton charge in the ISIS synchrotron: sample 1 was measured for a total of 6.0 mAh in the closed-cycle refrigerator (CCR); and sample 2 was measured for a total of 1.8 mAh in the vacuum tank, after removing the CCR. Sample 1 was measured in helium atmosphere at 60 mBar, and sample 2 in vacuum at 10^{-6} mBar.

IOP Conf. Series: Journal of Physics: Conf. Series **1055** (2018) 012016 doi:10.1088/1742-6596/1055/1/012016

3. Reduction of NCS data

The reduction of NCS data is generally performed in the Time-of-Flight (TOF) domain. After the differences of the six periods of data acquisition are taken, NCS spectra are composed of a collection of peaks, each corresponding to an isotope in the sample:

$$C_\theta(t) = \eta(E_1)\frac{E_0 I(E_0)}{Q}\sum_M M N_M b_M^2 J_M(x_M) \star R_M(x_M), \tag{1}$$

where E_0 and E_1 are the initial and final energies of the neutron, $I(E_0)$ is the incident flux at the sample position, Q is the momentum transfer, $J_M(x_M)$ is the neutron Compton profile (NCP) for an isotope of mass M, N_M is the number of isotopes under the beam, b_M is the scattering length, and $4\pi b_M^2$ is its bound scattering cross section [25]. Each NCP is a function of a scaling variable, generally referred to as the West-scaling variable, defined as:

$$y_M = \sqrt{2}\sigma_M x_M = \frac{M}{\hbar Q}\left(E - \frac{\hbar Q^2}{2M}\right) = \frac{M}{\hbar Q}(E - E_r), \tag{2}$$

with $E = E_0 - E_1$ the energy transfer, E_r the recoil energy of the nucleus of mass M, and σ_M the second moment of the NCP, defined as:

$$\sigma_M^2 = \int_{-\infty}^{\infty} y_M^2 J(y_M) dy_M. \tag{3}$$

In its most general form, the NCP can be expressed as a series of Gauss-Hermite polynomials, in a series often referred to as a Gram-Charlier expansion, of the form:

$$J_M(y_M) = \frac{1}{\sqrt{2\pi}\sigma_M}\exp\left(-\frac{y_M^2}{2\sigma_M^2}\right)\sum_n c_{2n}H_{2n}\left(\frac{y_M}{\sqrt{2}\sigma_M}\right) = \frac{\exp(-x_M^2)}{\sqrt{2\pi}\sigma_M}\sum_n c_{2n}H_{2n}(x_M), \tag{4}$$

where $n = 0, 1, 2, \ldots$, and the two conditions $c_0 = 1$ and $c_2 = 0$ are required so as to interpret the NCP as a probability function. Under these conditions, σ_M corresponds to the standard deviation of the leading Gaussian term in Eq. 4. This equation provides the most used model to fit experimental NCPs in NCS experiments. The NCP is expected to be a simple Gaussian profile when the nucleus experiences a harmonic and isotropic potential. Terms related to $n > 3$ are seldom expected or required in the fit, and it is common practice to assume the NCP of heavy elements as simple Gaussian functions during the data reduction stages.

One should notice that Eq. 4 only holds within the IA, *i.e.*, when both E and Q are very large compared to the characteristic binding energies and inverse inter-atomic distances in the system, respectively. In an experiment, corrections for the finite value of Q are generally required, referred to as Final-State Effects (FSEs). These are commonly assumed to be of the form [26]:

$$\Delta J(y_M, Q) = A_3(Q)\frac{\partial^3}{\partial y_M^3}J(y_M) = \frac{\sqrt{2}\sigma_M}{12Q}J(y_M, Q). \tag{5}$$

The proportionality of the FSE correction term upon σ_M in Eq. 5 is a consequence of the harmonic approximation, a condition that can be relaxed in the actual fit by including an additional factor as a free parameter [1].

According to the definition of TOF for a spectrometer with incident and final flight paths L_0 and L_1, respectively, and the condition that, at the peak centre, the energy transfer equals the nuclear recoil energy, $E = E_r$, one has:

$$t = \frac{L_0}{v_0} + \frac{L_1}{v_1} = \sqrt{\frac{m}{2E_1}}\left(\sqrt{\frac{E_1}{E_1 + E_r}}L_0 + L_1\right). \tag{6}$$

As $E_r = \hbar^2 Q^2/2M$ is inversely proportional to the mass of the scattering nucleus, light-weight elements contribute to the scattering intensity at low TOF values, as opposed to heavy-weight elements that, approaching the limit of infinite mass, have an upper bound to their NCP peak centre at $t_\infty = \sqrt{m/2E_1}(L_0 + L_1) \simeq 390$ μs on VESUVIO. As a consequence of this dependence upon the recoil energy, NCS is a MAss-selective Neutron SpEctroscopy (MANSE) [27]. Figure 2 (top-left) shows the forward-scattering intensity as a function of TOF and spectrum number (see Figure 1 for details on the numbering of spectra) for polyethylene sample 1. One can easily notice the change in peak position for the hydrogen NCP, with the peak at lower TOF values for detectors at high scattering angles (higher values of Q). Finally, one should appreciate the fact that the dynamic variables in Eq. 4 are all expressed as explicit functions of the time channel t in the TOF spectrum from each detector at scattering angle θ: $E_0(t)$, $Q(t,\theta)$, and $y_M(t,\theta)$. The loading of TOF spectra into MANTID using the "LoadVesuvio" algorithm is described in Appendix A.

3.1. Sequential pre-fit of raw TOF spectra

The corrections to be performed on the raw NCS data are based on the defining parameters of the NCPs, *i.e.*, the width of the peak σ_M, the main observable of the measurement, and the intensity of the peak I_M. Therefore, a preliminary fit of the raw spectra is needed so as to obtain estimates of σ_M and I_M, and the procedure can be iterated after the corrections have been performed. The positions of the peaks are fixed, at this stage, by the *a priori* knowledge of the mass of the constituents of the investigated sample. Details on the way the pre-fit is implemented in MANTID using the "VesuvioTOFFit" algorithm can be found in Appendix B. One can apply Eq. 1 to the case of polyethylene by fitting two profiles, one for hydrogen and one for carbon. As the stoichiometry of the sample is well known, with two hydrogen atoms per carbon atom, a constraint can be applied to improve the quality of the fit, especially for noisy data. The relative intensities of the two profiles, as measured by each detector, are in a fixed relation defined, in Eq. 1, by the ratio of their stoichiometric weights and bound scattering cross sections. Assuming an expression for the intensity of a given mass to be $I_M = N_M b_M^2$, one has

$$I_H = \frac{N_H b_H^2}{N_C b_C^2} I_C. \tag{7}$$

The result of the pre fit on the raw spectra is shown in Figure 2 (top-right).

3.2. Correction for the sample-dependent γ-background

The fitted values of σ_M and I_M, together with other calibration parameters related to the detector positions, can be used to correct for the sample-dependent GB, as discussed in Ref. [28]. The correction has an oscillatory character around each NCP, and it is shown in the bottom-right panel of Figure 2. Additional details on how to perform the GB correction in MANTID using the "VesuvioCalculateGammaBackground" algorithm are discussed in Appendix C.

3.3. Correction for multiple scattering

The multiple-scattering correction is originated by the finite thickness of the sample, and the consequent finite probability of a neutron to scatter more than once within the sample, before being detected [29]. The scattering probability can be expressed, for epithermal neutrons, as

$$1 - T(E_0 > 1\text{ eV}) = 1 - \exp(-\rho\sigma_f d) = \sum_{n=1}^{\infty} \frac{(\rho\sigma_f d)^n}{n!} \simeq \rho\sigma_f d + \frac{1}{2}(\rho\sigma_f d)^2 \ldots, \tag{8}$$

where $T(E_0 > 1\text{ eV})$ is the transmission for epithermal neutrons, ρ the number density of the sample, d is the thickness of the sample, and σ_f the free scattering cross section of the constituent

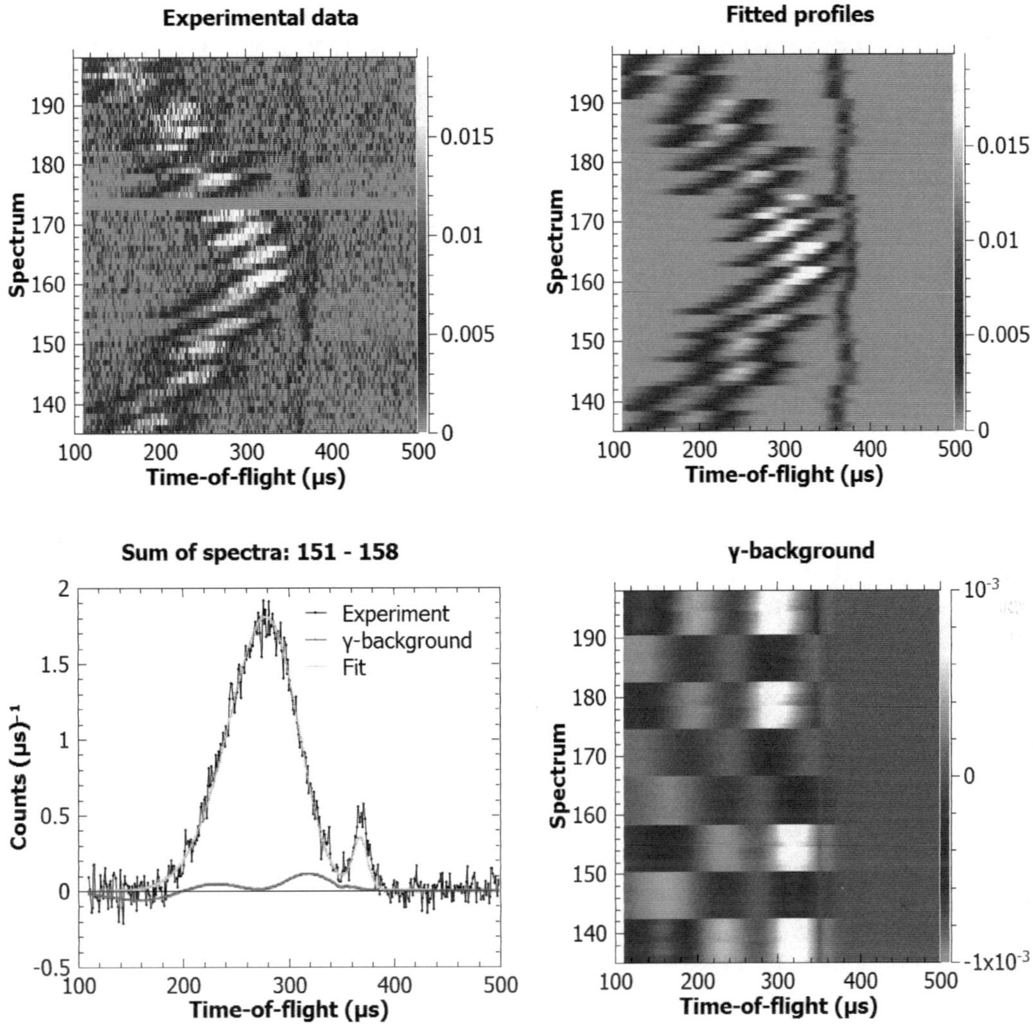

Figure 2. NCS spectra in the time-of-flight domain corresponding to a 0.25-mm-thick PE slab. Top-left: 2D representation of experimental data normalised to unit area. Detectors 173 and 174 have been masked; top-right 2D representation of fitted profiles normalised to unit area; bottom-right: 2D representation of the γ-background (GB) correction for spectra normalised to unit area; bottom-left: experimental data (black error bars), fitted profiles (green line), and GB correction for the bank of detectors 151 – 158, after normalisation to unit area.

elements of the sample [25, 30], CH_2 in the case of polyethylene, with stoichiometric coefficients $\tilde{N}_M$, and defined as:

$$\sigma_f = \sum_M \tilde{N}_M \frac{4\pi b_M^2}{\left(1+\frac{m}{M}\right)^2} \tag{9}$$

The product $\rho\sigma_f d$ is often referred to as the epithermal scattering power. A wise experimenter should chose d so as to minimise the contribution from multiple scattering aiming for a transmission of epithermal neutrons around 0.9, therefore making the terms with $n > 1$ in Eq. 8 negligible. The "VesuvioCanonicalThickness" algorithm can be used to define the best dimensions of a sample to be measured, and the "VesuvioTransmission' algorithm can be

IOP Conf. Series: Journal of Physics: Conf. Series **1055** (2018) 012016 doi:10.1088/1742-6596/1055/1/012016

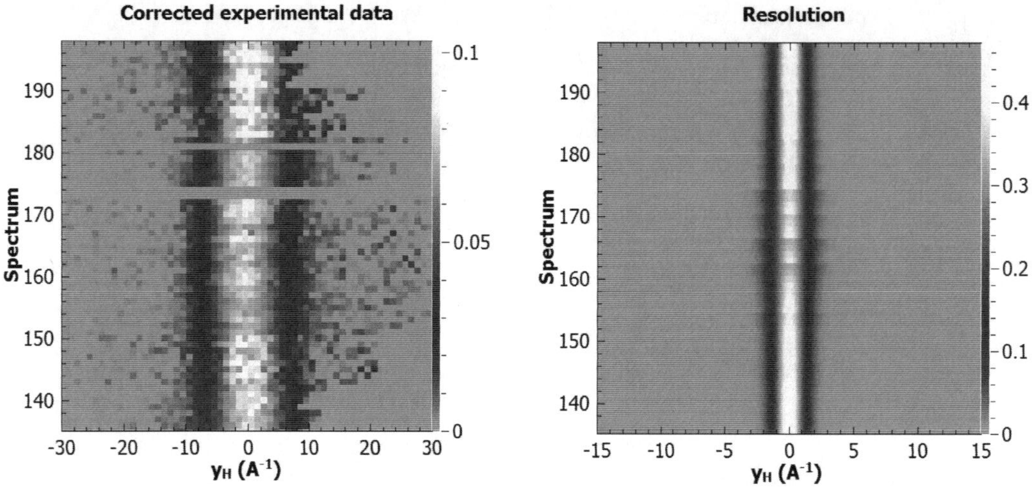

Figure 3. NCS spectra in the hydrogen West-scaling domain after corrections for MS, GB, and subtraction of NCPs from heavier masses. Left: 2D representation of experimental data normalised to unit area; Right 2D representation of resolution functions.

used to visualise the transmission spectra, provided a measurement of the empty container or instrument are available. Recent examples of the neutron-transmission capabilities on VESUVIO can be found in Refs. [31, 32]. The MS correction is evaluated using a Monte Carlo code [33] and the information about the fitted intensities $\{I_M\}$, and the experimental value of $T(E_0 > 1 \text{ eV})$. Additional details on how to perform the MS correction within MANTID using the "VesuvioCalculateMS" are discussed in Appendix D.

3.4. Sequential fit of corrected TOF spectra

An iterative procedure can be adopted, whereby the NCPs in the TOF domain are fitted after the corrections for GB and MS. The updated parameters for σ_M and I_M can be used to re-calculate the corrections until convergence is reached. The results of the sequential fit over all detectors for the width of the NCP, $\{\sigma_M\}$, represents a first, rough, estimate of this observable. The results for the polyethylene samples obtained from such statistical ensemble are reported in Table 1, and referred to as Mode 2, were obtained using the Simplex minimiser. The fit of TOF NCS data using a Levenberg-Marquardt minimiser was found to fail dramatically in the case of several experimental investigations, and its use at this stage is not recommended. The fit is also used to identify the NCPs from masses other than the mass of interest. These profiles should be subtracted from the experimental spectra before the analysis is continued in the domain of the West variable for the mass of interest. Additional details on the subtraction of NCPs from heavier masses are given in Appendix E. In the case of polyethylene, the fitted profile of the carbon NCP is subtracted from the data previously corrected for GB and MS, therefore eventually containing the hydrogen NCP alone. The fit over a bank of detectors corresponding to spectra 151 – 158, together with the GB correction is shown in Figure 2 (bottom-left).

4. Global fit of hydrogen NCP in the West-scaling domain

Corrected TOF spectra can be transformed to the hydrogen West-scaling variable, y_H, according to Eq. 2. One can appreciate the advantage of analysing the NCS spectra in the West-scaling domain from Figure 3 (left), where all the spectra share the same centre $y_H \simeq 0$, as opposed to the case in the TOF domain (Figure 2), where the peak position changes drastically. A number of strategies for the analysis of the hydrogen NCP are available at this juncture. All spectra are assumed to show the same physical information, *i.e.,* a probability function with standard deviation σ_M. However, two effects make the spectra dependent upon the scattering angle: i) the presence of FSE corrections, proportional to the inverse of the momentum transfer Q, slightly shifting the apparent centre of the NCP towards negative values of y_H; ii) the resolution function being broader for small scattering angles [34], as one can appreciate from Figure 3 (right).

At this stage, spectra are generally unit-area normalised, so as to satisfy the zero-order sum rule for probability distributions. Additional details on the data handling in the West-scaling domain are given in Appendix F

4.1. Global fit over all individual detectors

A number of scientific investigations in recent years [35–39] have been based on fits over all individual front-scattering spectra in the West-scaling domain, with the parameters defining the hydrogen NCP, namely σ_H and c_{2n}, being included in the fit as global parameters. Other variables, such as the amplitude of each experimental NCP and its peak centre, $y_0 \sim 0$, were considered spectrum-dependent. The number of parameters that were fitted in this way easily reached 100, and a robust and reliable framework needs to be available to undertake such a task. Traditionally, the analysis has been performed within the Minuit environment [40], used here as a benchmark to validate the results obtained within MANTID.

4.2. Fit of the spectral sum

Assuming that FSEs and the width of the resolution functions vary only slightly with the scattering angles considered, one can replace angle-dependent values with average ones. Consequently, by averaging all experimental spectra, resolution functions, and values of Q for every channel y_H, one can perform a single-spectrum fit using the function

$$\bar{F}(y_H, \bar{Q}) = J(y_H) \star \bar{R}(y_H) + \Delta J(y_H, \bar{Q}) \star \bar{R}(y_H), \tag{10}$$

where F, as opposed to J, traditionally denotes the experimental NCP, therefore including convolution with the resolution function and FSE corrections, and $\bar{F}(y_H, \bar{Q})$ is the average of all experimental spectra. The result of this procedure is reported in Table 1 and referred to as Mode 3, and the fitting curve is shown in Figure 4. The fitting procedure was tested using three minimisers: Simplex, Levenberg-Marquardt, and FABADA, yielding consistent results. Additional details on this fitting procedure are discussed in Appendix G.

4.3. Global fit over banks of detectors

A global fit over "banks" of detectors was implemented within the MANTID environment. The term "bank" here refers to a collection of detectors, not necessarily placed at similar scattering angles. In this procedure, the fit is performed over an ensemble of spectra with σ_M defined as a global parameter. The results from all banks are then considered as a statistical ensemble for which the average and standard deviations are evaluated. The results of this analysis using the Simplex minimiser are shown in Table 1, and referred to as modes 4 – 7, depending on the way each bank was defined. Additional details on the way the global fit over banks of detectors is implemented in MANTID can be found in Appendix H.

5. Discussion

As a number of procedures have been attempted to analyse the hydrogen NCP in polyethylene, we can now compare the obtained results. We comment below on each analysis mode:

Table 1. Results of the analysis of the width, σ_H, of the hydrogen NCP in polyethylene obtained for modes 1 – 8. The analysis has been performed on two measurements labelled Sample 1 (statistics ∼6 mAh) and Sample 2 (statistics ∼1.8 mAh). In the case of the results from mode 3 on Sample 2, values are reported in *italics* to stress their lack of accuracy.

Mode	Minimizer	banks	Sample 1 σ_H [Å^{-1}]	Sample 2 σ_H [Å^{-1}]
Minuit – Global fit (benchmark)				
1	Simplex	1	4.90 ± 0.03	4.87 ± 0.05
1	Migrad	1	4.90 ± 0.03	4.87 ± 0.05
MANTID – Sequential fit				
2	Simplex (TOF)	1	4.87 ± 0.25	4.85 ± 0.36
MANTID – Fit of the sum of spectra				
3	Levenberg-Marquardt	1	4.84 ± 0.02	*4.69 ± 0.04*
3	Simplex	1	4.83 ± 0.02	*4.70 ± 0.04*
3	FABADA	1	4.84 ± 0.04	*4.69 ± 0.06*
MANTID – Global fit				
4	Simplex	4	4.85 ± 0.19	4.83 ± 0.15
5	Simplex	8	4.82 ± 0.16	4.75 ± 0.22
6	Simplex	16	4.82 ± 0.16	4.76 ± 0.26
7	Simplex	8*	4.90 ± 0.08	4.85 ± 0.12
8	MANTID average		4.85 ± 0.03	4.81 ± 0.06

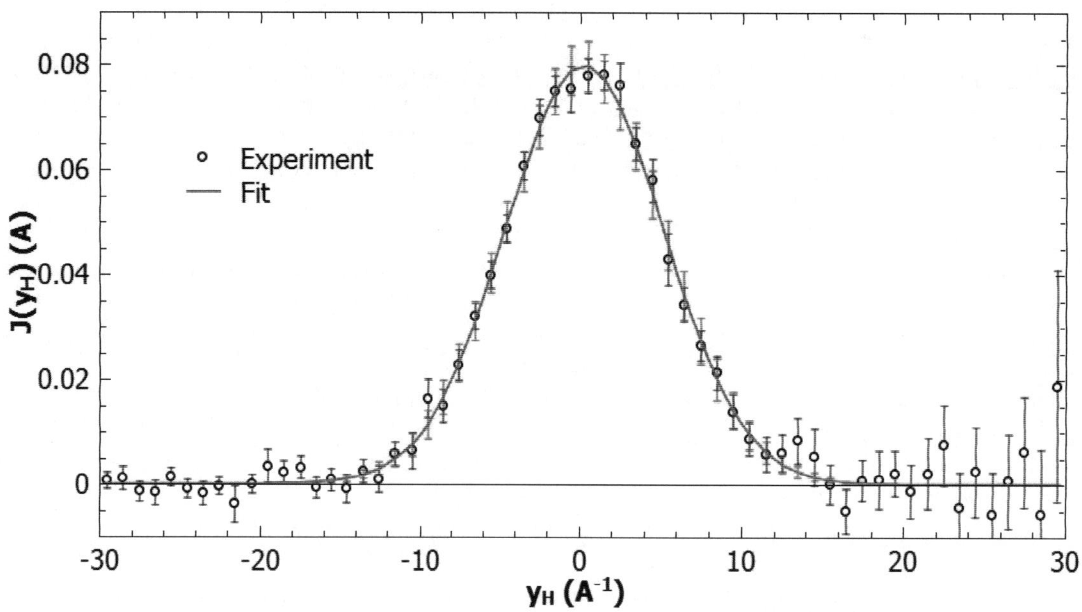

Figure 4. The sum of hydrogen NCPs in polyethylene from Sample 1, together with the fit obtained applying mode 3.

IOP Conf. Series: Journal of Physics: Conf. Series **1055** (2018) 012016 doi:10.1088/1742-6596/1055/1/012016

(1) The global fit of all individual spectra in the West-scaling domain. This is a common approach to the analysis of NCS data, that has been demonstrated to show reliability and reproducibility. At present, it is performed using FORTRAN routines based on the Minuit environment. Here, we take these results as a benchmark. The drawbacks of this approach lie in the time-consuming evaluation of the figure of merit, and the difficult implementation in environments where the available minimisers are not powerful enough.

(2) The sequential fit of TOF spectra. This is a necessary step in the data reduction. Due to the poor signal-to-noise ratio for individual spectra, the spread in values of $\{\sigma_M\}$ is generally large, which makes the results unusable for precise work. Yet, the average of the fitted values is relatively near the benchmark result, and can be used reliably for the GB and MS, and even FSE corrections.

(3) The fit on the sum of spectra in the West-scaling domain. This is a simple and quick approach allowing for a simple visualisation and interpretation of the results. The value of σ_H can be accessed with a high precision, however the result could lack accuracy whenever some spectra from faulty detectors are included in the sum and contribute with spurious intensity to the average NCP. This is the case, for example, of the results from the noisier Sample 2, reported in italics in Table 1. The values obtained from mode 3, in this case, provide a clear underestimate of the value of σ_H with respect to the other results in the table, an outcome that can be prevented by careful inspection, and possibly masking, of each individual spectrum prior to their sum. We notice that the task of recognising faulty spectra is generally more complex for noisier data.

(4) The global fit over 4 banks of 16 consecutive detectors. The four banks correspond to detectors: 135 – 150; 151 – 166; 167 – 182; and 183 – 198. σ_H is defined as a global parameter within the bank. Then, the results from each bank are averaged, and their standard deviation is used as the uncertainty on the result. The spread on the ensemble $\{\sigma_H\}$, is comparable with the uncertainty of each parameter from the global fit of individual banks. The result can be used as a more precise estimate with respect to mode 2, and a more accurate estimate than the one from mode 3.

(5) The global fit over 8 banks of 8 consecutive detectors. The same procedure as in mode 4, with the banks corresponding to the traditional grouping in Figure 1.

(6) The global fit over 16 banks of 4 consecutive detectors. The same procedure as in mode 4, with the banks defined as: 135 – 138; 139 – 142; and so on.

(7) The global fit over 8 banks of 8 non-consecutive detectors. Each group of detectors now includes a representative from all traditional banks from Figure 1. The analysis provides both precise and accurate results, yet the associated error-bars are a factor ca. 2.5 larger than what one would expect from mode 1.

As a number of strategies are available within the Mantid environment to analyse the NCS spectra, one can treat the many results as members of a statistical ensemble associated with a mean value and a standard deviation. We refer to this approach as mode 8 in Table 1. It is interesting to notice how the values of σ_H obtained in this way, as well as the magnitude of the associated uncertainty, are in remarkable agreement with the benchmark results. This suggests that more-robust minimisers could yield smaller uncertainties in the case of the global fits for modes 4 – 7.

6. Outlook and conclusions

We have presented an update on the implementation of the data reduction and analysis of NCS spectra within the MANTID environment. To this end, we have used a polyethylene standard sample to discuss several strategies to obtain values of the hydrogen NCP width, σ_H. As a result

IOP Conf. Series: Journal of Physics: Conf. Series **1055** (2018) 012016 doi:10.1088/1742-6596/1055/1/012016

of this analysis, we have provided a detailed Python script that can be readily used to analyse NCS data collected on VESUVIO. The present work suggests that the analysis of NCS data in MANTID can benefit from the inclusion of other, more robust, minimisers.

Acknowledgements
The authors gratefully acknowledge the UK Science & Technology Facilities Council for financial support and access to beam time at the ISIS Facility.

Appendix A. Initialisation procedure

The first step in order to analyse the NCS data in MANTID is to open a Python-script window and define the initialisation parameters. These include:

- the experimental run numbers to be analysed, according to the VESUVIO data-acquisition journal;
- the spectra corresponding to the detectors to be considered, according to the schematics in Figure 1;
- the differencing mode to combine the periods of data acquisition, in particular "SingleDifference" in the case of front-scattering detectors, and "DoubleDifference" in the case of backscattering detectors;
- the expected isotopes, and in particular their masses, in the sample or in the sample-plus-container system – noticing that the number and values of specified masses defines the number of NCPs to be fitted and the value of their peak centre;
- any expected constraint between intensities of pairs of NCPs, whenever an empirical formula is known, so as to facilitate the fitting procedure.

Additionally, one can declare faulty detectors or additional optional parameters, as in the listed code below.

```
#
import mantid
from mantid.simpleapi import *
import numpy as np

name='polyethylene_'          # A string for the name of the project
runs='17713-17779'            # The numbers of the runs to be analysed
empty_runs=None               # The numbers of the runs for the empty instrument
empty_scaling_factor=1.0      # Attenuation factor for the empty instrument spectra
spectra='135-198'             # Spectra to be analysed
# Optional instrument parameter file
ipfile='C:/MANTID/ipfiles/ip2018.par'
mode='SingleDifference'       # Differencing method
tof_binning='110.,1.,500'     # Binning of TOF spectra in us
masses=[1.0079,12.0]          # Array with the masses to be fitted in amu
guess_widths=[5,11]           # Array of widths (sigma) for the NCPs in A-1
# constraint on the intensities of fitted profiles, in the format [] or ([],[])
intensity_constraints=([1, -29.8])
detectors_masked='173,174,181'          # Optional detectors to be masked
transmission=0.9                        # Transmission value from VesuvioTransmission
verbose=False                           # Level of output from the script
correct_multiple_scattering=False       # Enable multiple scattering correction
correct_gamma_background=True           # Enable gamma-backtround correction
simple_gaussian_fit=True                # Simple Gaussian/ Gauss-Hermite profile for H
number_of_iterations=2                  # Iterations of the TOF data reduction
minimizer='Simplex'                     # Minimiser to be used
#
```

Once the basic variables have been defined, the data can be loaded into MANTID in the form of a "Workspace", a container for a collection of spectra, running the commands

```
# Loading of raw DINS data
LoadVesuvio(Filename=runs, SpectrumList=spectra, Mode=mode,
            InstrumentParFile=ipfile,OutputWorkspace=name+'raw')

# Subtraction by experimental background
if empty_runs:
    LoadVesuvio(Filename=empty_runs, SpectrumList=spectra, Mode=mode,
                InstrumentParFile=ipfile,OutputWorkspace=name+'empty')
```

IOP Conf. Series: Journal of Physics: Conf. Series **1055** (2018) 012016 doi:10.1088/1742-6596/1055/1/012016

```
    Scale(InputWorkspace=name+'empty',Factor=empty_scale_factor,
          Operation='Multiply', OutputWorkspace=name+'empty')
    Minus(LHSWorkspace=name+'raw',RHSWorkspace=name+'empty',
          OutputWorkspace=name)
else:
    RenameWorkspace(InputWorkspace=name+'raw',OutputWorkspace=name)

# Formatting of intensity constraints, if more than one
if intensity_constraints:
    if not isinstance(intensity_constraints[0], list):
        intensity_constraints = [intensity_constraints]
    intensity_constraints = [str(c) for c in intensity_constraints]
    intensity_constraints_str = ";".join(intensity_constraints)

# masking and rebinning of the TOF workspace
MaskDetectors(Workspace=name,SpectraList=detectors_masked)
Rebin(InputWorkspace=name,Params=tof_binning,OutputWorkspace=name)
```

At this stage it is possible to subtract experimental backgrounds to the data, merge non-sequential acquisition runs, or perform any needed operation on the TOF data prior to the correction and analysis directly related to NCS.

Appendix B. Sequential fit in TOF domain

A pre-fit of the NCS data is needed as both the γ-background (GB) and multiple-scattering (MS) corrections depend upon the NCP intensities, I_M and widths, σ_M. In a first step, one defines the number of expected profiles, together with their main properties. It is recommended to iterate over this procedure, so as to obtain precise values of the peak intensities after MS and GB corrections. In this script, the iteration procedure is iterated by the following piece of code:

```
# TOF Fit and corrections to be iterated
for iteration in range(number_of_iterations):
    # Definition of the workspace to be fitted
    if (iteration==0):
        CloneWorkspace(InputWorkspace=name,OutputWorkspace=name+'to_be_fitted')
    else:
        CloneWorkspace(InputWorkspace=name+'tof_corrected_iteration_'+str(iteration),
                       OutputWorkspace=name+'to_be_fitted')
```

Prior to the fitting procedure, one needs to define the workspaces where the total fit and each fitted profile will be guested, and the profile functions to be fitted:

```
    # creation of empty workspaces to guest the total fit and the fit of individual profiles
    CloneWorkspace(InputWorkspace=name,OutputWorkspace=name+'tof_fitted_profiles')
    for m in range(masses.__len__()):
        CloneWorkspace(InputWorkspace=name,OutputWorkspace=name+'tof_fitted_profile_'+str(m+1))

    # Definition of the string for the neutron Compton profiles to be fitted
    profiles=''
    if (simple_gaussian_fit):
        for m in range(masses.__len__()):
            width_string=[0.7*guess_widths[m],guess_widths[m],1.3*guess_widths[m]]
            width_string=str(width_string)
            profiles+='function=GramCharlier,width='+width_string+',hermite_coeffs=[1,0,0],k_free=0,
    sears_flag=1;'
    else:
        for m in range(masses.__len__()):
            width_string=[0.5*guess_widths[m],guess_widths[m],1.5*guess_widths[m]]
            width_string=str(width_string)
            if (m==0):
                profiles+= 'function=GramCharlier,width='+width_string+',hermite_coeffs=[1,0,1],k_free
    =0,sears_flag=1;'
            else:
                profiles+= 'function=GramCharlier,width='+width_string+',hermite_coeffs=[1,0,0],k_free
    =0,sears_flag=1;'
```

The sequential fit the TOF domain can be performed using the VesuvioTOFFit algorithm:

```
    for spec in range(ws.getNumberHistograms()):
        fit, par, chi_sq = VesuvioTOFFit(InputWorkspace=name+'to_be_fitted',
                                         Masses=masses,
                                         IntensityConstraints=intensity_constraints_str,
                                         MassProfiles=profiles,Minimizer=minimizer,
                                         WorkspaceIndex=spec,
                                         FitParameters=name+'parameters_'+str(spec))

        # Filling the workspaces for the total fit and individual fitted profiles
        for m in range(masses.__len__()):
            tmp=mtd[name+'tof_fitted_profile_'+str(m+1)]
            for bin in range(tmp.blocksize()):
```

```
            tmp.dataY(spec)[bin], tmp.dataE(spec)[bin] = fit.dataY(3+m)[bin], fit.dataE(3+m)[bin]
    tmp=mtd[name+'tof_fitted_profiles']
    for bin in range(ws.blocksize()):
        tmp.dataY(spec)[bin], tmp.dataE(spec)[bin] = fit.dataY(1)[bin], fit.dataE(1)[bin]
    if (not verbose):
        DeleteWorkspace('fit')
```

The relevant fitted parameters from the fit can be stored in a parameter workspace, including the spectrum number, and the values of the mass, width, and intensity for each fitted profile:

```
# Creation of a workspace with the fitted parameters
width, intensity=[None]*masses.__len__(),[None]*masses.__len__()
ws=mtd[name]
table = CreateEmptyTableWorkspace()
table.addColumn("int", "Spectrum")
for m in range(masses.__len__()):
    table.addColumn("double", "Mass"+str(m+1))
    table.addColumn("double", "Width"+str(m+1))
    table.addColumn("double", "Intensity"+str(m+1))
for spec in range(ws.getNumberHistograms()):
    par=mtd[name+'parameters_'+str(spec)]
    vector=[spec+3]
    for m in range(masses.__len__()):
        width[m], intensity[m] =par.cell(1+m*4,1), par.cell(3+m*4,1)
        vector.append(masses[m])
        vector.append(width[m])
        vector.append(intensity[m])
    table.addRow( vector)
    if (not verbose):
        DeleteWorkspace(name+'parameters_'+str(spec))
RenameWorkspace('table',name+'tof_fitted_parameters')
table=mtd[name+'tof_fitted_parameters']
```

As a result of the sequential TOF fit, the average values of σ_M for each fitted profile can be evaluated, corresponding to the analysis mode 2:

```
#DEFINITION OF THE WIDTHS (SIGMAS) OF FITTED NCPs, AVERAGED OVER ALL SPECTRA
print 'The fitted values for the standard deviations are:', '\n'
width=[None]*masses.__len__()
for m in range(masses.__len__()):
    width[m]=np.mean(table.column(2+3*m))
    print 'mass: ',masses[m], 'sigma (width): ', width[m], ' +/- ', np.std(table.column(2+3*m))
```

Appendix C. Subtraction of the γ-background

When enabled, the correction for GB can be performed using the code:

```
if (correct_gamma_background):
    # Create an empty workspace for the gamma correction
    CloneWorkspace(InputWorkspace=name, OutputWorkspace=name+'gamma')
    ws=mtd[name+'gamma']
    table=mtd[name+'tof_fitted_parameters']
    for spec in range(ws.getNumberHistograms()):
        profiles=''
        for m in range(masses.__len__()):
            mass,width,intensity=str(masses[m]), str(np.mean(table.column(2+3*m))),str(table.cell(
spec,3*(m+1)))
            profiles+= "name=GaussianComptonProfile,Mass="+mass+",Width="+width+",Intensity="+
intensity+';'
        background, corrected = VesuvioCalculateGammaBackground(InputWorkspace=name,
                                                   ComptonFunction=profiles,
WorkspaceIndexList=spec)
        for bin in range(ws.blocksize()):
            ws.dataY(spec)[bin],ws.dataE(spec)[bin]=background.dataY(0)[bin],background.dataE(0)[
bin]
    DeleteWorkspace('background')
    DeleteWorkspace('corrected')
    Minus(LHSWorkspace=name,RHSWorkspace=name+'gamma',OutputWorkspace=name+'tof_corrected')
```

Appendix D. Subtraction of the multiple scattering

In order to correct the experimental spectra for multiple scattering (MS), one first needs to define the geometry of the sample. THis can be done with the code:

```
if (correct_multiple_scattering):
    # DEFINITION OF THE SAMPLE GEOMETRY
    height, width, thick = 0.1, 0.1, 0.001  #(m)
    half_height, half_width, half_thick = 0.5*height, 0.5*width, 0.5*thick
    xml_str = \
    " <cuboid id=\"sample-shape\"> " \
    + "<left-front-bottom-point x=\"%f\" y=\"%f\" z=\"%f\" /> " % (half_width,-half_height,
half_thick) \
```

IOP Conf. Series: Journal of Physics: Conf. Series **1055** (2018) 012016 doi:10.1088/1742-6596/1055/1/012016

```
    + "<left-front-top-point x=\"%f\" y=\"%f\" z=\"%f\" />" % (half_width, half_height, half_thick
) \
    + "<left-back-bottom-point x=\"%f\" y=\"%f\" z=\"%f\" />" % (half_width, -half_height, -
half_thick) \
    + "<right-front-bottom-point x=\"%f\" y=\"%f\" z=\"%f\" />" % (-half_width, -half_height,
half_thick) \
    + "</cuboid>"
    CreateSampleShape(name, xml_str)
```

In a second step, the relative intensities of each NCP are obtained by averaging the values saved in the parameter workspace, output of the previous fit:

```
    print '\n', 'The relative intensities of the fitted masses are:', '\n'
    # DEFINITION OF THE RELATIVE INTENSITIES OF FITTED NCPs, AVERAGED OVER ALL SPECTRA
    amplitudes=[0]*masses.__len__()
    Ns=0
    ws=mtd[name]
    for spec in range(ws.getNumberHistograms()):
        sum=0
        det=ws.getDetector(spec)
        if det.isMasked():
            Ns+=1
        else:
            for m in range(masses.__len__()):
                sum+=table.cell(spec,3*(m+1))
            for m in range(masses.__len__()):
                if (sum != 0):
                    amplitudes[m]+=table.cell(spec,3*(m+1))/sum/ws.getNumberHistograms()
    for m in range(masses.__len__()):
        print 'mass: ',masses[m], 'relative intensity: ', amplitudes[m]
    amplitudes = map(lambda x: x*ws.getNumberHistograms()/(ws.getNumberHistograms()-Ns), amplitudes
)
    print 'Summing up to (expectec 1): ',np.sum(amplitudes)
```

At this point, the VesuvioThickness algorithm can be run to obtain the effective density of the sample, and the Monte Carlo simulation of the total and multiple scattering can be run using the VesuvioCalculateMS algorithm:

```
    MS_properties=[]
    for m in range(masses.__len__()):
        MS_properties.append(masses[m])
        MS_properties.append(amplitudes[m])
        MS_properties.append(width[m])

    NM=masses.__len__()
    dens, trans = VesuvioThickness(Masses=masses, Amplitudes=amplitudes, TransmissionGuess=
transmission, Thickness=0.1)
    dens10 = dens.cell(9,1)
    if (not verbose):
        DeleteWorkspace('dens')
        DeleteWorkspace('trans')

    sample_TS, sample_MS = VesuvioCalculateMS(name,
                                              NoOfMasses=NM, SampleDensity=dens10,
                                              AtomicProperties=MS_properties,
                                              BeamRadius=2.5,
                                              NumEventsPerRun=monte_carlo_events)
```

The total-scattering spectra obtained from the simulation can be scaled to the intensity of the experimental ones by multiplying for the ratio of integrated areas, and the same factor is used to scale the multiple scattering contribution, subsequently subtracted from the experimental spectra:

```
    tmp=Integration('sample_TS',250,400)
    sample_TS=Divide(LHSWorkspace='sample_TS',RHSWorkspace='tmp')
    sample_MS=Divide(LHSWorkspace='sample_MS',RHSWorkspace='tmp')
    tmp=Integration(name,250,400)
    sample_TS=Multiply(LHSWorkspace='sample_TS',RHSWorkspace='tmp')
    sample_MS=Multiply(LHSWorkspace='sample_MS',RHSWorkspace='tmp')
    DeleteWorkspace('tmp')
    RenameWorkspace('sample_TS',name+'tof_TotalScattering')
    RenameWorkspace('sample_MS',name+'tof_MultipleScattering')

    if (correct_gamma_background):
        Minus(LHSWorkspace=name+'tof_corrected',RHSWorkspace=name+'tof_MultipleScattering',
                                OutputWorkspace=name+'tof_corrected')
    else:
        Minus(LHSWorkspace=name,RHSWorkspace=name+'tof_MultipleScattering',
                                OutputWorkspace=name+'tof_corrected')
```

Appendix E. Subtraction of fitted NCPs from heavier masses

After a few iterations of the above corrections, the fitted NCPs from elements other than hydrogen can be subtracted from the experimental spectra:

```
for m in range(masses.__len__()-1):
    if (m==0):
        Minus(LHSWorkspace=name+'tof_corrected',RHSWorkspace=name+'tof_fitted_profile_'+str(m+2),
              OutputWorkspace=name+'tof_hydrogen')
    else:
        Minus(LHSWorkspace=name+'tof_hydrogen',RHSWorkspace=name+'tof_fitted_profile_'+str(m+2),
              OutputWorkspace=name+'tof_hydrogen')

CloneWorkspace(InputWorkspace=name+'tof_corrected',
               OutputWorkspace=name+'tof_corrected_iteration_'+str(iteration+1))
DeleteWorkspace(Workspace=name+'to_be_fitted')
```

Appendix F. Handling of the spectra in the West-scaling domain

The corrected spectra in the TOF domain can be converted to the hydrogen West-scaling variable using the ConvertToYSpace algorithm, and subsequently be unit-area normalised:

```
# Conversion to hydrogen West-scaling variable
rebin_params='-30,1,30'
ConvertToYSpace(InputWorkspace=name+'tof_hydrogen',Mass=1.0079,OutputWorkspace=name+'joy',
                QWorkspace=name+'q')
Rebin(InputWorkspace=name+'joy',Params=rebin_params,OutputWorkspace=name+'joy')
Rebin(InputWorkspace=name+'q',Params=rebin_params,OutputWorkspace=name+'q')
tmp=Integration(InputWorkspace=name+'joy',RangeLower='-30',RangeUpper='30')
Divide(LHSWorkspace=name+'joy',RHSWorkspace='tmp',OutputWorkspace=name+'joy')
```

When spectra are summed together, one should remember that detectors at different scattering angles probe different portions of the hydrogen West-scaling domain. In particular, spectra from detectors at high scattering angle display NCPs up to ca. 10 Å^{-1}, as opposed to spectra from low scattering angles that do not show this problem. To correct the sum of spectra in the West-scaling domain for this kinematic effect, a normalisation workspace can be defined as:

```
# Definition of a normalising workspace taking into consideration the kinematic constraints
ws=CloneWorkspace(InputWorkspace=name+'joy')
for j in range(ws.getNumberHistograms()):
    for k in range(ws.blocksize()):
        ws.dataE(j)[k] =0.
        if (ws.dataY(j)[k]!=0):
            ws.dataY(j)[k] =1.
ws=SumSpectra('ws')
RenameWorkspace('ws',name+'joy_sum_normalisation')
# Definition of the sum of all spectra
SumSpectra(name+'joy',OutputWorkspace=name+'joy_sum')
Divide(LHSWorkspace=name+'joy_sum',RHSWorkspace=name+'joy_sum_normalisation',
       OutputWorkspace=name+'joy_sum')
```

Finally, the resolution functions for each detector can be generated using the “VesuvioResolution” algorithm:

```
# Definition of the resolution functions
resolution=CloneWorkspace(InputWorkspace=name+'joy')
resolution=Rebin(InputWorkspace='resolution',Params='-15,0.125,15')
for i in range(resolution.getNumberHistograms()):
    VesuvioResolution(Workspace=name,WorkspaceIndex=str(i), Mass=1.0079, OutputWorkspaceYSpace='tmp')
    tmp=Rebin(InputWorkspace='tmp',Params='-15,0.125,15')
    for p in range (tmp.blocksize()):
        resolution.dataY(i)[p]=tmp.dataY(0)[p]
# Definition of the sum of resolution functions
resolution_sum=SumSpectra('resolution',StartWorkspaceIndex=0,EndWorkspaceIndex=63)
tmp=Integration('resolution_sum')
resolution_sum=Divide('resolution_sum','tmp')
DeleteWorkspace('tmp')
```

Appendix G. Fit of the average hydrogen NCP

The line-shape analysis of the average hydrogen NCP profile can be performed an average resolution function and an average correction for FSEs are provided. The stability of the fit procedure is checked against by using three different minimisers in the code below:

```
for minimizer_sum in ('Levenberg-Marquardt','Simplex','FABADA'):
    for iteration in range(2):
        if (iteration ==0):
            sigma, x0 =5, 0
        else:
            ws=mtd[name+'joy_sum_fitted_Parameters']
            sigma, x0 =ws.cell(2,1), ws.cell(1,1)
        factor=np.sqrt(2)*sigma/12.
        FSE=CloneWorkspace(name+'q')
        ws=mtd[name+'q']
        for i in range(ws.getNumberHistograms()):
            flag=True
            for p in range(ws.blocksize()):
                if (ws.dataY(i)[p] != 0):
                    x=(ws.dataX(i)[p]-x0)/np.sqrt(2)/sigma
                    FSE.dataY(i)[p] = np.exp(-x**2)/((2*3.1415*sigma**2)**0.5)*factor/ws.dataY(i)[p
]*(8*x**3-12*x)
                    FSE.dataE(i)[p] = 0
                if (ws.dataY(i)[p]==0):
                    if (flag):
                        FSE.dataY(i)[p-1]=0
                        flag=False
        SumSpectra('FSE',OutputWorkspace='FSE_sum')
        Divide(LHSWorkspace='FSE_sum',RHSWorkspace=name+'joy_sun_normalisation',OutputWorkspace='
FSE_sum')
        FSE_sum=ConvolveWorkspaces(Workspace1='FSE_sum', Workspace2='resolution_sum')
        Minus(LHSWorkspace=name+'joy_sum',RHSWorkspace='FSE_sum',OutputWorkspace=name+'joy_sum_fitted')

        if (simple_gaussian_fit):
            function='''composite=Convolution,FixResolution=true,NumDeriv=true;
                name=Resolution,Workspace=resolution_sum,WorkspaceIndex=0;
                name=UserFunction,Formula=A*exp( -(x-x0)^2/2/sigma^2)/(2*3.1415*sigma^2)^0.5,
                A=1,x0=0,sigma=5,    ties=()'''
        else:
            function='''composite=Convolution,FixResolution=true,NumDeriv=true;
                name=Resolution,Workspace=resolution_sum,WorkspaceIndex=0;
                name=UserFunction,Formula=A*exp( -(x-x0)^2/2/sigma^2)/(2*3.1415*sigma^2)^0.5
                *(1+c4/32*(16*((x-x0)/sqrt(2.)/sigma)^4-48*((x-x0)/sqrt(2.)/sigma)^2+12)),
                A=1,x0=0,sigma=5, c4=0,    ties=()'''

        if (minimizer_sum=='FABADA'):
            minimizer_string='FABADA, ChainLength=1000000,NumberBinsPDF=50,PDF='+name+'joy_sum_PDF'
            Fit(Function=function, MaxIterations=2000000 , InputWorkspace=name+'joy_sum_fitted',
                                        Output=name+'joy_sum_fitted',Minimizer=minimizer_string)
        else:
            Fit(Function=function, InputWorkspace=name+'joy_sum_fitted', Output=name+'joy_sum_fitted',
                                        Minimizer=minimizer_sum)
    ws=mtd[name+'joy_sum_fitted_Parameters']
    print 'Using the minimizer: ',minimizer_sum
    print 'Hydrogen standard deviation: ',ws.cell(2,1),' +/- ',ws.cell(2,2)
```

Appendix H. Global fit on banks of NCPs

The analysis of NCPs using σ_H as a global parameter over banks of spectra can be performed by running the following code:

```
# Definition of a workspace for the global fit with artificial error bars on the unphysical bins
CloneWorkspace(InputWorkspace=name+'joy', OutputWorkspace=name+'joy_global')
ws=mtd[name+'joy_global']
for j in range(ws.getNumberHistograms()):
    for k in range(ws.blocksize()):
        if (ws.dataE(j)[k]==0):
            ws.dataE(j)[k] =0.1


# Definition of the 1/Q workspace for correction for FSEs in the global fit
CloneWorkspace(InputWorkspace=name+'q',OutputWorkspace='one_over_q')
ws=mtd['one_over_q']
for j in range(ws.getNumberHistograms()):
    flag=True
    for k in range(ws.blocksize()):
        if (ws.dataY(j)[k]!=0):
            ws.dataY(j)[k] =1./ws.dataY(j)[k]
        if (ws.dataY(j)[k] == 0):
            if (flag):
                ws.dataY(j)[k-1] =0
                flag=False

# Global fit of the H J(y) with global values of Sigma, and FSE in the harmonic approximation
sample_ws = mtd[name+'joy_global']
resolution_ws = mtd['resolution']
if (simple_gaussian_fit):
    convolution_template = """
    (composite=Convolution,$domains=({0});
    name=Resolution,Workspace={1},WorkspaceIndex={0};
    (
      name=UserFunction,Formula=
      A*exp( -(x-x0)^2/2/Sigma^2)/(2*3.1415*Sigma^2)^0.5,
```

```
    A=1.,x0=0.,Sigma=5.0,   ties=();
    (
        composite=ProductFunction,NumDeriv=false;name=TabulatedFunction,Workspace=one_over_q,
  WorkspaceIndex={0},ties=(Scaling=1,Shift=0,XScaling=1);
        name=UserFunction,Formula=
        Sigma*1.4142/12.*exp( -(x)^2/2/Sigma^2)/(2*3.1415*Sigma^2)^0.5
        *((8*((x)/sqrt(2.)/Sigma)^3-12*((x)/sqrt(2.)/Sigma))),
        Sigma=6.0);ties=()
    ))"""
else:
    convolution_template = """
    (composite=Convolution,$domains=({0});
    name=Resolution,Workspace={1},WorkspaceIndex={0};
    (
      name=UserFunction,Formula=
      A*exp( -(x-x0)^2/2/Sigma^2)/(2*3.1415*Sigma^2)^0.5
      *(1+c4/32*(16*((x-x0)/sqrt(2.)/Sigma)^4-48*((x-x0)/sqrt(2.)/Sigma)^2+12)),
      A=1.,x0=0.,Sigma=5.0, c4=0, ties=();
      (
        composite=ProductFunction,NumDeriv=false;name=TabulatedFunction,Workspace=one_over_q,
  WorkspaceIndex={0},ties=(Scaling=1,Shift=0,XScaling=1);
        name=UserFunction,Formula=
        Sigma*1.4142/12.*exp( -(x)^2/2/Sigma^2)/(2*3.1415*Sigma^2)^0.5
        *((8*((x)/sqrt(2.)/Sigma)^3-12*((x)/sqrt(2.)/Sigma))),
        Sigma=6.0);ties=()
    ))"""


print '\n','Global fit in the West domain over banks','\n'
#   GLOBAL FIT
for nbanks in (4,8,16):
    print 'Number of grouping banks: ', nbanks
    w=[]
    for bank in range(nbanks):
        convolved_funcs = []
        ties = []
        datasets = {}
        counter = 0
        for i in range(64/nbanks*bank,64/nbanks*(bank+1)):
            spec_i = sample_ws.getSpectrum(i)
            det_i = sample_ws.getDetector(i)
            #print "Considering spectrum {0}".format(spec_i.getSpectrumNo())
            if (verbose):
                if det_i.isMasked():
                    print "Skipping masked spectrum {0}".format(spec_i.getSpectrumNo())
                    continue
            f1 = convolution_template.format(counter, resolution_ws.getName())
            convolved_funcs.append(f1)
            # Tie widths together for spectra together
            if counter > 0:
                ties.append('f{0}.f1.f0.Sigma= f{0}.f1.f1.f1.Sigma=f0.f1.f0.Sigma'.format(counter))
                if (not simple_gaussian_fit):
                    ties.append('f{0}.f1.f0.c4=f0.f1.f0.c4'.format(counter))
                #ties.append('f{0}.f1.f1.f1.c3=f0.f1.f1.f1.c3'.format(counter))
            # Attach datasets
            attr = 'InputWorkspace_{0}'.format(counter) if counter > 0 else 'InputWorkspace'
            datasets[attr] = sample_ws.getName()
            attr = 'WorkspaceIndex_{0}'.format(counter) if counter > 0 else 'WorkspaceIndex'
            datasets[attr] = i
            counter += 1
    # end
        multifit_func = 'composite=MultiDomainFunction;' + ';'.join(convolved_funcs) + ';ties=({0},f0.
  f1.f1.f1.Sigma=f0.f1.f0.Sigma)'.format(','.join(ties))
        minimizer_string=minimizer+',AbsError=0.00001,RealError=0.00001,MaxIterations=2000'
        Fit(multifit_func,Minimizer=minimizer_string, Output=name+'joy_nbanks_'+str(nbanks)+'_bank_'+
  str(bank)+'_fit', **datasets)
        ws=mtd[name+'joy_nbanks_'+str(nbanks)+'_bank_'+str(bank)+'_fit_Parameters']
        if (verbose):
            print 'bank: ', str(bank), ' -- sigma= ', ws.cell(2,1), ' +/- ', ws.cell(2,2)
        w.append(ws.cell(2,1))
        if (not verbose):
            DeleteWorkspace(name+'joy_nbanks_'+str(nbanks)+'_bank_'+str(bank)+'
  _fit_NormalisedCovarianceMatrix')
            DeleteWorkspace(name+'joy_nbanks_'+str(nbanks)+'_bank_'+str(bank)+'_fit_Workspaces')
    print 'Average hydrogen standard deviation: ',np.mean(w),' +/- ', np.std(w)


print '\n','Global fit in the West domain over 8 mixed banks','\n'
########################prova
#   GLOBAL FIT - MIXED BANKS
w=[]
for bank in range(8):
    dets=[ bank, bank+8, bank+16, bank+24, bank+32, bank+40, bank+48, bank+56]
    convolved_funcs = []
    ties = []
    datasets = {}
    counter = 0
    for i in dets:
        spec_i = sample_ws.getSpectrum(i)
        det_i = sample_ws.getDetector(i)
        #print "Considering spectrum {0}".format(spec_i.getSpectrumNo())
        if (verbose):
```

```
            if det_i.isMasked():
                #print "Skipping masked spectrum {0}".format(spec_i.getSpectrumNo())
                continue
        f1 = convolution_template.format(counter, resolution_ws.getName())
        convolved_funcs.append(f1)
        # Tie widths togethcr for spectra together
        if counter > 0:
            ties.append('f{0}.f1.f0.Sigma= f{0}.f1.f1.f1.Sigma=f0.f1.f0.Sigma'.format(counter))
            #ties.append('f{0}.f1.f0.c4=f0.f1.f0.c4'.format(counter))
            #ties.append('f{0}.f1.f1.f1.c3=f0.f1.f1.f1.c3'.format(counter))
        # Attach datasets
        attr = 'InputWorkspace_{0}'.format(counter) if counter > 0 else 'InputWorkspace'
        datasets[attr] = sample_ws.getName()
        attr = 'WorkspaceIndex_{0}'.format(counter) if counter > 0 else 'WorkspaceIndex'
        datasets[attr] = i
        counter += 1
# end
    multifit_func = 'composite=MultiDomainFunction;' + ';'.join(convolved_funcs) + ';ties=({0},f0.f1.
     f1.f1.Sigma=f0.f1.f0.Sigma)'.format(','.join(ties))
    minimizer_string=minimizer+',AbsError=0.00001,RealError=0.00001,MaxIterations=2000'
    Fit(multifit_func,Minimizer=minimizer_string, Output=name+'joy_mixed_banks_bank_'+str(bank)+'_fit',
      **datasets)
    ws=mtd[name+'joy_mixed_banks_bank_'+str(bank)+'_fit_Parameters']
    if (verbose):
        print 'bank: ', str(bank), ' -- sigma= ', ws.cell(2,1), ' +/- ', ws.cell(2,2)
    w.append(ws.cell(2,1))
    if (not verbose):
        DeleteWorkspace(name+'joy_mixed_banks_bank_'+str(bank)+'_fit_NormalisedCovarianceMatrix')
        DeleteWorkspace(name+'joy_mixed_banks_bank_'+str(bank)+'_fit_Workspaces')
print 'Average hydrogen standard deviation: ',np.mean(w),' +/- ', np.std(w)
```

References

[1] Andreani C, Krzystyniak M, Romanelli G, Senesi R and Fernandez-Alonso F 2017 *Advances in Physics* **66** 1–73

[2] Andreani C, Senesi R, Krzystyniak M, Romanelli G and Fernandez-Alonso F 2017 *Neutron Scattering - Applications in Biology, Chemistry, and Materials Science* (*Experimental Methods in the Physical Sciences* vol 49) ed Fernandez-Alonso F and Price D L (Academic Press) pp 403 – 457

[3] Last accessed on June 2018 URL `http://www.isis.stfc.ac.uk/Pages/VESUVIO.aspx`

[4] Romanelli G, Krzystyniak M, Senesi R, Raspino D, Boxall J, Pooley D, Moorby S, Schooneveld E, Rhodes N J, Andreani C and Fernandez-Alonso F 2017 *Measurement Science and Technology* **28** 095501

[5] Last accessed on June 2018 URL `http://www.isis.stfc.ac.uk`

[6] Krzystyniak M, Romanelli G, Fabian M, Gutmann M, Festa G, Arcidiacono L, Gigg M, Drubicki K, Andreani C, Senesi R and Fernandez-Alonso F 2018 *Journal of Physics: Conference Series* **1021** 012026

[7] Last accessed on June 2018 URL `http://download.mantidproject.org/`

[8] Arnold O, Bilheux J, Borreguero J, Buts A, Campbell S, Chapon L, Doucet M, Draper N, Leal R F, Gigg M, Lynch V, Markvardsen A, Mikkelson D, Mikkelson R, Miller R, Palmen K, Parker P, Passos G, Perring T, Peterson P, Ren S, Reuter M, Savici A, Taylor J, Taylor R, Tolchenov R, Zhou W and Zikovsky J 2014 *Nuclear Instruments and Methods in Physics Research Section A* **764** 156 – 166

[9] Romanelli G and Krzystyniak M 2016 *Nuclear Instruments and Methods in Physics Research Section A* **819** 84 – 88

[10] Krzystyniak M, Druzbicki K, Romanelli G, Gutmann M J, Rudic S, Imberti S and Fernandez-Alonso F 2017 *Physical Chemistry Chemical Physics* **19**(13) 9064–9074

[11] Monserrat D, Vispa A, Pardo L C, Tolchenov R, Mukhopadhyay S and Fernandez-Alonso F 2015 *Journal of Physics: Conference Series* **663** 012009

[12] Jackson S, Krzystyniak M, Seel A G, Gigg M, Richards S E and Fernandez-Alonso F 2014 *Journal of Physics: Conference Series* **571** 012009

[13] Nixon D 2015 *RAL Technical Reports* **07** URL `http://purl.org/net/epubs/work/22354868`

[14] Oram E 2016 *RAL Technical Reports* **11** URL `http://purl.org/net/epubs/manifestation/29541388`

[15] Oram E and McCann L 2016 *RAL Technical Reports* **19** URL `http://purl.org/net/epubs/work/30884953`

[16] West G B 1975 *Physics Reports* **18** 263–323

[17] Pietropaolo A and Senesi R 2011 *Physics Reports* **508** 45–90

[18] Schooneveld E M, Mayers J, Rhodes N J, Pietropaolo A, Andreani C, Senesi R, Gorini G, Perelli-Cippo E and Tardocchi M 2006 *Review Scientific Instruments* **77** 5103

[19] Pietropaolo A, Andreani C, Filabozzi A, Pace E and Senesi R 2007 *Nuclear Instruments and Methods in Physics Research Section A* **570** 498–510

[20] Andreani C, Colognesi D, Degiorgi E, Filabozzi A, Nardone M, Pace E, Pietropaolo A and Senesi R 2003 *Nuclear Instruments and Methods in Physics Research Section A* **497** 535–549

[21] Mayers J, Tomkinson J, Abdul-Redah T, Stirling W G, Andreani C, Senesi R, Nardone M, Colognesi D and Degiorgi E 2004 *Physica B Condensed Matter* **350** E659–E662

[22] Schooneveld E M, Pietropaolo A, Andreani C, Cippo E P, Rhodes N J, Senesi R, Tardocchi M and Gorini G 2016 *Reports on Progress in Physics* **79** 094301
[23] Pietropaolo A, Tardocchi M, Schooneveld E M and Senesi R 2006 *Nuclear Instruments and Methods in Physics Research Section A* **568** 826–838
[24] Mayers J and Reiter G 2012 *Measurement Science and Technology* **23** 045902
[25] Sears V F 1992 *Neutron News* **3** 26–37
[26] Pace E, Salmè G and West G 1991 *Physics Letters B* **273** 205 – 210
[27] Krzystyniak M, Seel A G, Richards S E, Gutmann M J and Fernandez-Alonso F 2014 *Journal of Physics: Conference Series* **571** 012002
[28] Mayers J 2011 *Measurement Science and Technology* **22** 015903
[29] Sears V 1975 *Advances in Physics* **24** 1–45
[30] Fermi E 1936 *Il Nuovo Cimento (1924-1942)* **13** 434–439
[31] Rodríguez Palomino L A, Dawidowski J, Márquez Damián J I, Cuello G J, Romanelli G and Krzystyniak M 2017 *Nuclear Instruments and Methods in Physics Research Section A* **870** 84 – 89
[32] Romanelli G, Rudić S, Zanetti M, Andreani C, Fernandez-Alonso F, Gorini G, Krzystyniak M and Škoro G 2018 *Nuclear Instruments and Methods in Physics Research Section A* **888** 88 – 95
[33] Andreani C, Merlo V and Ricci M A 1989 *Nuclear Instruments and Methods in Physics Research B* **36** 216–221
[34] Imberti S, Andreani C, Garbuio V, Gorini G, Pietropaolo A, Senesi R and Tardocchi M 2005 *Nuclear Instruments and Methods in Physics Research Section A* **522** 463
[35] Romanelli G, Ceriotti M, Manolopoulos D E, Pantalei C, Senesi R and Andreani C 2013 *The Journal of Physical Chemistry Letters* **4** 3251–3256
[36] Parmentier A, Shephard J J, Romanelli G, Senesi R, Salzmann C G and Andreani C 2015 *The Journal of Physical Chemistry Letters* **6** 2038–2042
[37] Romanelli G, Senesi R, Zhang X, Loh K P and Andreani C 2015 *Physical Chemistry Chemical Physics* **17** 31680–31684
[38] Romanelli G, Liscio A, Senesi R, Zamboni R, Treossi E, Liscio F, Giambastiani G, Palermo V, Fernandez-Alonso F and Andreani C 2016 *Carbon* **108** 199–203
[39] Andreani C, Romanelli G and Senesi R 2016 *The Journal of Physical Chemistry Letters* **7** 2216–2220
[40] James F 1994 *MINUIT Function Minimization and Error Analysis: Reference Manual Version 94.1 – CERN-D-506*

IOP Conf. Series: Journal of Physics: Conf. Series **1055** (2018) 012017 doi:10.1088/1742-6596/1055/1/012017

The road to a station for epithermal and thermal neutron analysis

G Romanelli[a], M Krzystyniak[a,b], G Festa[c], C Andreani[c,d], F Fernandez-Alonso[a,e], and R Senesi[d,f]

[a] ISIS Facility, Rutherford Appleton Laboratory, Chilton, Didcot, Oxfordshire, OX11 0QX, UK

[b] School of Science and Technology, Nottingham Trent University, Clifton Campus, Nottingham, NG11 8NS, UK

[c] Museo Storico della Fisica e Centro Studi e Ricerche "Enrico Fermi", Piazza del Viminale 1, Rome 00184 Italy

[d] Università degli Studi di Roma "Tor Vergata", Dipartimento di Fisica and Centro NAST, Via della Ricerca Scientifica 1, Rome , 00133 Italy

[e] Department of Physics and Astronomy, University College London, Gower Street, London, WC1E 6BT, UK

[f] CNR-IPCF Sezione di Messina, Viale Ferdinando Stagno d'Alcontres 37, Messina, 98158 Italy

E-mail: `Matthew.krzystyniak@stfc.ac.uk`

Abstract. Despite the large variety of research interests and themes motivating the current neutron research included in this collection, we have found common denominators characterising the manner in which the chosen research methodology tries to tackle the envisaged scientific questions. This article attempts to characterise those trends in current research with the aim of identifying the main mid-to-long term opportunities faced by electron-Volt neutron spectroscopy. The main realisation from this exercise is that the scientific community seems eager to combine neutron-based techniques over a broad energy range. To this end, the most natural choice seems to be to resort to neutron instruments where such capabilities are already present from the outset, with the most prominent example being the VESUVIO spectrometer at the ISIS pulsed neutron and muon source in the UK. However broad the operational basis of the existing neutron beamline infrastructure may be, progress, achievable only through further instrument upgrades, is the only way forward. The need to move forward is clearly seen within the community and is well documented by the research presented in this collection. This need for a substantial upgrade has crystallised in the form of a proposition to build a station rather than a conventional beamline, for Epithermal and Thermal Neutron Analysis station, hereafter ETNA.

1. Introduction

The application of electron-Volt (eV) neutron spectroscopy to scientific investigations on condensed-matter systems has witnessed, over the last few years, an impressive development. Looking at the proceedings of the VI (2014) [1] and VII (2017) international workshops on eV neutron spectroscopy [2], one can notice how the use of Deep Inelastic Neutron Scattering (DINS) to characterise Nuclear Quantum Effects (NQEs) mostly on hydrogen dynamics [3] has lead to MAss-selective Neutron SpEctroscopy (MANSE), whereby the Nuclear Momentum Distributions (NMDs) of several elements in a given sample are accessed at the same time [4, 5, 6].

Furthermore, a number of recent investigations carried out on the VESUVIO spectrometer [7], the flagship instrument for eV neutron spectroscopy at the ISIS Facility [8], successfully combined DINS with concurrent Neutron Diffraction (ND) measurements [9, 10], or employed broad-range Neutron Transmission (NT) [11, 12]. Moreover, a novel trend of concurrent characterisation of condensed-matter systems by means of DINS, ND, and Neutron Resonance Capture Analysis (NRCA) is proving to be a potent strategy and thus is gaining momentum [13].

Here, we summarise the most prominent and successful examples of the currently prevailing trends in eV neutron spectroscopy, with the aim to identify the main mid-to-long term objectives faced by the techniques employed as drivers for much-needed upgrade of the existing beamline infrastructure.

2. Neutron spectroscopy and modelling

Inelastic Neutron Scattering (INS) can be considered as the neutron-based technique nearest to DINS. On the one hand, mass-resolved DINS spectra can be obtained from measured vibrational densities of states by evaluating the multi-phonon expansion well beyond the first term, and in the limit of large energy and momentum transfers [14], *i.e.*, within the celebrated Impulse Approximation (IA) [15]. On the other hand, the widths of NMDs accessed through DINS can be related via harmonic models to the frequencies of molecular vibrations [16], or to simulated atom-projected vibrational densities of states [17].

Several articles in this collection have explored the link between vibrational spectroscopy and epithermal neutrons. Parker *et al.* give an example of how thermal and epithermal neutrons at spallation neutron sources can be used to study the stretching mode in H_2, with future applications to characterise the molecule's interaction with materials – a simpler task from the simulation point of view than the modelling of the 0→1 rotational transition. The article provides an instructive example on how the INS signal from light atoms, investigated with epithermal neutrons, follows the nuclear recoil line while approaching the IA. These results gain a particular interest when considered in the framework of a previous study by Senesi *et al.* [18] discussing the DINS signal obtained on direct-geometry INS spectrometers, particularly in the case of the SEQUOIA instrument at SNS, where the energy of the incident neutrons was as high as 6 eV.

The contribution by Kolensikov *et al.* is an exquisite example of how the understanding of a given system, in particular ultra-confined water, can be enhanced by comparing INS and DINS data. In this case, INS peaks interpreted as tunnelling modes affecting the confined water molecule were directly related to the delocalisation of water's hydrogen accessed through the analysis of its NMD [19]. Moreover, the comparison of INS and DINS data at several temperatures allows for a deeper discussion on the origin of hydrogen bonding and its evolution with the thermodynamic phase.

The attempt to reproduce both INS and DINS spectra by first-principles simulations can be regarded as a stringent selection protocol for a given theoretical model, especially for those systems particularly affected by NQEs or anharmonic local potentials, as in the case of formic acid [10]. State-of-the-art computer simulations including NQEs, based on the path integral formalism [20], have found in DINS experiments a much-needed benchmark, such as in [21, 9]. However, the time-consuming nature of first-principles simulations has pushed the community to look for new strategies [22]. In this regard, Druzbiki *et al.* show how to successfully convert the results from a *per se* classical computer simulation based on molecular dynamics so as to account for NQEs by using a coloured-noise thermostat previously suggested by Ceriotti [23].

The relation between NMDs and the ordering degree of the local potential affecting carbon in several allotropes was tackled by Armstrong *et al.* The authors replaced the concept of bond strength with the Hausdorff-Besicovitch fractal dimension, a concept borrowed from mathematical topology, deriving a relation between the fractal dimension of a given system and the carbon nuclear kinetic energy, supporting the experimental results with simulated atom-

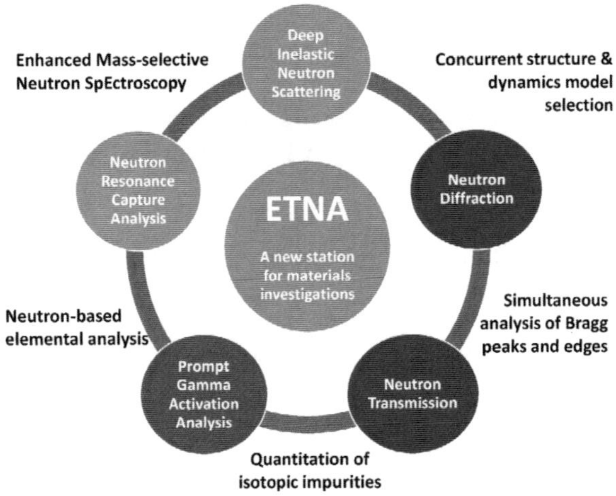

Figure 1. Experimental possibilities available on ETNA.

projected vibrational densities of states.

3. Concurrent investigation of structure and dynamics

Building upon the synergy between INS and DINS, and taking advantage of the possibility to concurrently measure DINS spectra and ND data, recent investigations were able to model and optimise the structure and dynamics of complex systems over a uniquely large time- and space-window [24, 25]. Following this strategy, Krzystyniak *et al.* tackle the challenging problem of glassy-versus-polycrystalline structures in Zr–Be binary mixtures (see also Ref. [26]).

Another contribution by Krzystyniak *et al.*, applies concurrent ND and DINS to nitrogen-doped superconducting radio-frequency niobium cavities. The conditioning procedure was found to lead to systematically larger Nb lattice parameters, also providing an estimate of nitrogen concentration. Moreover, DINS was found particularly sensitive to the broadening of the niobium NMD in the doped sample as compared to a standard one, that was related to an increased degree of ordering and binding in the niobium local potential.

4. Neutron-induced photon spectroscopies

Epithermal neutrons are particularly suitable for applications to elemental analysis of bulk materials, in scattering experiments based on DINS and MANSE, as well as in absorption experiments based on NRCA, neutron resonance transmission analysis, and Prompt-Gamma Activation Analysis (PGAA). The effort of the community towards this end is well represented in the collection of articles presented here, as well as in Ref. [13]. For example, Nardini *et al.* present an interesting combination of NRCA and other chemometric tools, in particular principal-component analysis, applied to archaeological artefacts so as to identify trace elements, and thereby their origin and history.

The contribution by Romanelli *et al.* continues along the lines of previous studies by Mayers *et al.* [27] and Postma and Schillebeeckx [28] to use NRCA to analyse the resonance line shape to obtain information about the local quantum temperature affecting an atom, and the associated NQEs. This approach is suggested as a high-throughput technique whereby relatively fast measurements can be applied to parametric studies, for example as a function of temperature.

The technique is able to tackle the dynamics of heavy-weight elements for which the DINS resolution would otherwise be a limiting factor.

Recent studies made use of a high-purity Ge detector to acquire high-resolution photon-energy and time-of-flight spectra, whereby the photon peaks can be correlated to the arrival time of neutrons in the sample area. The contribution by Onorati *et al.*(see also Ref. [29]) discusses the origin of the γ-background levels on the VESUVIO spectrometer, as well as other beam-lines at ISIS, including the effects due to a recent change to the water moderator serving the instrument [30, 31]. Details on the absolute calibration of the high-purity Ge detector are given in the contribution by Parmentier *et al.* The characterisation and indexing of the γ-ray spectra is the first step to design low-background measurements and increase the detection and quatitation thresholds for elemental analysis.

5. Detection and instrument modelling

As already mentioned in the conclusions of the proceedings of the VI workshop [32], the long acquisition times to collect DINS data with good statistics is one of the limiting factors preventing parametric studies on the VESUVIO spectrometer. The preliminary work by Ulpiani *et al.* shows how it is possible to increase the count rate for the front-scattering YAP detectors by fine-tuning the electronics parameters underpinning the detection procedure. In addition to shorter measurements, such an upgrade could improve the detection limits of hydrogen in materials, paving the way for a new application of the technique in industrial problems.

More generally, there is a need to optimise the geometry of the VESUVIO spectrometer so as to increase the beam flux without sensibly worsening the instrument's resolution. The modelling of the instrument using McStas, presented by Di Giulio *et al.*, goes in this direction. Codes for the simulation of neutron transport are not optimised for epithermal neutrons, therefore the development of new components, such as those simulating energy filters based on nuclear resonances, are much needed and will speed-up any foreseeable instrument-development project.

6. Strategies for data analysis

Following a previous update [33], Romanelli *et al.* have presented a new script to run the analysis of DINS spectra in MANTID. The script makes use of the algorithms included in the software to load and visualise the data, perform preliminary fits on the raw data, and run corrections for multiple scattering and γ backgrounds. Moreover, a number of consistent analysis strategies are presented, including sequential or global fits of hydrogen NMDs from different detectors in the West-scaling variable. A detailed and independent approach to the analysis of DINS data is presented by Dawidowski *et al.* as well. The combination of several procedures to reduce and fit DINS data is the strongest proof of consistency in the data analysis, and the implementation of these strategies is being implemented progressively in the new MANTID releases.

The novelty in the handling of DINS data in MANTID includes the implementation of Bayesian tools for the selection of the fitting model, also discussed in Ref. [34], here stressing non-negativity and other physical constraints to fitted NMDs. Krzystyniak *et al.* also show how to extract value of the mean kinetic energy from the final-state corrections to the IA, a procedure that can provide an estimation of the width of experimental NMDs at an early stage of data reduction, to be used as a starting parameter for subsequent data analysis.

7. Outlook and ways forward

Looking at these proceedings and at a recent review [4], the scientific community using eV neutron spectroscopy on the VESUVIO spectrometer has broadened the area of investigated subjects towards, for example, energy materials [35] and real-life systems [36]. Such an evolution in the scientific programme has been catalysed, amongst other reasons, by an increased attention to the several techniques simultaneously available during a DINS experiment. Although the instrument was designed originally for applications in neutron spectroscopy, it is now considered a good-resolution diffractometer [30], as well as a station where neutron cross sections can be accurately measured over a broad energy range, from a fraction of meV up to tens of keV. Additional attention is now paid to the NRCA capabilities of the photon-sensitive front-scattering detectors, and PGAA experiments are becoming more regular.

Driven by the needs the community has expressed, there are a number of upgrades that could pave the way to even more challenging investigations. The fine tuning of the detection electronic chain has already showed promising results [37], with saturation problems arising from the increase in count rates being mitigated by the recent installation of last-generation data-acquisition electronics, DAE3, now in its test period on VESUVIO. The optimisation of the detection procedures should be reinforced by an increase of the angular coverage of scattered neutrons, a limiting factor at present for acquisition times, as well as a rearrangement of the existing detectors onto Debye-Scherrer cones at constant scattering angle, a way to increase the signal-to-noise ratio at the data-analysis stage.

Cleaner measurements of NRCA and PGAA require a suppression of the background within the VESUVIO blockhouse, a task addressed by additional Pb shielding towards the beam stop and the moderator, under installation at present, and the possibility to cycle in the primary path a Cd foil to attenuate the neutron flux at thermal energies. The foil changer, in its design stage at present, will allow additional tests of direct-geometry DINS spectra by cycling a resonant foil in the incident beam, with the possibility to improve the MANSE capabilities of the instrument. Also, the instrument is at present collecting data with a fine binning of 0.1 μs, instead of the traditional 0.5 μs binning, so as to improve the resolution both for NRCA and DINS, at low time-of-flight values.

Furthermore, the capabilities available with neutron transmission can be increased by including a monitor with both spatial- and energy-resolution, therefore making neutron imaging available, with possible applications to the characterisation of catalytic processes as in the case of [12]. Such a detector would serve as a monitor for a more robust correction of DINS data, and as a diagnostic tool.

The evolution of eV neutron spectroscopy has proved to be successful and exciting. Driven by this well-documented and highly exciting upsurge in research activity, a clearly emerging community-driven need for a substantial upgrade of the VESUVIO instrument will pave the way to tackle challenging investigations yet to come. The main result is the awareness of the need for a station for Epithermal and Thermal Neutron Analysis, a future instrument already referred to as ETNA.

As organisers, we have enjoyed the VII international workshop on electron-Volt neutron spectroscopy, and the related proceedings collected in this volume. We look forward to the next workshop, eager to understand, from this prolific scientific community, what are the best ways forward. As always, the sky is the limit.

Acknowledgements
We would like to thank the Science & Technology Facilities Council for the support in the organisation of the workshop and the publication of these proceedings. We acknowledge the financial support and effort of the Museo Storico della Fisica e Centro Studi e Ricerche "Enrico Fermi", the School of Neutron Scattering "F. P. Ricci", the Consiglio Nazionale delle Ricerche, and the British Embassy in Rome. Finally, a special thanks to the international advisory committee, and to those who attended the workshop and who participated in this volume with exquisite examples of their research.

References
[1] Seel A G, Senesi R and Fernandez-Alonso F 2014 *Journal of Physics: Conference Series* **571** 011001

[2] Romanelli G, Krzystyniak M, Festa G, Andreani C, Senesi R and Fernandez-Alonso F 2018 *Neutron News* **29** 4–6
[3] Andreani C, Colognesi D, Mayers J, Reiter G F and Senesi R 2005 *Advances in Physics* **54** 377–469
[4] Andreani C, Krzystyniak M, Romanelli G, Senesi R and Fernandez-Alonso F 2017 *Advances in Physics* **66** 1–73
[5] Andreani C, Senesi R, Krzystyniak M, Romanelli G and Fernandez-Alonso F 2017 *Neutron Scattering - Applications in Biology, Chemistry, and Materials Science* (*Experimental Methods in the Physical Sciences* vol 49) ed Fernandez-Alonso F and Price D L (Academic Press) pp 403 – 457
[6] Andreani C, Senesi R, Krzystyniak M, Romanelli G and Fernandez-Alonso F 2018 **41**
[7] Last accessed on June 2018 URL `http://www.isis.stfc.ac.uk/Pages/VESUVIO.aspx`
[8] Last accessed on June 2018 URL `http://www.isis.stfc.ac.uk`
[9] Andreani C, Romanelli G and Senesi R 2016 *The Journal of Physical Chemistry Letters* **7** 2216–2220
[10] Krzystyniak M, Druzbicki K, Romanelli G, Gutmann M J, Rudic S, Imberti S and Fernandez-Alonso F 2017 *Physical Chemistry Chemical Physics* **19**(13) 9064–9074
[11] Rodríguez Palomino L A, Dawidowski J, Márquez Damián J I, Cuello G J, Romanelli G and Krzystyniak M 2017 *Nuclear Instruments and Methods in Physics Research Section A* **870** 84 – 89
[12] Romanelli G, Rudić S, Zanetti M, Andreani C, Fernandez-Alonso F, Gorini G, Krzystyniak M and Škoro G 2018 *Nuclear Instruments and Methods in Physics Research Section A* **888** 88 – 95
[13] Krzystyniak M, Romanelli G, Fabian M, Gutmann M, Festa G, Arcidiacono L, Gigg M, Drubicki K, Andreani C, Senesi R and Fernandez-Alonso F 2018 *Journal of Physics: Conference Series* **1021** 012026
[14] Watson G I 1996 *Journal of Physics Condensed Matter* **8** 5955–5975
[15] Sears V F 1984 *Physical Review B* **30** 44–51
[16] Moreh R, Sellyey W C, Sutton D and Vodhanel R 1985 *Physical Review C* **31** 2314–2316
[17] Krzystyniak M, Richards S E, Seel A G and Fernandez-Alonso F 2013 *Physical Review B* **88**(18) 184304
[18] Senesi R, Kolesnikov A I and Andreani C 2014 *Journal of Physics: Conference Series* **571** 012007
[19] Kolesnikov A I, Reiter G F, Choudhury N, Prisk T R, Mamontov E, Podlesnyak A, Ehlers G, Seel A G, Wesolowski D J and Anovitz L M 2016 *Phys. Rev. Lett.* **116**(16) 167802
[20] Car R and Parrinello M 1985 *Phys. Rev. Lett.* **55**(22) 2471–2474
[21] Cheng B, Behler J and Ceriotti M 2016 *The Journal of Physical Chemistry Letters* **7** 2210–2215
[22] Ceriotti M, Fang W, Kusalik P G, McKenzie R H, Michaelides A, Morales M A and Markland T E 2016 *Chemical Reviews* **116** 7529–7550
[23] Ceriotti M, Bussi G and Parrinello M 2009 *Physical Review Letters* **103** 030603
[24] Krzystyniak M and Fernandez-Alonso F 2011 *Phys. Rev. B* **83**(13) 134305
[25] Krzystyniak M, Druzbicki K and Fernandez-Alonso F 2015 *Physical Chemistry Chemical Physics* **17**(46) 31287–31296
[26] Syrykh G F, Stolyarov A A, Krzystyniak M, Romanelli G and Sadykov R A 2017 *JETP Letters* **105** 591–594
[27] Mayers J, Baciocco G and Hannon A 1989 *Nuclear Instruments and Methods in Physics Research Section A: Accelerators, Spectrometers, Detectors and Associated Equipment* **275** 453 – 459
[28] Postma H and Schillebeeckx P 2006 *Neutron Resonance Capture and Transmission Analysis* (John Wiley & Sons, Ltd) ISBN 9780470027318
[29] Onorati D 2017 *Optimization of γ detection methods for neutron energy analysis on eV neutron spectrometers* (M.Sc Thesis - Università degli Studi di Roma "Tor Vergata")
[30] Romanelli G, Krzystyniak M, Senesi R, Raspino D, Boxall J, Pooley D, Moorby S, Schooneveld E, Rhodes N J, Andreani C and Fernandez-Alonso F 2017 *Measurement Science and Technology* **28** 095501
[31] Škoro G, Bewley R, Lilley S, Ewings R, Romanelli G, Gutmann M, Smith R, Rudi S and Ansell S 2018 *Journal of Physics: Conference Series* **1021** 012039
[32] Seel A G, Krzystyniak M, Senesi R and Fernandez-Alonso F 2014 *Journal of Physics: Conference Series* **571** 012014
[33] Jackson S, Krzystyniak M, Seel A G, Gigg M, Richards S E and Fernandez-Alonso F 2014 *Journal of Physics: Conference Series* **571** 012009
[34] Romanelli G and Krzystyniak M 2016 *Nuclear Instruments and Methods in Physics Research Section A* **819** 84 – 88
[35] Druźbicki K, Pinna R S, Rudić S, Jura M, Gorini G and Fernandez-Alonso F 2016 *The Journal of Physical Chemistry Letters* **7** 4701–4709
[36] Tian K V, Yang B, Yue Y, Bowron D T, Mayers J, Donnan R S, Dobó-Nagy C, Nicholson J W, Fang D C, Greer A L, Chass G A and Greaves G N 2015 *Nature Communications* **6** 8631
[37] Ulpiani P 2017 *Inelastic neutron scattering at eV energies with enhanced counting statistics and noise reduction* (M.Sc Thesis - Università degli Studi di Roma "Tor Vergata")